STUDENT PRACTICE WORKBOOK

PRECALCULUS
GRAPHICAL, NUMERICAL, ALGEBRAIC
SEVENTH EDITION

Franklin D. Demana
The Ohio State University

Bert K. Waits
The Ohio State University

Gregory D. Foley
Liberal Arts and Science Academy of Austin

Daniel Kennedy
Baylor School

PEARSON

Addison
Wesley

Boston San Francisco New York
London Toronto Sydney Tokyo Singapore Madrid
Mexico City Munich Paris Cape Town Hong Kong Montreal

Pearson Prentice Hall™ is a trademark of Pearson Education, Inc.
Pearson® is a registered trademark of Pearson plc.
Prentice Hall® is a registered trademark of Pearson Education, Inc.

0-13-198580-9
1 2 3 4 5 6 7 8 9 10 BB 10 09 08 07 06

Contents

Chapter P — Prerequisites

P.1 Real Numbers

Alternate Example 4 — Converting Between Intervals and Inequalities

Convert interval notation to inequality notation or vice versa. Find the endpoints and state whether the interval is bounded, its type, and graph the interval.

(a) $(-3, 2]$ (b) $(2, \infty)$ (c) $-1 \leq x < 4$

SOLUTION

(a) The interval $(-3, 2]$ corresponds to $-3 < x \leq 2$ and is bounded and half-open (see Figure P.1a). The endpoints are -3 and 2.

(b) The interval $(2, \infty)$ corresponds to $x > 2$ and is unbounded and open (see Figure P.1b). The only endpoint is 2.

(c) The inequality $-1 \leq x < 4$ corresponds to the half-open, bounded interval $[-1, 4)$ (see Figure P.1c). The endpoints are -1 and 4.

(a)

```
    ←+——+——(——+——+——+——]——+——+——+——→ x
     -5  -4  -3  -2  -1   0   1   2   3   4   5
```

(b)

```
    ←+——+——+——+——+——+——(——+——+——+——→ x
     -5  -4  -3  -2  -1   0   1   2   3   4   5
```

(c)

```
    ←+——+——+——[——+——+——+——+——)——+——→ x
     -5  -4  -3  -2  -1   0   1   2   3   4   5
```

Figure P.1 Graphs of the intervals of real numbers in Alternate Example 4.

Exercises for Alternate Example 4

In Exercises 1–6, convert interval notation to inequality notation or vice versa. Find the endpoints and state whether the interval is bounded and its type. Then graph the interval.

1. $[-2, 3]$ 2. $(-\infty, 0)$ 3. $-4 \leq x \leq 4$ 4. $2 < x < 5$ 5. $-3 < x < 5$ 6. $[3, \infty)$

Alternate Example 5 — Using the Distributive Property

(a) Write the expanded form of $2b(x - 3)$.

(b) Write the factored form of $5x + 10xy$.

SOLUTION

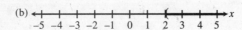

(a) $2b(x - 3) = 2b(x) - 2b(3)$
$= 2bx - 6b$

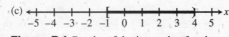

(b) $5x + 10xy = 5x + 2 \cdot 5xy$
$= 5x \cdot 1 + 5x(2y)$
$= 5x(1 + 2y)$

Exercises for Alternate Example 5

In Exercises 7–14, use the distributive property to expand or factor.

7. $b(x^2 - 3)$ **8.** $2x(3 - c)$ **9.** $(2x^2 + 5)n$ **10.** $3r(4 + 2n)$ **11.** $3n + 6mn$ **12.** $14c + 7bc$

13. $15gh + 7gz$ **14.** $n^3p + 2n^3t$

Alternate Example 7 **Simplifying Expressions Involving Powers**

Write each expression in simplified form with positive exponents.

(a) $(7nm^3)(3n^2m^2)$ **(b)** $\dfrac{n^3z^{-1}}{n^2z^{-2}}$ **(c)** $\left(\dfrac{m^4}{3}\right)^{-2}$

SOLUTION

(a) $(7nm^3)(3n^2m^2) = (7 \cdot 3)(n \cdot n^2)(m^3 \cdot m^2)$
$$= 21n^3m^5$$

(b) $\dfrac{n^3z^{-1}}{n^2z^{-2}} = \dfrac{n^3}{n^2} \cdot \dfrac{z^{-1}}{z^{-2}}$
$$= nz^{-1-(-2)}$$
$$= nz$$

(c) $\left(\dfrac{m^4}{3}\right)^{-2} = \dfrac{(m^4)^{-2}}{3^{-2}}$
$$= \dfrac{m^{-8}}{3^{-2}}$$
$$= \dfrac{3^2}{m^8}$$
$$= \dfrac{9}{m^8}$$

Exercises for Alternate Example 7

In Exercises 15–20, simplify the expression. Assume that the variables in denominators are nonzero.

15. $(7x^2y^3)(4x^3y^4)$ **16.** $(3a^4b^4)(-5a^3b^6)$ **17.** $\dfrac{c^4d^{-2}}{c^5d^3}$ **18.** $\dfrac{x^3y^2}{x^2y^4}$ **19.** $\left(\dfrac{k^{-2}}{4}\right)^3$

20. $\left(\dfrac{h^{-2}}{-3}\right)^{-2}$

Alternate Example 9 **Using Scientific Notation**

Simplify $\dfrac{(2{,}400)(130)}{12{,}000}$.

SOLUTION
Using Mental Arithmetic

$$\dfrac{(2{,}400)(130)}{12{,}000} = \dfrac{(2.4)10^3 \times (1.3)10^2}{(1.2)10^4} \qquad \text{Write in scientific notation.}$$

$$= \dfrac{(2.4)(1.3)}{1.2} \cdot \dfrac{10^3 \times 10^2}{10^4} \qquad \text{Apply laws of exponents.}$$

$$= 2.6 \cdot 10^1$$

$$= 26$$

Using a Calculator

Figure P.2 shows two ways to perform the computation. In the first, the numbers are entered in decimal form. In the second, the numbers are entered in scientific notation. The calculator uses "2.6E1" to stand for 2.6×10^1.

Figure P.2 Be sure you understand how your calculator displays scientific notation.

Exercises for Alternate Example 9

In Exercises 21–26, simplify using scientific notation.

21. $\dfrac{3.2 \times 10^7}{1.6 \times 10^5}$ **22.** $\dfrac{98{,}000 \times 50}{14{,}000}$ **23.** $(3.2 \times 10^{-2})(1.5 \times 10^3)$ **24.** $(6.0 \times 10^{-1})(1.8 \times 10^{-2})$

25. $\dfrac{3{,}600 \times 180}{720}$ **26.** $\dfrac{(4.5)10^{-4} \times (3.0)10^{-1}}{9 \times 10^2}$

P.2   Cartesian Coordinate System

Alternate Example 6   Finding Standard Form Equations of Circles

Find the standard form equation of the circle.

(a) Center $(-3, -2)$, radius 6 **(b)** Center $(0, 0)$, radius 7

SOLUTION

(a) $(x - h)^2 + (y - k)^2 = r^2$ Standard form equation
$(x - (-3))^2 + (y - (-2))^2 = 6^2$ Substitute values for h, k, and r. $h = -3$, $k = -2$, $r = 6$
$(x + 3)^2 + (y + 2)^2 = 36$

(b) $(x - h)^2 + (y - k)^2 = r^2$ Standard form equation
$(x - 0)^2 + (y - 0)^2 = 7^2$ Substitute values for h, k, and r. $h = 0$, $k = 0$, $r = 7$
$x^2 + y^2 = 49$

Exercises for Alternate Example 6

In Exercises 1–6, find the standard form equation of each circle.

1. Center $(0, 4)$, radius 4 **2.** Center $(1, 1)$, radius 10 **3.** Center $(0, 0)$, radius 1 **4.** Center $(-4, 0)$, radius 2.5

5. Center $\left(\dfrac{1}{2}, \dfrac{1}{2}\right)$, radius 4 **6.** Center $(-3, -3)$, radius 8

Alternate Example 8   Verifying Right Triangles

Use the converse of the Pythagorean theorem and the distance formula to show that the points $(-6, 8)$, $(8, 6)$, and $(2, 0)$ determine a right triangle.

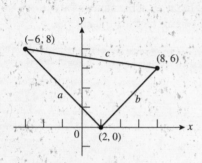

Figure P.3 The triangle in Alternate Example 8.

SOLUTION

The three points are plotted in Figure P.3. We need to show that the lengths of the sides of the triangle satisfy the Pythagorean relationship $a^2 + b^2 = c^2$. Applying the distance formula, we find that

$$a = \sqrt{(-6-2)^2 + (8-0)^2} = \sqrt{128}$$

$$b = \sqrt{(8-2)^2 + (6-0)^2} = \sqrt{72}$$

$$c = \sqrt{(-6-8)^2 + (8-6)^2} = \sqrt{200}$$

The triangle is a right triangle because

$$a^2 + b^2 = \left(\sqrt{128}\right)^2 + \left(\sqrt{72}\right)^2 = 128 + 72 = 200 = c^2$$

Exercises for Alternate Example 8

In Exercises 7–12, use the converse of the Pythagorean theorem and the distance formula to show that the given points determine a right triangle.

7. $(0, 0), (0, 1), (1, 0)$ **8.** $(1, 1), (13, 1), (1, 6)$ **9.** $(-4, 0), (-4, -3), (0, 0)$ **10.** $(-3, 4), (2, 4), (2, 10)$

11. $(0, 0), (-1, 2), (6, 3)$ **12.** $(4, 0), (6, 2), (6, -2)$

Alternate Example 9 **Using the Midpoint Formula**

It is a fact from geometry that the midpoints of the diagonals of a rectangle bisect each other. Prove this with a midpoint formula.

SOLUTION

We can position a rectangle in the rectangular coordinate plane as shown in Figure P.4. Applying the midpoint formula for the coordinate plane to segments OC and AB, we find that

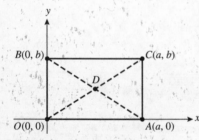

Figure P.4 The coordinates of C must be (a, b) in order for quadrilateral $OBCA$ to be a rectangle.

$$\text{midpoint of segment } OC = \left(\frac{0+a}{2}, \frac{0+b}{2}\right) = \left(\frac{a}{2}, \frac{b}{2}\right)$$

$$\text{midpoint of segment } AB = \left(\frac{0+a}{2}, \frac{0+b}{2}\right) = \left(\frac{a}{2}, \frac{b}{2}\right)$$

The midpoints of segments OC and AB are the same, so the diagonals of the rectangle $OBCA$ meet at their midpoints and thus bisect each other.

Exercises for Alternate Example 9

In Exercises 13–18, prove that the diagonals of the figure determined by the points bisect each other.

13. Square $(-1, 1), (2, 1), (2, 4), (-1, 4)$ **14.** Rectangle $(-2, 2), (-2, 6), (8, 6), (8, 2)$

15. Parallelogram $(0, 0), (-2, 0), (-4, 6), (-2, 6)$ **16.** Parallelogram $(-6, 6), (1, 6), (2, 3), (-5, 3)$

17. Rectangle $(-2, 8), (-2, 2), (2, 2), (2, 8)$ **18.** Square whose sides have length a

P.3 Linear Equations and Inequalities

Alternate Example 2 **Solving a Linear Equation**

Solve $-3(x + 2) + 2(x - 1) = -4x + 4$. Support the result with a calculator.

SOLUTION

$$-3(x + 2) + 2(x - 1) = -4x + 4$$

$-3x - 6 + 2x - 2 = -4x + 4$	Distributive property
$-x - 8 = -4x + 4$	Combine like terms.
$3x = 12$	Add 8, and add $4x$.
$x = 4$	Divide by 3.

```
4 → X
                    4
-3(X+2)+2(X-1)
                  -12
-4X+4
                  -12
```

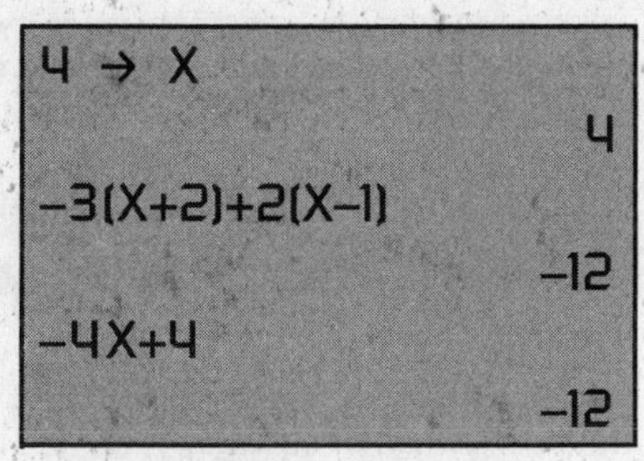

Figure P.5 The top line stores the number 4 into the variable x.

To support our algebraic work we can use a calculator to evaluate the original equation for $x = 4$. Figure P.5 shows that each side of the original equation is equal to -12 if $x = 4$.

Exercises for Alternate Example 2

In Exercises 1–6, solve the equation. Support the result with a calculator.

1. $-(x - 5) - 2(x - 7) = -2x + 10$

2. $5(2x) - 2(4x - 3) = -x - 9$

3. $x + 2 + 5(x - 1) = 7x + 3$

4. $6(5x + 1) - 2(x - 2) = 5x + 10$

5. $-7(2x - 2) + 5(x - 1) = -10x$

6. $2(3x + 6) + 2(8x - 7) = 25x - 14$

Alternate Example 3 **Solving a Linear Equation Involving Fractions**

Solve $\dfrac{5y + 2}{3} = 3 + \dfrac{y}{2}$.

SOLUTION

The denominators are 3, 1, and 2. The LCD of the fractions is 6. (See Appendix A.3 if necessary.)

$$\frac{5y + 2}{3} = 3 + \frac{y}{2}$$

$6\left(\dfrac{5y + 2}{3}\right) = 6\left(3 + \dfrac{y}{2}\right)$	Multiply by the LCD 6.
$6 \cdot \dfrac{5y + 2}{3} = 6 \cdot 3 + 6 \cdot \dfrac{y}{2}$	Distributive property
$2(5y + 2) = 18 + 3y$	
$10y + 4 = 18 + 3y$	Simplify.
$10y = 14 + 3y$	Subtract 4.
$7y = 14$	Subtract $3y$.
$y = 2$	Divide by 7.

We leave it to you to check the solution either using pencil and paper or a calculator.

Exercises for Alternate Example 3

In Exercises 7–12, solve the equation.

7. $\dfrac{5x + 2}{3} = 3 + \dfrac{x}{2}$

8. $\dfrac{5x - 1}{2} = 4 - \dfrac{x}{2}$

9. $\dfrac{x}{3} - 1 = 1 - \dfrac{x - 1}{2}$

10. $\dfrac{-x + 7}{3} = 1 + \dfrac{x}{5}$

11. $\dfrac{4x - 5}{4} = -2 - \dfrac{x}{2}$

12. $\dfrac{2x}{3} = 1 + \dfrac{x}{4}$

Alternate Example 4 **Solving a Linear Inequality**

Solve $4(x - 2) + 5 \geq 7x - 6$.

SOLUTION

$$4(x - 2) + 5 \geq 7x - 6$$
$$4x - 8 + 5 \geq 7x - 6 \qquad \text{Distributive property}$$
$$4x - 3 \geq 7x - 6 \qquad \text{Simplify.}$$
$$4x \geq 7x - 3 \qquad \text{Add 3.}$$
$$-3x \geq -3 \qquad \text{Subtract } 7x.$$
$$x \leq 1 \qquad \text{Multiply by } -\frac{1}{3}. \text{ (The inequality reverses.)}$$

The solution set of the inequality is the set of all real numbers less than or equal to 1. In interval notation, the solution set is $(-\infty, 1]$.

Exercises for Alternate Example 4

In Exercises 13–18, solve the inequality.

13. $-2(2x - 1) - 1 > 3x - 6$
14. $4(2x - 1) + 5 \leq x - 13$
15. $-x - 15 < 3(x - 6) - 5$
16. $3(-2x + 2) \geq -4x$

17. $2(2x - 1) + 3x \leq 8x - 7$
18. $-3(2x - 1) + 2(x - 1) < -3x + 6$

Alternate Example 6 **Solving a Double Inequality**

Solve the inequality and graph its solution set.

$$1 \leq \frac{2x - 1}{3} < 5$$

SOLUTION

$$1 \leq \frac{2x - 1}{3} < 5$$
$$3 \leq 2x - 1 < 15 \qquad \text{Multiply by 3.}$$
$$4 \leq 2x < 16 \qquad \text{Add 1.}$$
$$2 \leq x < 8 \qquad \text{Divide by 2.}$$

The solution set is the set of all real numbers greater than or equal to 2 and less than 8. In interval notation, the solution set is $[2, 8)$. Its graph is shown in Figure P.6.

Figure P.6 The graph of the solution set of the double inequality in Alternate Example 6.

Exercises for Alternate Example 6

In Exercises 19–24, solve the inequality and graph the solution set.

19. $-2 < \frac{3x - 2}{4} < 4$
20. $4 \leq \frac{5x + 2}{3} \leq 9$
21. $-3 < \frac{5x + 2}{6} \leq 2$
22. $-1 \leq \frac{4x - 7}{7} < 3$
23. $3 < \frac{2x + 3}{3} \leq 5$

24. $-1 < \frac{2x - 1}{5} < 3$

P.4 Lines in the Plane

Alternate Example 2 **Using the Point-Slope Form**

Use the point-slope form to find an equation for the line that passes through $(-5, 2)$ and has slope -3.

SOLUTION

Substitute $x_1 = -5$ and $y_1 = 2$, and $m = -3$ into the point-slope form, and simplify the resulting equation.

$$y - y_1 = m(x - x_1)$$ Point-slope form

$$y - 2 = -3(x - (-5))$$ $x_1 = -5,\ y_1 = 2,\ m = -3$

$$y - 2 = -3x - (-3)(-5)$$ Distributive property

$$y - 2 = -3x - 15$$

$$y = -3x - 13$$ A common simplified form

Exercises for Alternate Example 2

In Exercises 1–6, find a point-slope form equation for the line through the point with given slope.

1. $(4, 6)$; slope -2 **2.** $(0, -2)$; slope 2 **3.** $(4, 4)$; slope 1.5

4. $(6, 0)$; slope 5 **5.** $(-3, -2)$; slope -1 **6.** $(5, -2)$; slope -10

Alternate Example 4 **Use a Graphing Utility**

Draw the graph of $3x - 2y = 6$.

SOLUTION

First we solve for y.

$$3x - 2y = 6$$

$$-2y = -3x + 6$$ Solve for y.

$$y = \frac{3}{2}x - 3$$ Divide by -2.

Figure P.7 shows the graph of $y = \frac{3}{2}x - 3$, or equivalently, the graph of the linear equation $3x - 2y = 6$ in the $[-5.1, 5.1]$ by $[-4.7, 4.7]$ viewing window.

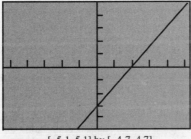

[-5.1, 5.1] by [-4.7, 4.7]

Figure P.7 The graph of $3x - 2y = 6$. The points $(0, -3)$ (y-intercept) and $(2, 0)$ (x-intercept) appear to lie on the graph and, as pairs, are solutions of the equation, providing visual support that the graph is correct.

Exercises for Alternate Example 4

In Exercises 7–12, graph the linear equation on a grapher.

7. $2x - y = 3$ **8.** $-2x + 2y = 4$ **9.** $3x - 2y = 4$ **10.** $3x - y = 4$ **11.** $-x + 2y = 6$ **12.** $-4x - 4y = 12$

Alternate Example 5 Finding an Equation of a Parallel Line

Find an equation of the line through $P(2, -1)$ that is parallel to the line L with equation $3x + 2y = 5$.

SOLUTION

We find the slope of L by writing its equation in slope-intercept form.

$$3x + 2y = 5 \qquad \text{Equation for } L$$
$$2y = -3x + 5 \qquad \text{Subtract } 3x.$$
$$y = -\frac{3}{2}x + \frac{5}{2} \qquad \text{Divide by 2.}$$

The slope of L is $-\frac{3}{2}$.

The line whose equation we seek has slope $-3/2$ and contains the point $(x_1, y_1) = (2, -1)$. Thus, the point-slope form equation for the line we seek is

$$y - (-1) = -\frac{3}{2}(x - 2)$$
$$y + 1 = -\frac{3}{2}x + 3 \qquad \text{Distributive property}$$
$$y = -\frac{3}{2}x + 2$$

Exercises for Alternate Example 5

In Exercises 13–18, find an equation for the line passing through the point and parallel to the given line.

13. $P(-2, -2)$; $3x - 2y = 4$ **14.** $P(0, -3)$; $-2x + 2y = 5$ **15.** $P(-3, 0)$; $-x + 3y = 4$ **16.** $P(-5, 1)$; $3x + 4y = -2$

17. $P(2, 5)$; $2x - y = 0$ **18.** $P(3, -4)$; $-4x + 8y = -2$

Alternate Example 6 Finding an Equation of a Perpendicular Line

Find an equation of the line through $P(2, -3)$ that is perpendicular to the line L with equation $3x - y = 3$. Support the result with a grapher.

SOLUTION

We find the slope of L by writing its equation in slope-intercept form.

$$3x - y = 3 \qquad \text{Equation for } L$$
$$y = 3x - 3 \qquad \text{Subtract } 3x. \text{ Multiply by } -1.$$

The slope of L is 3.

The line whose equation we seek has slope $-1/3$ and passes through $(x_1, y_1) = (2, -3)$. Thus, the point-slope form equation for the line we seek is

$$y - (-3) = -\frac{1}{3}(x - 2)$$
$$y + 3 = -\frac{1}{3}x + \frac{2}{3} \qquad \text{Distributive property}$$
$$y = -\frac{1}{3}x - \frac{7}{3}$$

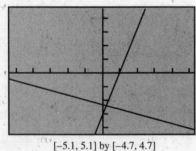

[−5.1, 5.1] by [−4.7, 4.7]

Figure P.8 The graphs of $y = 3x - 3$ and $y = (-1/3)x - 7/3$ in this square viewing window appear to intersect at a right angle.

Figure P.8 shows the graphs of the two equations in a square viewing window and suggests that the graphs are perpendicular.

Exercises for Alternate Example 6

In Exercises 19–24, find an equation for the line passing through the point and perpendicular to the given line. Support your work graphically.

19. $P(-3, -4); -3x - 2y = -3$ **20.** $P(2, 1); -x + 2y = -3$ **21.** $P(-5, -2); 5x + 2y = 6$ **22.** $P(3, 0); x + 4y = 6$

23. $P(0, 4); -3x + 3y = 1$ **24.** $P(5, 7); 5x - 10y = -7$

P.5 Solving Equations Graphically, Numerically, and Algebraically

Alternate Example 1   Solving by Finding *x*-Intercepts

Solve the equation $2x^2 - 7x + 3 = 0$ graphically.

SOLUTION

Solve Graphically

Find the x-intercepts of the graph of $y = 2x^2 - 7x + 3$ (Figure P.9). We use TRACE to see that $(0.5, 0)$ and $(3, 0)$ are x-intercepts of this graph. Thus, the solutions of this equation are $x = 0.5$ and $x = 3$. Answers obtained graphically are really approximations, although in general they are very good approximations.

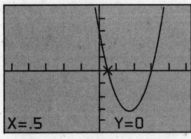

[−5.1, 5.1] by [−4.7, 4.7]

Figure P.9 It appears that $(0.5, 0)$ and $(3, 0)$ are x-intercepts of the graph of $y = 2x^2 - 7x + 3$.

Solve Algebraically

In this case, we can use factoring to find exact values.

$$2x^2 - 7x + 3 = 0$$
$$(2x - 1)(x - 3) = 0 \qquad \text{Factor.}$$

We can conclude that

$$2x - 1 = 0 \qquad \text{or} \qquad x - 3 = 0$$

$$x = \frac{1}{2} \qquad \text{or} \qquad x = 3$$

So, $x = 1/2$ and $x = 3$ are the exact solutions of the original equation.

Exercises for Alternate Example 1

In Exercises 1–6, solve the equation graphically. Confirm by using factoring to solve the equation.

1. $2x^2 - 8 = 0$ **2.** $3x^2 - 12x = 0$ **3.** $x^2 - x - 6 = 0$ **4.** $2x^2 - x - 3 = 0$ **5.** $2x^2 - 7x + 3 = 0$

6. $5x^2 + 19x + 12 = 0$

Alternate Example 4 **Solving Using the Quadratic Formula**

Solve the equation $2x^2 - 5x = 2$.

SOLUTION

First we subtract 2 from each side of the equation to put it in the form $ax^2 + bx + c = 0$: $2x^2 - 5x - 2 = 0$. We can see that $a = 2$, $b = -5$, and $c = -2$.

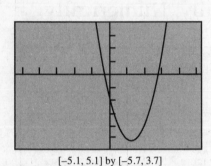

[−5.1, 5.1] by [−5.7, 3.7]

Figure P.10 The graph of $y = 2x^2 - 5x - 2$.

$$x = \frac{-b \pm \sqrt{b^2 - 4ac}}{2a}$$ Quadratic formula

$$x = \frac{-(-5) \pm \sqrt{(-5)^2 - 4(2)(-2)}}{2(2)}$$ $a = 2$, $b = -5$, $c = -2$

$$x = \frac{5 \pm \sqrt{41}}{4}$$ Simplify.

$$x = \frac{5 + \sqrt{41}}{4} \approx 2.85 \quad \text{or} \quad x = \frac{5 - \sqrt{41}}{4} \approx -0.35$$

The graph of $y = 2x^2 - 5x - 2$ in Figure P.10 supports that the x-intercepts are approximately −0.35 and 2.85.

Exercises for Alternate Example 4

In Exercises 7–12, solve the equation by using the quadratic formula.

7. $2x^2 - 5x = 0$ **8.** $3x^2 - x = 2$ **9.** $3x^2 - 7x + 1 = 0$ **10.** $4x^2 - 6x = 3$ **11.** $x^2 - 6x = 5$ **12.** $10x^2 - x = 2$ ■

Alternate Example 5 **Solving Graphically**

Solve the equation $x^3 + 2x - 1 = 0$ graphically.

SOLUTION

Figure P.11a suggests that $x = 0.45339765$ is the solution we seek. Figure P.11b provides numerical support that $x = 0.45339765$ is a close approximation to the solution because, when $x = 0.45339765$, $x^3 + 2x - 1 = -3.96799 \times 10^{-9}$ or −0.00000000396779, which is nearly zero.

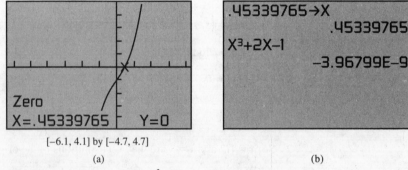

[−6.1, 4.1] by [−4.7, 4.7]

(a)   (b)

Figure P.11 The graph of $y = x^3 + 2x - 1$. (a) shows that (0.4533976, 0) is an approximation to the x-intercept of the graph. (b) supports this conclusion.

Exercises for Alternate Example 5

In Exercises 13–18, solve the equation graphically.

13. $x^3 - x + 2 = 0$ **14.** $2x^3 + x = 0$ **15.** $x^3 + x - 4 = 0$ **16.** $x^3 + 6x - 1 = 0$ **17.** $-x^3 + 2x + 3 = 0$

18. $x^3 + 6x + 2 = 0$ ■

Alternate Example 6   Solving Using Tables

Solve the equation $x^3 + 2x - 1 = 0$ using grapher tables.

SOLUTION

From Figure P.11a, we know that the solution we seek is between $x = 0$ and $x = 1$. Figure P.12a sets the starting point of the table (TblStart = 0) at $x = 0$ and increments the numbers in the table (ΔTbl = 0.1) by 0.1. Figure P.12b shows that the zero of $x^3 + 2x - 1$ is between $x = 0.4$ and $x = 0.5$.

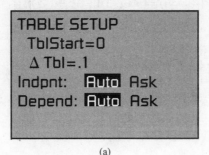

(a)

X	Y1
0	−1
.1	−.799
.2	−.592
.3	−.373
.4	−136
.5	.125
.6	.416

Y1 ◼ X³+2X−1

(b)

Figure P.12 (a) gives the setup that produces the table in (b).

Exercises for Alternate Example 6

In Exercises 19–24, find the real solutions of the equation to two decimal places by using grapher tables.

19. $x^3 - 2x - 2 = 0$ **20.** $-x^3 + 2x^2 - 6 = 0$ **21.** $-x^2 + 4x - 2 = 0$ **22.** $\dfrac{x^2}{2} + 4x - 2 = 0$

23. $x^3 + 4x^2 - 2 = 0$ **24.** $x^3 - 4x^2 - 2 = 0$

Alternate Example 7   Solving by Finding Intersections

Solve the equation $|2x + 3| = 5$.

SOLUTION

Figure P.13 suggests that the V-shaped graph of $y = |2x + 3|$ intersects the graph of the horizontal line $y = 5$ twice. We can use TRACE or the intersection feature of our grapher to see that the two points of intersection have coordinates $(-4, 5)$ and $(1, 5)$. This means that the original equation has two solutions: -4 and 1.

We can use algebra to find the exact solutions. The only two real numbers with absolute value 5 are 5 itself and -5. So, if $|2x + 3| = 5$, then

$$2x + 3 = 5 \quad \text{or} \quad 2x + 3 = -5$$
$$x = 1 \quad \text{or} \quad x = -4$$

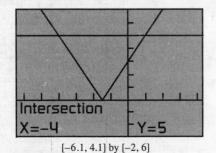

[−6.1, 4.1] by [−2, 6]

Figure P.13 The graphs of $y = |2x + 3|$ and $y = 5$ intersect at $(-4, 5)$ and $(1, 5)$.

Exercises for Alternate Example 7

In Exercises 25–30, solve each equation by finding intersections.

25. $|x - 2| = 3$ **26.** $|2x| = 6$ **27.** $|x + 3| = 0$ **28.** $|2x - 2| = 4$ **29.** $|2x - 1| = 2x$

30. $|2x + 5| = -3x + 5$

P.6 Complex Numbers

Alternate Example 1 **Adding and Subtracting Complex Numbers**

Add or subtract.

(a) $(6 - 5i) + (-3 - 7i)$ 　　　　　　(b) $(2 + 11i) - (-8 - 2i)$

SOLUTION

(a) $(6 - 5i) + (-3 - 7i) = (6 + (-3)) + (-5 + (-7))i = 3 + (-12)i = 3 - 12i$

(b) $(2 + 11i) - (-8 - 2i) = (2 - (-8)) - (11 - (-2))i = 10 + 13i$

Exercises for Alternate Example 1

In Exercises 1–8, write the sum or difference in the standard form $a + bi$.

1. $(-1 - 3i) + (-11 + i)$ 　　2. $(-10 + 2i) - (-1 - i)$ 　　3. $(7 + 7i) + (2 + 2i)$ 　　4. $(18 - 3i) - (1 - 9i)$

5. $(4 + 30i) + (-11 - 29i)$ 　　6. $(-8 + 18i) - (21 + 12i)$ 　　7. $(1 - 7i) + (12 + 32i)$ 　　8. $(7 + 13i) - (17 - 79i)$

Alternate Example 3 **Raising a Complex Number to a Power**

If $z = \dfrac{\sqrt{2}}{2} + \dfrac{\sqrt{2}}{2}i$, find z^2 and z^3.

SOLUTION

$$z^2 = \left(\frac{\sqrt{2}}{2} + \frac{\sqrt{2}}{2}i\right)\left(\frac{\sqrt{2}}{2} + \frac{\sqrt{2}}{2}i\right)$$

$$= \frac{2}{4} + \frac{2}{4}i + \frac{2}{4}i + \frac{2}{4}i^2$$

$$= \frac{2}{4} + \frac{2}{4}i + \frac{2}{4}i + \frac{2}{4}(-1)$$

$$= i$$

$$z^3 = z^2 \cdot z$$

$$= i\left(\frac{\sqrt{2}}{2} + \frac{\sqrt{2}}{2}i\right)$$

$$= \frac{\sqrt{2}}{2}i + \frac{\sqrt{2}}{2}i^2$$

$$= \frac{\sqrt{2}}{2}i + \frac{\sqrt{2}}{2}(-1)$$

$$= -\frac{\sqrt{2}}{2} + \frac{\sqrt{2}}{2}i$$

Exercises for Alternate Example 3

In Exercises 9–14, write the complex number in standard form.

9. $(1 + 2i)^2$ 　　10. $(-3 + 4i)^3$ 　　11. $(6 - 8i)^3$ 　　12. $(-3 + i)^2$ 　　13. $(-2 - 3i)^3$ 　　14. $(-3 - 5i)^2$

Alternative Example 4 **Dividing Complex Numbers**

Write the complex number in standard form.

(a) $\dfrac{3}{2+i}$

(b) $\dfrac{3+3i}{2-i}$

SOLUTION

(a) $\dfrac{3}{2+i} = \dfrac{3}{2+i} \cdot \dfrac{2-i}{2-i}$

$= \dfrac{3(2-i)}{(2+i)(2-i)}$

$= \dfrac{6-2i}{2^2-i^2}$

$= \dfrac{6-2i}{4-(-1)}$

$= \dfrac{6-2i}{4+1}$

$= \dfrac{6-2i}{5}$

$= \dfrac{6}{5} - \dfrac{2i}{5}$

(b) $\dfrac{3+3i}{2-i} = \dfrac{3+3i}{2-i} \cdot \dfrac{2+i}{2+i}$

$= \dfrac{(3+3i)(2+i)}{(2-i)(2+i)}$

$= \dfrac{6+3i+6i+3i^2}{2^2-i^2}$

$= \dfrac{6+9i+3(-1)}{4-(-1)}$

$= \dfrac{6+9i-3}{4+1}$

$= \dfrac{3+9i}{5}$

$= \dfrac{3}{5} + \dfrac{9i}{5}$

Exercises for Alternate Example 4

In Exercises 15–20, write the complex number in standard form.

15. $\dfrac{3}{2i}$

16. $\dfrac{-2+i}{2i}$

17. $\dfrac{3-2i}{1+2i}$

18. $\dfrac{3-i}{3+i}$

19. $\dfrac{4+3i}{3+4i}$

20. $\dfrac{6-8i}{3-4i}$

Alternate Example 5 **Solving a Quadratic Equation**

Solve $2x^2 + 2x + 1 = 0$.

SOLUTION

Solve Algebraically

Using the quadratic formula with $a = 2$, $b = 2$, and $c = 1$, we obtain

$$x = \dfrac{-2 \pm \sqrt{2^2 - 4(2)(1)}}{2(2)}$$

$$= \dfrac{-2 \pm \sqrt{-4}}{4}$$

$$= \dfrac{-2 \pm 2i}{4}$$

$$= -\dfrac{1}{2} \pm \dfrac{1}{2}i$$

So, the solutions are $-\dfrac{1}{2} + \dfrac{1}{2}i$ and $-\dfrac{1}{2} - \dfrac{1}{2}i$, a complex conjugate pair.

Confirm Numerically

Substituting $-\frac{1}{2} + \frac{1}{2}i$ into the original equation, we obtain

$$2\left(-\frac{1}{2} + \frac{1}{2}i\right)^2 + 2\left(-\frac{1}{2} + \frac{1}{2}i\right) + 1 = (-i) + (-1 + i) + 1 = 0$$

By a similar computation we can confirm the second solution.

Exercises for Alternate Example 5

In Exercises 21–28, solve the equation.

21. $x^2 + 5x + 7 = 0$ **22.** $3x^2 - x + 4 = 0$ **23.** $4x^2 - 3x + 5 = 0$ **24.** $4x^2 + 3 = 0$

25. $x^2 - x + 13 = 0$ **26.** $2x^2 - 10x + 13 = 0$ **27.** $x^2 + 3x + 3 = 0$ **28.** $5x^2 + 20 = 0$

P.7 Solving Inequalities Algebraically and Graphically

> **Alternate Example 2** **Solving Another Absolute Value Inequality**
>
> Solve $|2x + 1| > 4$.

SOLUTION

The solution of this absolute value inequality consists of the solutions of both these inequalities.

$$
\begin{array}{llll}
2x + 1 < -4 & \text{or} & 2x + 1 > 4 & \\
2x < -5 & \text{or} & 2x > 3 & \text{Subtract 1.} \\
x < -\dfrac{5}{2} & \text{or} & x > \dfrac{3}{2} & \text{Divide by 2.}
\end{array}
$$

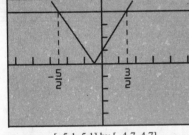

[–5.1, 5.1] by [–4.7, 4.7]

Figure P.14 The graphs of $y = |2x + 1|$ and $y = 4$.

The solution consists of all numbers that are either one of the two intervals $\left(-\infty, -\dfrac{5}{2}\right)$ or $\left(\dfrac{3}{2}, \infty\right)$, which may be written as $\left(-\infty, -\dfrac{5}{2}\right) \cup \left(\dfrac{3}{2}, \infty\right)$. The notation "$\cup$" is read as "union." Figure P.14 shows that points on the graph of $y = |2x + 1|$ are above the points on the graph of $y = 4$ for values of x to the left of $-\dfrac{5}{2}$ and to the right of $\dfrac{3}{2}$.

Exercises for Alternate Example 2

In Exercises 1–6, solve the absolute value inequality.

1. $|x| > 1$ **2.** $|x + 2| \geq 2$ **3.** $|x - 2| \leq 3$ **4.** $|x - 3| > 3$ **5.** $|2x - 5| \leq 1$ **6.** $|3x + 1| \geq 2$

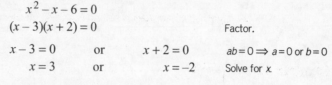

Alternate Example 3 **Solving a Quadratic Inequality**

Solve $x^2 - x - 6 \geq 0$.

SOLUTION

First we solve the corresponding equation.

$$x^2 - x - 6 = 0$$
$$(x - 3)(x + 2) = 0 \qquad\qquad\qquad \text{Factor.}$$
$$x - 3 = 0 \qquad \text{or} \qquad x + 2 = 0 \qquad ab = 0 \Rightarrow a = 0 \text{ or } b = 0$$
$$x = 3 \qquad \text{or} \qquad x = -2 \qquad \text{Solve for } x.$$

The solutions of the corresponding quadratic equation are –2 and 3, and they are solutions of the original inequality because $0 \geq 0$ is true. Figure P.15 shows that the points on the graph of $y = x^2 - x - 6$ are on or above the x-axis for values of x at or to the left of –2 and at or to the right of 3.

[−5.1, 5.1] by [−6.7, 2.7]

Figure P.15 The graph of $y = x^2 - x - 6$ appears to cross the x-axis at $x = -2$ and $x = 3$.

Exercises for Alternate Example 3

In Exercises 7–12, solve the inequality.

7. $x^2 - 9 \geq 0$ **8.** $x^2 - 5x + 4 < 0$ **9.** $x^2 + 3x - 4 \geq 0$ **10.** $x^2 - 6x + 9 > 0$ **11.** $x^2 + 7x + 12 \geq 0$

12. $2x^2 + 7x + 3 \leq 0$

Alternate Example 7 **Solving a Cubic Inequality**

Solve $x^3 - 3x^2 + 1 \geq 0$ graphically.

SOLUTION

We can use the graph of $y = x^3 - 3x^2 + 1$ in Figure P.16 to show that the solutions of the corresponding equation $x^3 - 3x^2 + 1 = 0$ are approximately –0.53, 0.65, and 2.88. The points on the graph of $y = x^3 - 3x^2 + 1$ are above the x-axis for values of x between –0.53 and 0.65, and for values of x to the right of 2.88.

The solution of the inequality is $[-0.53, 0.65] \cup [2.88, \infty)$. We use square brackets because the zeros of $x^3 - 3x^2 + 1 = 0$ are also solutions of the inequality.

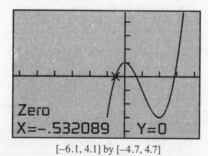

[−6.1, 4.1] by [−4.7, 4.7]

Figure P.16 The graph of $y = x^3 - 3x^2 + 1$ appears to be on or above the x-axis between the negative x-intercept and the smaller of the two positive x-intercepts and to the right of the larger x-intercept.

Exercises for Alternate Example 7

In Exercises 13–18, solve the cubic inequality graphically.

13. $x^3 + x - 3 \geq 0$ **14.** $x^3 - 3x > 0$ **15.** $x^3 + 5x^2 + x > 0$ **16.** $-x^3 + x^2 + x \geq 0$ **17.** $-x^3 + 4x^2 + 2x > 0$

18. $x^3 - 5x^2 + 4 \geq 0$

Chapter 1 Functions and Graphs

1.1 Modeling and Equation Solving

Alternate Example 3 Comparing Pizzas

A pizzeria sells a rectangular pizza 10″ by 12″ for the same price as its large round pizza (12″ diameter). If both pizzas are of the same thickness, which option gives the most pizza for the money?

SOLUTION

We need to compare the *areas* of the pizzas. Fortunately, geometry has provided algebraic models that allow us to compare the areas from the given information. For the rectangular pizza:

$Area = l \times w = 10 \times 12 = 120$ square inches.

For the circular pizza:

$$Area = \pi r^2 = \pi \left(\frac{12}{2}\right)^2 = 36\pi \approx 113.1 \text{ square inches.}$$

The rectangular pizza is larger and therefore gives more for the money.

Exercises for Alternate Example 3

In Exercises 1–6, the two pizzas are of the same thickness. Which option gives the most pizza for the money?

1. square pizza 10 inches on a side
   rectangular pizza 10 inches long and 8 inches wide

2. square pizza 10 inches on a side
   circular pizza with a 6-inch diameter

3. rectangular pizza 12 inches long and 7 inches wide
   rectangular pizza 10 inches long and 8 inches wide

4. circular pizza with an 8-inch diameter
   circular pizza with a 9-inch diameter

5. square pizza 15 inches on a side
   square pizza 14 inches on a side

6. circular pizza with a 10-inch diameter
   rectangular pizza 10 inches long and 8 inches wide

Alternate Example 6 Solving an Equation Algebraically

Find all real numbers x for which $2x^3 = -3x^2 + 2x$.

SOLUTION

We begin by changing the form of the equation to $2x^3 + 3x^2 - 2x = 0$.

We can then solve this equation algebraically by factoring.

$$2x^3 + 3x^2 - 2x = 0$$
$$x(2x^2 + 3x - 2) = 0$$
$$x(x + 2)(2x - 1) = 0$$

$x = 0$ or $x + 2 = 0$ or $2x - 1 = 0$

$x = 0$ or $x = -2$ or $x = \frac{1}{2}$

Exercises for Alternate Example 6

In Exercises 7–12, solve the equation algebraically.

7. $x^3 = 4x^2 - 3x$ **8.** $x^3 + 2x^2 = 8x$ **9.** $x^3 + 16x = 8x^2$ **10.** $2x^3 = 13x^2 - 20x$ **11.** $2x^3 = 5x^2$

12. $6x^3 = 17x^2 - 5x$

Alternate Example 7   Solving an Equation: Comparing Methods

Solve the equation $x^2 = 3x - 1$.

SOLUTION

Solve Algebraically

The given equation is equivalent to $x^2 - 3x + 1 = 0$. This quadratic equation has irrational roots that can be found by the quadratic formula.

$$x = \frac{3 + \sqrt{5}}{2} \approx 2.61803$$

and

$$x = \frac{3 - \sqrt{5}}{2} \approx 0.381966$$

While the decimal answers are certainly accurate enough for all practical purposes, it is important to note that only the expressions found by the quadratic formula give the *exact* real number answers. The tidiness of exact answers is a worthy mathematical goal. Realistically, however, exact answers are often impossible to obtain, even with the most sophisticated mathematical tools.

Solve Graphically

We first find an equivalent equation with 0 on the right-hand side: $x^2 - 3x + 1 = 0$. We then graph the equation $y = x^2 - 3x + 1$, as shown in Figure 1.1.

We then use the grapher to locate the *x*-intercepts of the graph:

$x \approx 0.381966$ and $x \approx 2.61803$.

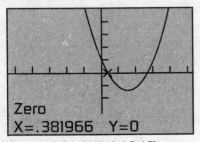

Zero
X=.381966   Y=0

[–5.1, 5.1] by [–4.7, 4.7]

Figure 1.1 The graph of $y = x^2 - 3x + 1$.

Exercises for Alternate Example 7

In Exercises 13–18, solve the equation.

13. $x^2 - 4x + 2 = 0$ **14.** $x^2 - 5x + 2 = 0$ **15.** $x^2 = x + 7$ **16.** $x^2 + 3x = 1$ **17.** $x^2 + 4x = 6$

18. $x^2 = 4x + 7$

Alternate Example 8   Applying the Problem-Solving Process

The engineers at an auto manufacturer pay students $0.10 per mile plus $20 per day to road test their new vehicles.

(a) How much did the auto manufacturer pay Dennis to drive 200 miles in one day?

(b) Erma earned $30 test-driving a new car in one day. How far did she drive?

SOLUTION

Model

A picture of a car or of Dennis or Erma would not be helpful, so we go directly to designing the model. Both Dennis and Erma earned $20 for one day, plus $0.10 per mile. Multiply dollars/miles to get dollars.

So, if p represents the pay for driving x miles in one day, our algebraic model is

$p = 20 + 0.10x$.

Solve Algebraically

(a) To get Dennis's pay we let $x = 200$ and solve for p:

$$p = 20 + 0.10(200)$$
$$= 40$$

(b) To get Erma's mileage we let $p = 30$ and solve for x.

$$30 = 20 + 0.10x$$
$$10 = 0.10x$$
$$x = \frac{10}{0.10}$$
$$x = 100$$

Support Graphically

Figure 1.2a shows that the point (200, 40) is on the graph of $y = 20 + 0.1x$, supporting our answer to (a). Figure 1.2b shows that the point (100, 30) is on the graph of $y = 20 + 0.1x$, supporting out answer to (b). (We could also have supported our answer numerically by simply substituting in for each x and confirming the value of p.)

Interpret

Dennis earned $40 for driving 200 miles in one day. Erma drove 300 miles in one day to earn $30.

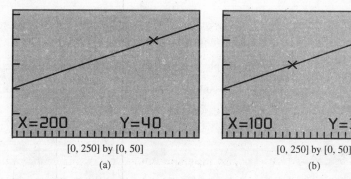

[0, 250] by [0, 50]   [0, 250] by [0, 50]

(a)   (b)

Figure 1.2 Graphical support for the algebraic solution in Alternate Example 8.

Exercises for Alternate Example 8

In Exercises 19–21, the engineers at an auto manufacturer pay students $0.20 per mile plus $50 per day to road test their new vehicles. Answer each question.

19. How much did the auto manufacturer pay Jacob to drive 250 miles in one day?

20. Maria earned $85 test-driving a new car in one day. How far did she drive?

21. Kyle earned $75 test-driving a new car in one day and $90 on a second day. How far did he drive altogether?

In Exercises 22–24, engineers have a tank that contains 50 gallons of water. They let in 10 gallons of water per minute. Answer each question.

22. How many gallons of water will be in the tank after 12 minutes?

23. How many minutes will it take for the tank to contain 85 gallons in all?

24. How many minutes will it take for the tank to contain 700 gallons in all?

1.2 Functions and Their Properties

Alternate Example 3 **Finding the Domain of a Function**

Find the domain of each of these functions:

(a) $f(x) = \sqrt{x + 4}$

(b) $g(x) = \dfrac{\sqrt{x}}{x - 3}$

(c) $A(x) = x^2$, where $A(x)$ is the area of a square with sides of length x.

SOLUTION

Solve Algebraically

(a) The expression under a radical may not be negative. We set $x + 4 \geq 0$ and solve to find $x \geq -4$. The domain of f is the interval $[-4, \infty)$.

(b) The expression under a radical may not be negative; therefore $x \geq 0$. Also, the denominator of a fraction may not be zero; therefore $x \neq 3$. The domain of g is the interval $[0, \infty)$ with the number 3 removed, which we can write as the union of two intervals: $[0, 3) \cup (3, \infty)$.

(c) The algebraic expression has domain all real numbers, but the behavior being modeled restricts s from being negative. The domain of A is the interval $[0, \infty)$.

Support Graphically

We can support our answers in (a) and (b) graphically, as the calculator should not plot points where the function is undefined.

(a) Notice that the graph of $y = \sqrt{x + 4}$ (Figure 1.3a) shows points only for $x \geq -4$, as expected.

(b) The graph of $y = \dfrac{\sqrt{x}}{x - 3}$ (Figure 1.3b) shows points only for $x \geq 0$ as expected, but shows an unexpected line through the x-axis at $x = 3$. This line, a form of grapher failure described on page 78 of the text, should not be there. Ignoring it, we see that 3, as expected, is not in the domain.

(c) The graph of $y = x^2$ (Figure 1.3c) shows the unrestricted domain of the algebraic expression; all real numbers. The calculator has no way of knowing that x is the length of a side of a square.

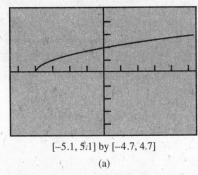

[−5.1, 5.1] by [−4.7, 4.7]

(a)

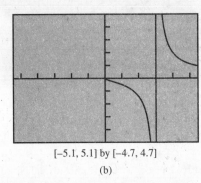

[−5.1, 5.1] by [−4.7, 4.7]

(b)

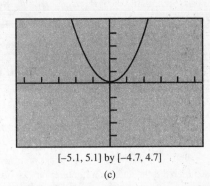

[−5.1, 5.1] by [−4.7, 4.7]

(c)

Figure 1.3 Graphical support of the algebraic solutions in Alternate Example 3. The vertical line in (b) should be ignored because it results from grapher failure. The points in (c) with negative x-coordinates should be ignored because the calculator does not know that x is a length (but we do).

Exercises for Alternate Example 3

In Exercises 1–8, find the domain of each of these functions.

1. $f(x) = 2\sqrt{x - 1}$

2. $g(x) = \dfrac{\sqrt{2x}}{x + 2}$

3. $P(s) = 4s$, where $P(s)$ is the perimeter of a square with sides of length s.

4. $h(x) = \dfrac{(x + 1)\sqrt{x}}{x}$

5. $f(x) = \sqrt{2x - 1}$

6. $V(r) = \dfrac{4}{3}\pi r^3$, $V(r)$ the volume of a sphere with radius r.

7. $g(x) = \dfrac{x - 3}{\sqrt{x}}$

8. $m(x) = \sqrt{x}$

Alternate Example 6 **Analyzing a Function for Increasing-Decreasing Behavior**

For each function, tell the intervals on which it is increasing and the intervals on which it is decreasing.

(a) $f(x) = (x - 2)^2$

(b) $g(x) = \dfrac{x^2}{x^2 - 4}$

SOLUTION

Solve Graphically

(a) We can see from the graph in Figure 1.4 that f is decreasing on $(-\infty, 2]$ and increasing on $[2, \infty)$. (Notice that we include 2 in both intervals. Don't worry that this sets up some contradiction about what happens at 2, because we only talk about functions increasing or decreasing on intervals, and 2 is not an interval.)

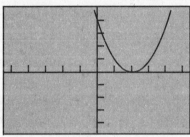

[−5.1, 5.1] by [−4.7, 4.7]

Figure 1.4 The function $f(x) = (x - 2)^2$ decreases on $(-\infty, 2]$ and increases on $[2, \infty)$.

(b) We see from the graph in Figure 1.5 that g is increasing on $(-\infty, -2)$, increasing again on $(-2, 0]$, decreasing on $[0, 2)$, and decreasing again on $(2, \infty)$.

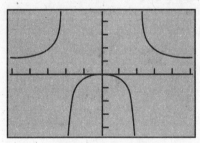

[−5.1, 5.1] by [−4.7, 4.7]

Figure 1.5 The function $g(x) = \dfrac{x^2}{x^2 - 4}$ increases on $(-\infty, -2)$ and $(-2, 0]$; the function decreases on $[0, 2)$ and $(2, \infty)$.

Exercises for Alternate Example 6

In Exercises 9–14, identify the intervals on which the function is increasing and decreasing.

9. $f(x) = 3(x + 2)^2$

10. $f(x) = -(x + 3)^2$

11. $g(x) = \dfrac{x^2}{x^2 - 16}$

12. $g(x) = \dfrac{x^2}{x^2 - 25}$

13. $f(x) = |x - 2| + 1$

14. $f(x) = \dfrac{x^3}{3} - x^2 + 1$

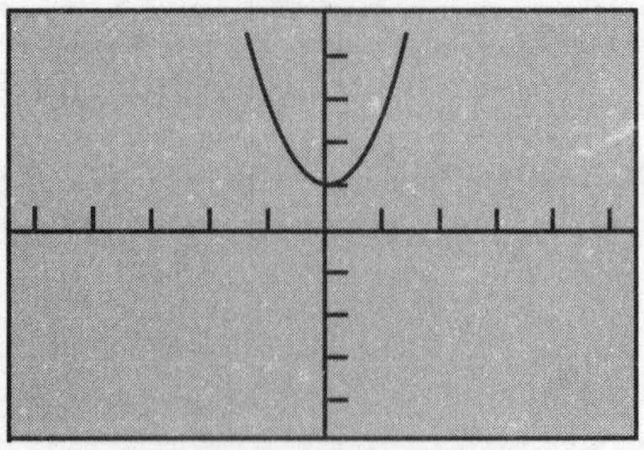

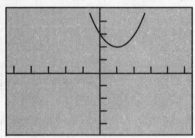

[−5.1, 5.1] by [−4.7, 4.7]

Figure 1.6 This graph appears to be symmetric with respect to the *y*-axis, so we conjecture that *f* is an even function.

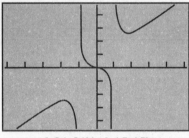

[−5.1, 5.1] by [−4.7, 4.7]

Figure 1.7 This graph does not appear to be symmetric with respect to either the *y*-axis or the origin, so we conjecture that *g* is neither even nor odd.

[−5.1, 5.1] by [−4.7, 4.7]

Figure 1.8 This graph appears to be symmetric with respect to the origin, so we conjecture that *h* is odd.

Alternate Example 9 **Checking Functions for Symmetry**

Tell whether each of the following functions is odd, even, or neither.

(a) $f(x) = 2x^2 + 1$ **(b)** $g(x) = x^2 - 2x + 3$ **(c)** $h(x) = \dfrac{x^3}{x^2 - 1}$

SOLUTION

(a) Solve Graphically

The graphical solution is shown in Figure 1.6.

Confirm Algebraically

We need to verify that $f(-x) = f(x)$ for all x in the domain of f.

$f(-x) = 2(-x)^2 + 1 = 2x^2 + 1 = f(x)$

Since this identity is true for all x, the function f is indeed even.

(b) Solve Graphically

The graphical solution is shown in Figure 1.7.

Confirm Algebraically

We need to verify that $g(-x) \neq g(x)$ and $g(-x) \neq -g(x)$

$$g(-x) = (-x)^2 - (-2x) + 3$$
$$= x^2 + 2x + 3$$
$$g(x) = x^2 - 2x + 3$$
$$-g(x) = -x^2 + 2x - 3$$

So, $g(-x) \neq g(x)$ and $g(-x) \neq -g(x)$.

We conclude that g is neither odd nor even.

(c) Solve Graphically

The graphical solution is shown in Figure 1.8.

Confirm Algebraically

We need to verify that

$h(-x) = -h(x)$

for all x in the domain of h.

$$h(-x) = \frac{(-x)^3}{(-x)^2 - 1}$$
$$= \frac{-x^3}{x^2 - 1}$$
$$= -h(x)$$

Since this identity is true for all x except ± 1 (which is not in the domain of h), the function is odd.

Exercises for Alternate Example 9

In Exercises 15–20, state whether each of the following functions is odd, even, or neither.

15. $f(x) = 3x^2 + 2$ **16.** $f(x) = x^2 - 2x$ **17.** $m(x) = \dfrac{2x^2}{x^2 - 16}$ **18.** $c(x) = \dfrac{x^2}{x^2 + 4}$

19. $f(x) = (x - 5)^2 - 2x$ **20.** $g(x) = \dfrac{x^3}{2x^2 - 9}$

Alternate Example 10 **Identifying the Asymptotes of a Graph**

Identify any horizontal or vertical asymptotes of the graph of $y = \dfrac{x}{x^2 - 4x + 3}$.

SOLUTION

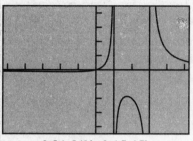

[−5.1, 5.1] by [−4.7, 4.7]

Figure 1.9 The graph of

$y = \dfrac{x}{x^2 - 4x + 3}$ has vertical

asymptotes of $x = 1$ and $x = 3$ and a horizontal asymptote of $y = 0$.

The quotient $\dfrac{x}{x^2 - 4x + 3} = \dfrac{x}{(x-1)(x-3)}$ is undefined at $x = 1$ and $x = 3$, which makes them likely sites for vertical asymptotes. The graph (Figure 1.9) provides support, showing vertical asymptotes of at $x = 1$ and $x = 3$.

For large values of x, the numerator (a large number) is dwarfed by the denominator (a product of two large numbers), suggesting that $\displaystyle\lim_{x \to \infty} \dfrac{x}{(x-1)(x-3)} = 0$. This would indicate a horizontal asymptote of $y = 0$ as $x \to \infty$. Similar logic suggests that

$\displaystyle\lim_{x \to -\infty} \dfrac{x}{(x-1)(x-3)} = -0 = 0$, indicating the same horizontal asymptote as $x \to -\infty$.

Again, the graph provides support for this.

Exercises for Alternate Example 10

In Exercises 21–26, find all horizontal or vertical asymptotes of the function.

21. $y = \dfrac{1}{x^2 - 4}$ **22.** $y = \dfrac{5}{5 - x}$ **23.** $y = \dfrac{3}{x - 5}$ **24.** $y = \dfrac{x}{x^2 - 4x + 4}$ **25.** $y = \dfrac{x}{x^2 + 1}$

26. $y = \dfrac{x}{x^2 - 3x - 4} + 2$

1.3 Twelve Basic Functions

Alternate Example 5 **Analyzing a Function Graphically**

Graph the function $y = (x - 3)^2$. Then answer the following questions:

(a) On what interval is the function increasing? On what interval is it decreasing?

(b) Is the function odd, even, or neither?

(c) Does the function have any extrema?

(d) How does the graph relate to the graph of the basic function $y = x^2$?

SOLUTION

The graph is shown in Figure 1.10.

(a) The function is increasing if its graph is headed upward as it moves from left to right. We see that it is increasing on the interval $[3, \infty)$. The function is decreasing if its graph is headed downward as it moves from left to right. We see that it is decreasing on the interval $(-\infty, 3]$.

(b) The graph is not symmetric with respect to the y-axis, nor is it symmetric with respect to the origin. The function is neither.

(c) Yes, we see that the function has a minimum value of 0 at $x = 3$. (This is easily confirmed by the algebraic fact that $(x - 3)^2 \geq 0$ for all x.)

(d) We see that the graph of $y = (x - 3)^2$ is just the graph of $y = x^2$ moved three units to the right.

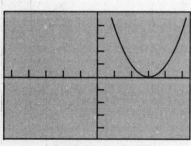

[−5.1, 5.1] by [−4.7, 4.7]

Figure 1.10 The graph of $y = (x - 3)^2$

Exercises for Alternate Example 5

For Exercises 1–6, answer the following questions: **(a)** On what interval is the function increasing? On what interval is it decreasing? **(b)** Is the function odd, even, or neither? **(c)** Does the function have any extrema? **(d)** How does the graph relate to the graph of the basic function $y = x^2$?

1. $y = -x^2$ **2.** $y = (x+1)^2$ **3.** $y = -(x-3)^2$ **4.** $y = x^2 + 1$ **5.** $y = x^2 - 1$ **6.** $y = x^2 - 4x + 4$

Alternate Example 7 Defining a Function Piecewise

Using basic functions from this section, construct a piecewise definition for the function whose graph is shown in Figure 1.11. Is your function continuous?

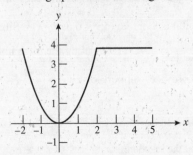

Figure 1.11 A piecewise defined function.

SOLUTION

This appears to be the graph of $y = x^2$ to the left of $x = 2$ and the graph of $y = 4$ to the right of 2. We can therefore define it piecewise as

$$f(x) = \begin{cases} x^2 & \text{if } x \le 2 \\ 4 & \text{if } x \ge 2 \end{cases}$$

The function is continuous.

Exercises for Alternate Example 7

In Exercises 7–12, using basic functions from this section, construct a piecewise definition for the function whose graph is shown.

7.

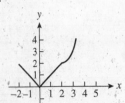

8.

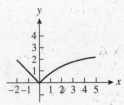

9.

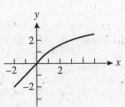

10.

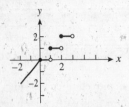

11.

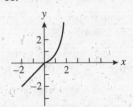

12.

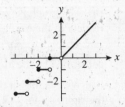

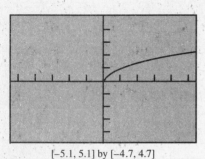

[−5.1, 5.1] by [−4.7, 4.7]

Figure 1.12 The graph of $f(x) = \sqrt{x}$.

Alternate Example 9 Analyzing a Function

Give a complete analysis of the basic function $f(x) = \sqrt{x}$.

SOLUTION

BASIC FUNCTION The Square Root Function

$$f(x) = \sqrt{x} \ \text{(Figure 1.12)}$$

Domain: all nonnegative real numbers
Range: $[0, \infty)$
Continuous
Increasing on $[0, \infty)$
Not symmetric with respect to the *y*-axis or the origin: neither even nor odd
Bounded below
Local minimum at $(0, 0)$
No horizontal asymptotes
No vertical asymptotes
End behavior: $\lim\limits_{x \to \infty} \sqrt{x} = \infty$

Exercises for Alternate Example 9

In Exercises 13–18, give a complete analysis of each basic function.

13. $f(x) = x^3$ **14.** $f(x) = \ln x$ **15.** $f(x) = e^x$ **16.** $f(x) = \text{int}\,(x)$ **17.** $f(x) = \dfrac{1}{x}$ **18.** $f(x) = |x|$

1.4 Building Functions from Functions

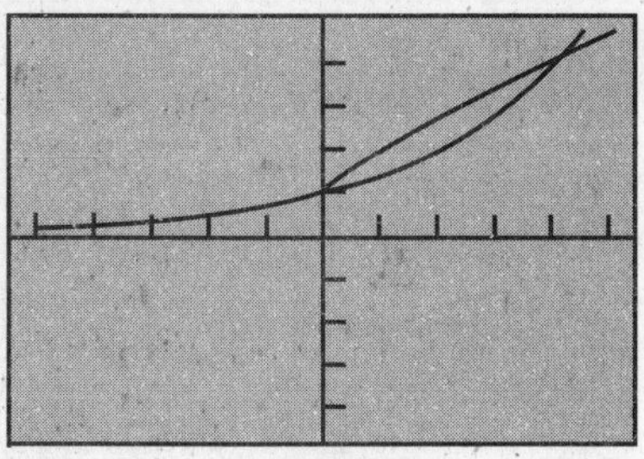

[−5.1, 5.1] by [−4.7, 4.7]

Figure 1.13 The graphs of $y = 2^{\sqrt{x}}$ and $y = \sqrt{2^x}$ are not the same.

Alternate Example 2 Composing Functions

Let $f(x) = 2^x$ and $g(x) = \sqrt{x}$. Find $(f \circ g)(x)$ and $(g \circ f)(x)$ and verify that the functions $f \circ g$ and $g \circ f$ are not the same.

SOLUTION

$$(f \circ g)(x) = f(g(x)) = f(\sqrt{x}) = 2^{\sqrt{x}}$$

$$(g \circ f)(x) = g(f(x)) = g(2^x) = \sqrt{2^x}$$

One verification that these functions are not the same is that they have different domains: $f \circ g$ is defined only for $x \geq 0$, while $g \circ f$ is defined for all real numbers. We could also consider their graphs (Figure 1.13), which agree only at $x = 0$ and $x = 4$.

Exercises for Alternate Example 2

In Exercises 1–6, find $(f \circ g)(x)$ and $(g \circ f)(x)$.

1. $f(x) = 2x^2$ and $g(x) = \dfrac{1}{x}$ **2.** $f(x) = 2x$ and $g(x) = x^2$ **3.** $f(x) = x + 1$ and $g(x) = \sqrt{x}$

4. $f(x) = x^3$ and $g(x) = \sqrt{x} - 1$ **5.** $f(x) = x^2$ and $g(x) = \dfrac{1}{x+1}$ **6.** $f(x) = 2^x$ and $g(x) = x^2$

Alternate Example 4   Decomposing Functions

For each function h, find functions f and g such that $h(x) = f(g(x))$.

(a) $h(x) = 2(x - 5)^2 + 2(x - 5)$

(b) $h(x) = \sqrt[3]{x^2 - 3}$

SOLUTION

(a) We can see that h is quadratic in $x - 5$. Let $f(x) = 2x^2 + 2$ and let $g(x) = x - 5$. Then

$$h(x) = f(g(x)) = f(x - 5) = 2(x - 5)^2 + 2(x - 5).$$

(b) We can see that h is the cube root of the function $x^2 - 3$. Let $f(x) = \sqrt[3]{x}$ and let $g(x) = x^2 - 3$. Then

$$h(x) = f(g(x)) = f(x^2 - 3) = \sqrt[3]{x^2 - 3}.$$

Exercises for Alternate Example 4

In Exercises 7–12, for each function h, find functions f and g such that $h(x) = f(g(x))$. (Note: There may be more than one possible decomposition.)

7. $h(x) = 2(x^2 + 5)^2$

8. $h(x) = \dfrac{1}{x^2 + 1}$

9. $h(x) = \left(\dfrac{1}{x^2 + 1}\right)^3$

10. $h(x) = -2(x^2 - 1)^3 + x^2 - 1$

11. $h(x) = \dfrac{1}{x^2 - 9} + 3(x^2 - 9)^3$

12. $h(x) = \sqrt[3]{x^2} + x^2$

Alternate Example 7   Using Implicitly Defined Functions

Describe the graph of the relation $4x^2 + 4xy + y^2 = 4$.

SOLUTION

This looks like a difficult task at first, but notice that the expression on the left of the equal sign is a factorable trinomial. This enables us to split the relation into two implicitly defined functions as follows.

$$4x^2 + 4xy + y^2 = 4$$
$$(2x + y)^2 = 4 \qquad \text{Factor.}$$
$$2x + y = \pm 2 \qquad \text{Extract square roots.}$$
$$2x + y = 2 \text{ or } 2x + y = -2$$
$$y = -2x + 2 \text{ or } y = -2x - 2 \qquad \text{Solve for } y.$$

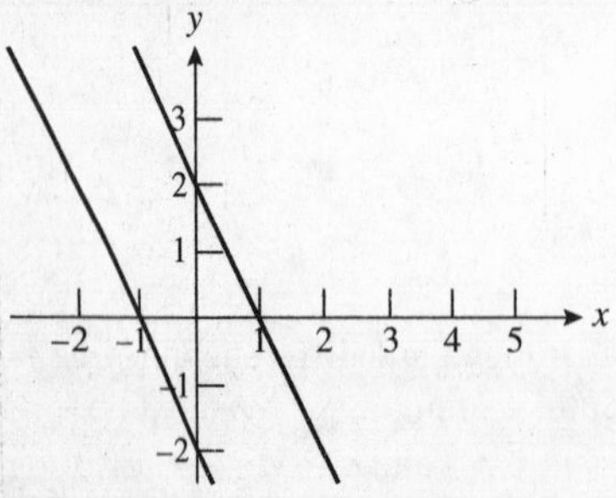

Figure 1.14 The graph of the relation $4x^2 + 4xy + y^2 = 4$

The graph consists of two parallel lines (Figure 1.14), each the graph of one of the implicitly defined functions.

Exercises for Alternate Example 7

In Exercises 13–18, describe the graph of the relation.

13. $(x - y)(x - y) = 9$

14. $x^2 - y^2 = 0$

15. $x^2 - 3xy - 4y^2 = 0$

16. $x^2 - 5xy + 6y^2 = 0$

17. $9x^2 - 12xy + 4y^2 = 16$

18. $3x^2 + 2xy - y^2 = 0$

1.5 Parametric Relations and Inverses

Alternate Example 1 **Defining a Function Parametrically**

Consider the set of all ordered pairs (x, y) defined by the equations

$$x = t - 1$$
$$y = t^2 - 1$$

where t is any real number.

(a) Find the points determined by $t = -3, -2, -1, 0, 1, 2,$ and 3.

(b) Find an algebraic relationship between x and y. Is y a function of x?

(c) Graph the relation in the (x, y) plane.

SOLUTION

(a) Substitute each value of t into the formulas for x and y to find the point that it determines parametrically:

t	$x = t - 1$	$y = t^2 - 1$	(x, y)
-3	-4	8	(-4, 8)
-2	-3	3	(-3, 3)
-1	-2	0	(-2, 0)
0	-1	-1	(-1, -1)
1	0	0	(0, 0)
2	1	3	(1, 3)
3	2	8	(2, 8)

(b) We can find the relationship between x and y algebraically by the method of substitution. First we solve for t in terms of x to obtain $x + 1$.

$$y = t^2 - 1 \qquad \text{Given}$$
$$y = (x + 1)^2 - 1 \qquad t = x + 1$$
$$= x^2 + 2x + 1 - 1 \qquad \text{Expand.}$$
$$= x^2 + 2x \qquad \text{Simplify.}$$

This is consistent with the ordered pairs we had found in the table. As t varies over all the real numbers, we will get all the ordered pairs in the relation $y = x^2 + 2x$, which does indeed define y as a function of x.

(c) Since the parametrically defined relation consists of all ordered pairs in the relation $y = x^2 + 2x$, we can get the graph by simply graphing the parabola $y = x^2 + 2x$. See Figure 1.15.

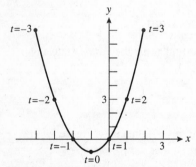

Figure 1.15 (Alternate Example 1)

Exercises for Alternate Example 1

In Exercises 1–6, consider the set of all ordered pairs (x, y) defined by the equations. **(a)** Find the points determined by $t = -3, -2, -1, 0, 1, 2,$ and 3. **(b)** Find an algebraic relationship between x and y. Is y a function of x? **(c)** Graph the relation in the (x, y) plane.

1. $x = 2t$ and $y = t - 1$ **2.** $x = t + 2$ and $y = 2t - 1$ **3.** $x = t - 1$ and $y = t^2$ **4.** $x = 2t$ and $y = t^2 - 1$

5. $x = 2t$ and $y = t^2 + 2$ **6.** $x = t + 1$ and $y = t^2 - 3$

Alternate Example 4   **Finding an Inverse Function Algebraically**

Find an equation for $f^{-1}(x)$ if $f(x) = \dfrac{2x}{2x-1}$.

SOLUTION

The graph of f in Figure 1.16 suggests that f is one-to-one. The original function satisfies the equation $y = \dfrac{2x}{2x-1}$. If f is truly one-to-one, the inverse function f^{-1} will satisfy the equation $x = \dfrac{2y}{2y-1}$. (Note that we just switch the x and y.)

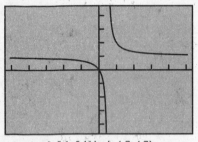

[-5.1, 5.1] by [-4.7, 4.7]

Figure 1.16   The graph of $f(x) = \dfrac{2x}{2x-1}$.

$$x = \frac{2y}{2y-1}$$
$$x(2y-1) = 2y \qquad \text{Multiply by } 2y-1.$$
$$2xy - x = 2y \qquad \text{Distributive property}$$
$$2xy - 2y = x \qquad \text{Isolate the } y \text{ terms.}$$
$$y(2x-2) = x \qquad \text{Factor out } y.$$
$$y = \frac{x}{2x-2} \qquad \text{Divide by } 2x-2.$$

Therefore $f^{-1}(x) = \dfrac{x}{2x-2}$.

Exercises for Alternate Example 4

In Exercises 7–12, find an equation for $f^{-1}(x)$.

7. $f(x) = \dfrac{x}{x+1}$ 8. $f(x) = \dfrac{x}{x-1}$ 9. $f(x) = \dfrac{2x}{3x-4}$ 10. $f(x) = \dfrac{x+1}{x-1}$ 11. $f(x) = \dfrac{2x+3}{x-3}$

12. $f(x) = \sqrt[3]{x-2}$

Alternate Example 7   **Finding an Inverse Function**

Show that $f(x) = \sqrt{2x-3}$ has an inverse function and find a rule for $f^{-1}(x)$. State any restrictions on the domains of f and f^{-1}.

SOLUTION

Solve Algebraically

The graph of f passes the horizontal line test, so f has an inverse function (Figure 1.17). Note that f has domain $[1.5, \infty)$ and range $[0, \infty)$.

To find f^{-1} we write

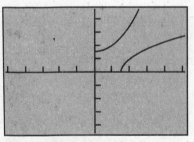

[-5.1, 5.1] by [-4.7, 4.7]

Figure 1.17   The graph of $f(x) = \sqrt{2x-3}$ and its inverse, a restricted $y = \dfrac{1}{2}x^2 + \dfrac{3}{2}$.

$$y = \sqrt{2x-3} \qquad \text{where } x \geq 1.5, y \geq 0$$
$$x = \sqrt{2y-3} \qquad \text{where } y \geq 1.5, x \geq 0 \qquad \text{Interchange } x \text{ and } y.$$
$$x^2 = 2y - 3 \qquad \text{Square.}$$
$$y = \frac{1}{2}x^2 + \frac{3}{2} \qquad \text{Solve for } y.$$

Thus, $f^{-1}(x) = \dfrac{1}{2}x^2 + \dfrac{3}{2}$, with an inherited domain restriction of $x \geq 0$. Figure 1.17 shows the two functions. Note the domain restriction of $x \geq 0$ imposed on the parabola $y = \dfrac{1}{2}x^2 + \dfrac{3}{2}$.

Support Graphically

Use a grapher in parametric mode and compare the graphs of the two sets of parametric equations with Figure 1.17:

$$x = t \qquad \text{and} \qquad x = \sqrt{2x - 3}$$
$$y = \sqrt{2x - 3} \qquad \qquad y = t$$

Exercises for Alternate Example 7

In Exercises 13–18, find an equation for $f^{-1}(x)$. State any restrictions on the domains of f and f^{-1}.

13. $f(x) = 2x + 3$

14. $f(x) = \dfrac{1}{2x + 5}$

15. $f(x) = \dfrac{1}{-3x + 2}$

16. $f(x) = \sqrt{5x + 1}$

17. $f(x) = \sqrt{4x - 8}$

18. $f(x) = 2\sqrt{x - 1}$

1.6 Graphical Transformations

> **Alternate Example 3** **Finding Equations for Reflections**

Find an equation for the reflection of $f(x) = \dfrac{3x + 2}{x^2 + 1}$ across each axis.

SOLUTION

Solve Algebraically

Across the x-axis: $y = -f(x) = -\dfrac{3x + 2}{x^2 + 1} = \dfrac{-3x - 2}{x^2 + 1}$

Across the y-axis: $y = f(-x) = \dfrac{3(-x) + 2}{(-x)^2 + 1} = \dfrac{-3x + 2}{x^2 + 1}$

Support Graphically

The graphs in Figure 1.18 support our algebraic work.

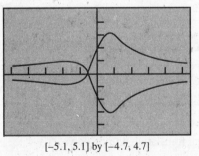 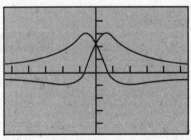

[−5.1, 5.1] by [−4.7, 4.7]   [−5.1, 5.1] by [−4.7, 4.7]
(a)   (b)

Figure 1.18 Reflections of $f(x) = \dfrac{3x + 2}{x^2 + 1}$ across (a) the x-axis and (b) the y-axis.

Exercises for Alternate Example 3

In Exercises 1–6, find an equation for the reflection of each function across each axis.

1. $f(x) = \dfrac{3x + 2}{3x^2 - 5}$

2. $f(x) = \dfrac{1}{2x + 5}$

3. $f(x) = \dfrac{1}{-3x + 2}$

4. $f(x) = \sqrt{x - 2} + 1$

5. $f(x) = \dfrac{x}{x^2 + 7}$

6. $f(x) = \dfrac{3x^2}{4x^2 - 1}$

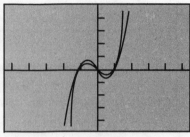

[−5.1, 5.1] by [−4.7, 4.7]

(a)

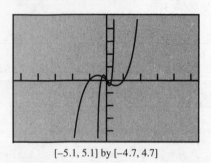

[−5.1, 5.1] by [−4.7, 4.7]

(b)

Figure 1.19 The graph of $y_1 = x^3 - x$ shown with (a) a vertical stretch and (b) a horizontal shrink.

Alternate Example 5 **Finding Equations for Stretches and Shrinks**

Let C_1 be the curve defined by $y_1 = f(x) = x^3 - x$. Find equations for the following non-rigid transformations of C_1:

(a) C_2 is a vertical stretch of C_1 by a factor of 2.

(b) C_3 is a horizontal shrink of C_1 by a factor of $1/4$.

SOLUTION

Solve Algebraically

(a) Denote the equation for C_2 by y_2. Then

$$y_2 = 2 \cdot y_1$$
$$= 2(x^3 - x)$$
$$= 2x^3 - 2x$$

(b) Denote the equation for C_3 by y_3. Then

$$y_3 = f\left(\frac{x}{1/4}\right)$$
$$= f(4x)$$
$$= (4x)^3 - 4x$$
$$= 64x^3 - 4x$$

Support Graphically

The graphs in Figure 1.19 support our algebraic work.

Exercises for Alternate Example 5

In Exercises 7–12, find equations for the following non-rigid transformations of the given function.

7. $f(x) = x^2 + 2x$; vertical stretch by a factor of 3

8. $f(x) = 2x^2 - 3x$; horizontal shrink by a factor of 0.5

9. $f(x) = 2x^3 + 4x^2$; vertical shrink by a factor of 0.5

10. $f(x) = -x^3 + 5x$; horizontal stretch by a factor of 2

11. $f(x) = -2x^3 + x^2$; vertical stretch by a factor of 3

12. $f(x) = x^2 - x$; horizontal stretch by a factor of 2

Alternate Example 6 **Combining Transformations in Order**

(a) The graph of $y = x^2$ undergoes the following transformations, in order. Find the equation of the graph that results.

- a horizontal shift 3 units to the left
- a vertical stretch by a factor of 2
- a vertical translation 4 units down.

(b) Apply the transformations in (a) in the opposite order and find the equation of the graph that results.

SOLUTION

(a) Applying the transformations in order, we have

$$x^2 \Rightarrow (x+3)^2 \Rightarrow 2(x+3)^2 \Rightarrow 2(x+3)^2 - 4$$

Expanding the final expression, we get the function $y = 2x^2 + 12x - 14$.

(b) Applying the transformations in the opposite order, we have

$$x^2 \Rightarrow x^2 - 4 \Rightarrow 2(x^2 - 4) \Rightarrow 2((x + 3)^2 - 4)$$

Expanding the final expression, we get the function $y = 2x^2 + 12x + 10$.

The second graph is 24 units higher than the first because the vertical stretch lengthens the vertical translation when the translation occurs first. Order often matters when stretches, shrinks, or reflections are involved.

Exercises for Alternate Example 6

In Exercises 13–18, the graph of $y = x^2$ undergoes the following transformations, in order. **(a)** Find the equation of the graph that results. **(b)** Apply the transformations in **(a)** in the opposite order and find the equation of the graph that results.

13. a horizontal shift 2 units to the right
a vertical shrink by a factor of 1/2
a vertical translation 2 units up.

14. a horizontal shift 1 unit to the right
a vertical stretch by a factor of 3
a vertical translation 1 unit down.

15. no horizontal shift
a horizontal stretch by a factor of 2
a vertical translation 3 units up.

16. a horizontal shift 3 units to the left
no horizontal or vertical stretch or shrink
a vertical translation 1 unit down.

17. a horizontal shift 4 units to the right
a vertical stretch by a factor of 3
a vertical translation 4 units up.

18. a horizontal shift 2 units to the left
a horizontal shrink by a factor of 1/4
a vertical translation 5 units down.

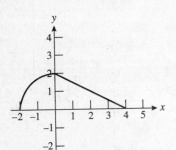

Figure 1.20 The graph of the function $y = f(x)$ in Alternate Example 7.

Alternate Example 7 **Transforming a Graph Geometrically**

The graph of $y = f(x)$ is shown in Figure 1.20. Determine the graph of the composite function $y = 2f(x - 1) - 2$ by showing the effect of a sequence of transformations on the graph of $y = f(x)$.

SOLUTION

The graph of $y = 2f(x - 1) + 2$ can be obtained from the graph of $y = f(x)$ by the following sequence of transformations:

(a) a vertical stretch by a factor of 2 to get $y = 2f(x)$ (Figure 1.21a)

(b) a horizontal translation 1 unit to the right to get $y = 2f(x - 1)$ (Figure 1.21b)

(c) a vertical translation 2 units down to get $y = 2f(x - 1) - 2$ (Figure 1.21c)

(The order of the first two transformations can be reversed without changing the final graph.)

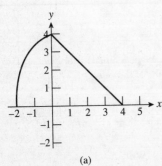

(a)

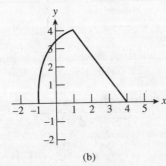

(b)

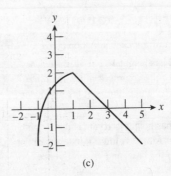

(c)

Figure 1.21 The graph of the function $y = f(x)$ in Alternate Example 7.

Exercises for Alternate Example 7

In Exercises 19–24, the graph of $y = f(x)$ is shown. Sketch the graph of the composite function.

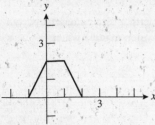

19. $y = 2f(x) - 1$ **20.** $y = 0.5f(x) + 1$ **21.** $y = 2f(x - 2)$ **22.** $y = 0.5f(x + 1)$ **23.** $y = 2f(x - 1) - 1$

24. $y = 1.5f(x - 1) + 1$

1.7 Modeling with Functions

Alternate Example 1 Finding Functions from Formulas

A right circular cylinder has height 8 inches. Write the volume V of the cylinder as a function of its base

(a) radius   **(b)** diameter   **(c)** circumference

SOLUTION

(a) The familiar formula from geometry gives V as a function of r with $h = 8$:

$$V = 8\pi r^2$$

(b) This formula is not so familiar. However we know that $r = \dfrac{d}{2}$, so we can substitute that expression for r in the volume formula:

$$V = 8\pi r^2 = 8\pi \left(\frac{d}{2} \right)^2 = 2\pi d^2.$$

(c) Since $C = 2\pi r$, we can solve for r to get $r = \dfrac{C}{2\pi}$. Then we substitute to get V:

$$V = 8\pi r^2 = 8\pi \left(\frac{C}{2\pi} \right)^2 = \frac{2C^2}{\pi}$$

Exercises for Alternate Example 1

In Exercises 1–6, write a function for each situation using known formulas.

1. A right rectangular prism has height 10 cm. Its base is a square. Write functions for the volume V in terms of the length s of a side of the base and in terms of the perimeter P of the base.

2. A right circular cylinder has height is 10 cm and base with radius r. Write a functions for the volume V in terms of the radius r, the diameter d of the base, and circumference C of the base.

3. A circle is inscribed in a square. Write a function for the area of the circle in terms of the side of the square.

4. A cube is contained in a sphere. Write a function for the surface area of the sphere in terms of the side s of the cube.

5. A rectangle has length 4 and width w. Write functions for the area A in terms of length and width and in terms of the perimeter.

6. A right circular cone has height 9. Write functions for the volume V in terms of the radius r, the diameter d of the base, and circumference C of the base.

Alternate Example 3 Protecting an Antenna

A small satellite dish is packaged with a cardboard cylinder for protection. The parabolic dish is 16 in. in diameter and 4 in. deep, and the diameter of the cardboard cylinder is 8 in. How tall must the cylinder be to fit in the middle of the dish and be flush with the top of the dish? (See Figure 1.22.)

SOLUTION

The diagram in Figure 1.22a showing the cross section of this 3-dimensional problem is also a 2-dimensional graph of a quadratic function. We can transform our basic function $y = x^2$ with a vertical shrink so that it goes through the points $(-8, 4)$ and $(8, 4)$, thereby producing a graph of the parabola in the coordinate plane (Figure 1.22b).

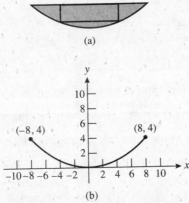

(a)

(b)

Figure 1.22 (a) A parabolic satellite dish with a protective cardboard cylinder in the middle for packaging. (b) The parabola in the coordinate plane.

$$y = kx^2 \qquad \text{Vertical shrink}$$
$$4 = k(\pm 8)^2 \qquad \text{Substitute } x = \pm 8, y = 4.$$
$$k = \frac{4}{64} = \frac{1}{16} \qquad \text{Solve for } k.$$

Thus, $y = \frac{1}{16}x^2$.

To find the height of the cardboard cylinder, we first find the y-coordinate of the parabola 4 inches from the center, that is, when $x = 4$:

$$y = \frac{1}{16}(4)^2 = 1$$

From that point to the top of the dish is $4 - 1 = 3$ in.

Exercises for Alternate Example 3

In Exercises 7–12, a small satellite dish is packaged with a cardboard cylinder for protection. How tall must the cylinder be to fit in the middle of the dish and be flush with the top of the dish?

7. The parabolic dish is 32 in. in diameter and 8 in. deep, and the diameter of the cardboard cylinder is 16 in.

8. The parabolic dish is 36 in. in diameter and 9 in. deep, and the diameter of the cardboard cylinder is 18 in.

9. The parabolic dish is 30 in. in diameter and 7.5 in. deep, and the diameter of the cardboard cylinder is 15 in.

10. The parabolic dish is 40 in. in diameter and 10 in. deep, and the diameter of the cardboard cylinder is 20 in.

11. The parabolic dish is 20 in. in diameter and 5 in. deep, and the diameter of the cardboard cylinder is 10 in.

12. The parabolic dish is 50 in. in diameter and 12.5 in. deep, and the diameter of the cardboard cylinder is 25 in.

Alternate Example 4 **Finding the Model and Solving**

Grain is leaking through a hole in a storage bin at a constant rate of 5 cubic inches per minute. The grain forms a cone-shaped pile on the ground below. As it grows, the height of the cone always remains equal to its radius. If the cone is one foot tall now, how tall will it be in two hours?

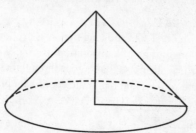

Figure 1.23 A cone with equal height and radius.

SOLUTION

Reading the problem carefully, we realize that the formula for the volume of the cone is needed. (Figure 1.23). From memory or by looking it up, we get the formula $V = \frac{1}{3}\pi r^2 h$. A careful reading also reveals that the height and the radius are always equal, so we can get volume directly as a function of height: $V = \frac{1}{3}\pi h^3$.

When $h = 12$ in., the volume is $V = \frac{1}{3}\pi(12)^3 = 576\pi$ in.^{3}.

Two hours later, the volume will have grown by $120 \text{ min} \cdot \frac{5 \text{ in.}^3}{\text{min}} = 600$ in.^{3}. The total volume of the pile at that point will be $(576\pi + 600)$ in.^{3}. Finally, we use the volume formula once again to solve for h:

$$V = \frac{1}{3}\pi h^3$$
$$V = 576\pi + 600$$
$$h^3 = \frac{3(576\pi + 600)}{\pi}$$
$$h = \sqrt[3]{\frac{3(576\pi + 600)}{\pi}}$$
$$h \approx 12.43 \text{ inches}$$

Exercises for Alternate Example 4

In Exercises 13–18, grain is leaking through a hole in a storage bin at the given constant rate. The grain forms a cone-shaped pile on the ground below. As it grows, the height of the cone always remains equal to its radius. Given the height of the cone at the present time, how tall will it be after the given elapsed time?

13. rate of leakage: 10 cubic inches per minute height at the present time: 15 inches elapsed time: 45 minutes

14. rate of leakage: 12 cubic inches per minute height at the present time: 12 inches elapsed time: 12 minutes

15. rate of leakage: 20 cubic inches per minute height at the present time: 10 inches elapsed time: 1 hour

16. rate of leakage: 6 cubic inches per minute height at the present time: 4 inches elapsed time: 2 hours

17. rate of leakage: 15 cubic inches per minute height at the present time: 2 feet elapsed time: 45 minutes

18. rate of leakage: 10 cubic inches per minute height at the present time: 2 feet elapsed time: 2 hours

Chapter 2

Polynomial, Power, and Rational Functions

2.1 Linear and Quadratic Functions and Modeling

Alternate Example 2 Finding an Equation of a Linear Function

Write an equation for the linear function f such that $f(3) = 3$ and $f(5) = 8$.

SOLUTION

Solve Algebraically

We seek a line through the points $(3, 3)$ and $(5, 8)$. The slope is

$$m = \frac{y_2 - y_1}{x_2 - x_1} = \frac{8 - 3}{5 - 3} = \frac{5}{2}$$

Using this slope and the coordinates of $(3, 3)$ with the point-slope formula, we have

$$y - y_1 = m(x - x_1)$$
$$y - 3 = \frac{5}{2}(x - 3)$$
$$y - 3 = \frac{5}{2}x - \frac{15}{2}$$
$$y = \frac{5}{2}x - \frac{9}{2}$$

Converting to function notation gives us the desired form:

$$f(x) = \frac{5}{2}x - \frac{9}{2}$$

Support Graphically

We can graph $y = \frac{5}{2}x - \frac{9}{2}$ and see that it includes the points $(3, 3)$ and $(5, 8)$. (See Figure 2.1.)

Confirm Numerically

Using $f(x) = \frac{5}{2}x - \frac{9}{2}$, we prove that $f(3) = 3$ and $f(5) = 8$:

$$f(3) = \frac{5}{2}(3) - \frac{9}{2} = \frac{15}{2} - \frac{9}{2} = 3 \text{ and } f(5) = \frac{5}{2}(5) - \frac{9}{2} = \frac{25}{2} - \frac{9}{2} = 8.$$

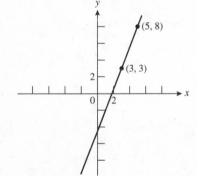

Figure 2.1 The graph of $y = \frac{5}{2}x - \frac{9}{2}$ passes through $(3, 3)$ and $(5, 8)$.

Exercises for Alternate Example 2

In Exercises 1–6, write an equation for the linear function f that satisfies the given conditions.

1. $f(2) = 3$ and $f(6) = 3$

2. $f(0) = 2$ and $f(3) = 4$

3. $f(-1) = 0$ and $f(2) = 4$

4. $f(-2) = -3$ and $f(2) = -4$

5. $f(-3) = 6$ and $f(2) = -1$

6. $f(1) = -2$ and $f(3) = 9$

Alternate Example 5 **Finding the Vertex and Axis of a Quadratic Function**

Use the vertex form of a quadratic function to find the vertex and axis of the graph of $f(x) = -2x + x^2 - 3$. Rewrite the equation in vertex form.

SOLUTION

Solve Algebraically

The standard polynomial form of f is

$$f(x) = x^2 - 2x - 3.$$

So, $a = 1$, $b = -2$, and $c = -3$, and the coordinates of the vertex are

$$h = -\frac{b}{2a} = -\frac{-2}{2(1)} = 1 \quad \text{and} \quad k = f(h) = f(1) = 1^2 - 2(1) - 3 = -4.$$

The equation of the axis is $x = 1$, the vertex is $(1, -4)$, and the vertex form of f is

$$f(x) = (x - 1)^2 + (-4).$$

Exercises for Alternate Example 5

In Exercises 7–12, find the vertex and axis of symmetry of the graph of each function. Rewrite the equation of the function in vertex form.

7. $f(x) = x^2 + 2x$ **8.** $f(x) = 2 - 6x + x^2$ **9.** $f(x) = x^2 + 6x + 5$ **10.** $f(x) = x^2 + 10x + 25$

11. $f(x) = x^2 - 3$ **12.** $f(x) = -x^2 - 12x - 43$

Alternate Example 6 **Using Algebra to Describe the Graph of a Quadratic Function**

Use completing the square to describe the graph of $f(x) = 2x^2 + 4x - 1$. Support your answer graphically.

SOLUTION

Solve Algebraically

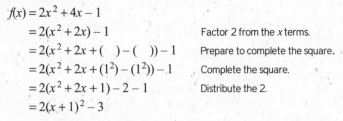

$f(x) = 2x^2 + 4x - 1$	
$= 2(x^2 + 2x) - 1$	Factor 2 from the x terms.
$= 2(x^2 + 2x + (\quad) - (\quad)) - 1$	Prepare to complete the square.
$= 2(x^2 + 2x + (1^2) - (1^2)) - 1$	Complete the square.
$= 2(x^2 + 2x + 1) - 2 - 1$	Distribute the 2.
$= 2(x + 1)^2 - 3$	

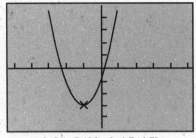

[−5.1, 5.1] by [−4.7, 4.7]

Figure 2.2 The graphs of $f(x) = 2x^2 + 4x - 1$ and $y = 2(x + 1)^2 - 3$ appear to be identical. The vertex $(-1, -3)$ is highlighted.

The graph of f is an upward-opening parabola with vertex $(-1, -3)$, axis of symmetry $x = -1$ and intersects the x-axis at about -2.225 and 0.225. The exact values of x are

$$-1 - \frac{\sqrt{6}}{2} \quad \text{and} \quad -1 + \frac{\sqrt{6}}{2}.$$

Support Graphically

The graph in Figure 2.2 supports these results.

Exercises for Alternate Example 6

In Exercises 13–18, use completing the square to describe the graph of each function. Support your answer graphically.

13. $f(x) = x^2 - 2x - 6$ **14.** $f(x) = x^2 + 6x - 2$ **15.** $f(x) = -x^2 - 2x + 3$ **16.** $f(x) = x^2 - 8x + 12$

17. $f(x) = 2x^2 - 8x + 5$ **18.** $f(x) = 3x^2 + 36x + 96$

2.2 Power Functions with Modeling

Alternate Example 2 Analyzing Power Functions

State the power and constant of variation for the function, graph it, and analyze it.

(a) $f(x) = \sqrt[4]{x}$

(b) $g(x) = \dfrac{1}{x^3}$

SOLUTION

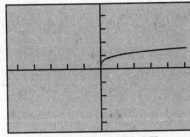

[−5.1, 5.1] by [−4.7, 4.7]

(a)

(a) Because $f(x) = \sqrt[4]{x} = x^{1/4} = 1 \cdot x^{1/4}$, its power is 1/4, and its constant of variation is 1. The graph of f is shown in Figure 2.3a.

Domain: all nonnegative real numbers
Range: $[0, \infty)$
Continuous
Increasing on $[0, \infty)$
Not symmetric with respect to the y-axis or the origin: neither even nor odd
Bounded below
Local minimum at $(0, 0)$
No horizontal asymptotes
No vertical asymptotes
End behavior: $\displaystyle\lim_{x \to \infty} \sqrt[4]{x} = \infty$

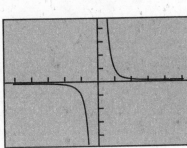

[−5.1, 5.1] by [−4.7, 4.7]

(b)

Figure 2.3 The graphs of
(a) $f(x) = \sqrt[4]{x} = x^{1/4}$ and
(b) $g(x) = \dfrac{1}{x^3} = x^{-3}$.

(b) Because $g(x) = \dfrac{1}{x^3} = x^{-3} = 1 \cdot x^{-3}$, its power is −3, and its constant of variation is 1.

The graph of g is shown in Figure 2.3b.

Domain: all real numbers except 0
Range: $(-\infty, 0) \cup (0, \infty)$
Continuous except at $x = 0$
Decreasing on $(-\infty, 0) \cup (0, \infty)$
Symmetric with respect to origin
Not bounded below or above
No local extrema
Horizontal asymptote: $y = 0$
Vertical asymptote: $x = 0$
End behavior: $\displaystyle\lim_{x \to -\infty} x^{-3} = 0$ and $\displaystyle\lim_{x \to \infty} x^{-3} = 0$

Exercises for Alternate Example 2

In Exercises 1–6, state the power and constant of variation for the function, graph it, and analyze it.

1. $f(x) = 2\sqrt{x}$ **2.** $g(x) = -\dfrac{1}{2\sqrt[3]{x}}$ **3.** $g(x) = \dfrac{1}{x^4}$ **4.** $f(x) = \sqrt[5]{x}$ **5.** $f(x) = \sqrt[6]{x}$

6. $g(x) = \dfrac{1}{x^5}$

Alternate Example 3 Graphing Monomial Functions

Describe how to obtain the graph of the given function from the graph of $g(x) = x^n$ with the same power n. Sketch the graph by hand and support your answer with a grapher.

(a) $f(x) = 3x^3$ **(b)** $f(x) = -0.5x^4$

SOLUTION

(a) We obtain the graph of $f(x) = 3x^3$ by vertically stretching the graph of $g(x) = x^3$ by a factor of 3. Both are odd functions. See Figure 2.4a.

(b) We obtain the graph of $f(x) = -0.5x^4$ by vertically shrinking the graph of $g(x) = x^4$ by a factor of 0.5 and then reflecting it across the x-axis. Both are even functions. See Figure 2.4b.

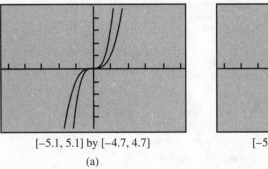
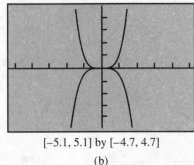

[−5.1, 5.1] by [−4.7, 4.7]   [−5.1, 5.1] by [−4.7, 4.7]
(a)   (b)

Figure 2.4 The graphs of (a) $f(x) = 3x^3$ with basic monomial $g(x) = x^3$, and (b) $f(x) = -0.5x^4$ with basic monomial $g(x) = x^4$.

Exercises for Alternate Example 3

In Exercises 7–12, describe how to obtain the graph of the given function from the graph of $g(x) = x^n$ with the same power n.

7. $f(x) = 2x^2$ **8.** $f(x) = -2x^4$ **9.** $f(x) = -x^3$ **10.** $f(x) = 2x^4$ **11.** $f(x) = 0.5x^6$ **12.** $f(x) = -2x^5$ ■

Alternate Example 4 Graphing Power Functions $f(x) = k \cdot x^a$

State the values of the constants k and a. Describe the portion of the curve that lies in Quadrant I or IV. Determine whether f is even, odd, or undefined for $x < 0$. Describe the rest of the curve if any. Graph the function to see whether it matches the description.

(a) $f(x) = 3x^{-3}$ **(b)** $f(x) = -x^{1.5}$ **(c)** $f(x) = -x^{0.8}$

SOLUTION

(a) Because $k = 3$ is positive and $a = -3$ is negative, the graph passes through $(1, 3)$ and is asymptotic to both axes. The graph is decreasing in the first quadrant. The function f is odd because

$$f(-x) = 3(-x)^{-3} = \frac{3}{(-x)^3} = -\frac{3}{x^3} = -3x^{-3} = -f(x).$$

So its graph is symmetric about the origin. The graph in Figure 2.5a supports all aspects of the description.

(b) Because $k = -1$ is negative and $a = 1.5 > 1$, the graph contains $(0, 0)$ and passes through $(1, -1)$. In the fourth quadrant, it is decreasing. The function f is undefined for $x < 0$ because

$$f(x) = -x^{1.5} = -x^{3/2} = -\left(\sqrt{x}\right)^3,$$

and the square-root function is undefined for $x < 0$. So the graph of f has no points in Quadrants II and III. The graph in Figure 2.5b matches the description.

(c) Because $k = -1$ is negative and $0 < a < 1$, the graph contains $(0, 0)$ and passes through $(1, -1)$. In the fourth quadrant, it is decreasing. The function f is even because

$$f(-x) = -(-x)^{0.8} = -(-x)^{4/5} = -\left(\sqrt[5]{-x}\right)^4 = -\left(-\sqrt[5]{x}\right)^4 = -\left(\sqrt[5]{x}\right)^4 = -x^{0.8} = f(x).$$

So, the graph of f is symmetric about the y-axis. The graph in Figure 2.5c fits the description.

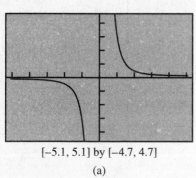

[−5.1, 5.1] by [−4.7, 4.7]

(a)

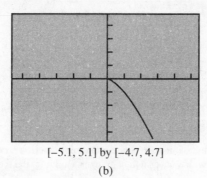

[−5.1, 5.1] by [−4.7, 4.7]

(b)

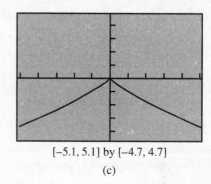

[−5.1, 5.1] by [−4.7, 4.7]

(c)

Figure 2.5 The graphs (a) $f(x) = 3x^{-3}$, (b) $f(x) = -x^{1.5}$, and (c) $f(x) = -x^{0.8}$.

Exercises for Alternate Example 4

In Exercises 13–18, state the values of the constants k and a. Describe the portion of the curve that lies in Quadrant I or IV. Determine whether f is even, odd, or undefined for $x < 0$. Describe the rest of the curve if any. Graph the function to see whether it matches the description.

13. $f(x) = 2x^{-2}$ **14.** $f(x) = 0.5x^{-5}$ **15.** $f(x) = -x^{2.5}$ **16.** $f(x) = -2x^{1.4}$ **17.** $f(x) = -x^{0.1}$

18. $f(x) = -x^{0.2}$

Polynomial Functions of Higher Degree with Modeling

2.3

Alternate Example 1   **Graphing Transformations of Monomial Functions**

Describe how to transform the graph of an appropriate monomial function $f(x) = a_n x^n$ into the graph of the given function. Sketch the transformed graph by hand and support your answer with a grapher. Compute the location of the y-intercept as a check on the transformed graph.

(a) $g(x) = 2(x - 1)^3$ **(b)** $h(x) = -(x + 1)^4 - 2$

SOLUTION

(a) You can obtain the graph of $g(x) = 2(x - 1)^3$ by shifting the graph of $f(x) = 2x^3$ one unit to the right, as shown in Figure 2.6a. The y-intercept of the graph of g is $g(0) = 2(0 - 1)^3 = -2$, which appears to agree with the transformed graph.

(b) You can obtain the graph of $h(x) = -(x + 1)^4 - 2$ by shifting the graph of $f(x) = -x^4$ one unit to the left and two units down, as shown in Figure 2.6b. The y-intercept of the graph of h is $h(0) = -(0 + 1)^4 - 2 = -3$, which appears to agree with the transformed graph.

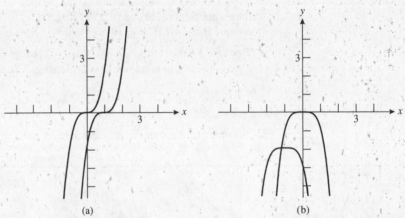

(a)   (b)

Figure 2.6 (a) The graphs of $g(x) = 2(x-1)^3$ and $f(x) = 2x^3$. (b) The graphs of $h(x) = -(x+1)^4 - 2$ and $f(x) = -x^4$.

Exercises for Alternate Example 1

In Exercises 1–6, describe how to transform the graph of an appropriate monomial function $f(x) = a_n x^n$ into the graph of the given function. Sketch the transformed graph by hand and support your answer with a grapher. Compute the location of the y-intercept as a check on the transformed graph.

1. $g(x) = 3(x+1)^3$ **2.** $h(x) = -2(x+1)^4$ **3.** $g(x) = 2(x-1)^3 - 1$ **4.** $h(x) = -3(x-1)^4 + 1$

5. $g(x) = 2x^3 - 1$ **6.** $h(x) = -(x+2)^4 + 2$

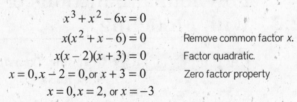 **Finding the Zeros of a Polynomial Function**

Find the zeros of $f(x) = x^3 + x^2 - 6x$.

SOLUTION
Solve Algebraically

We solve the related equation $f(x) = 0$ by factoring:

$$x^3 + x^2 - 6x = 0$$
$$x(x^2 + x - 6) = 0 \qquad \text{Remove common factor } x.$$
$$x(x-2)(x+3) = 0 \qquad \text{Factor quadratic.}$$
$$x = 0, x - 2 = 0, \text{or } x + 3 = 0 \qquad \text{Zero factor property}$$
$$x = 0, x = 2, \text{or } x = -3$$

So, the zeros of f are -3, 0, and 2.

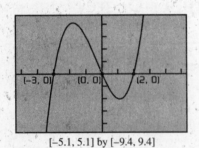

[−5.1, 5.1] by [−9.4, 9.4]

Figure 2.7 The graph of $f(x) = x^3 + x^2 - 6x$ showing the x-intercepts.

Solve Graphically

Use the features of your calculator to approximate the zeros of f. Figure 2.7 shows that there are three values. Based on your algebraic solution we can be sure that these values are correct.

Exercises for Alternate Example 5

In Exercises 7–12, find the zeros of each polynomial function.

7. $f(x) = x^3 - 16x$ **8.** $f(x) = x^3 - 9x^2 - 10x$ **9.** $f(x) = x^3 + 9x^2 + 8x$ **10.** $f(x) = x^3 - 8x^2 + 7x$

11. $f(x) = x^3 + 2x^2 - 35x$ **12.** $f(x) = x^3 + 15x^2 + 56x$

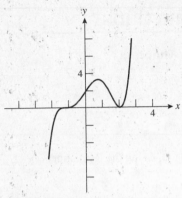

Figure 2.8 A sketch of the graph of $f(x) = (x + 1)^3(x - 2)^2$ showing the x-intercepts.

Alternate Example 6 **Sketching the Graph of a Factored Polynomial**

State the degree and list the zeros of the function $f(x) = (x + 1)^3(x - 2)^2$. State the multiplicity of each zero and whether the graph crosses the x-axis at the corresponding x-intercept. Then sketch the graph of f by hand.

SOLUTION

The degree of f is 5 and the zeros are $x = -1$ and $x = 2$. The graph cross the x-axis at $x = -1$ because the multiplicity 3 is odd. The graph does not cross the x-axis at $x = 2$ because the multiplicity 2 is even. Notice that the values of f are positive for $x > 2$, positive for $-1 < x < 2$, and negative for $x < -1$. Figure 2.8 shows a sketch of the graph of f.

Exercises for Alternate Example 6

In Exercises 13–18, state the degree and list the zeros of each function. State the multiplicity of each zero and whether the graph crosses the x-axis at the corresponding x-intercept. Then sketch the graph of f by hand.

13. $f(x) = x(x + 2)^2$ **14.** $f(x) = (x + 2)(x - 1)^3$ **15.** $f(x) = (x - 3)^2(x - 1)^2$ **16.** $f(x) = (x + 1)^3(x + 3)^3$

17. $f(x) = (x - 2)^2(x + 1)^3$ **18.** $f(x) = x(x + 2)(x - 3)^2$

Alternate Example 8 **Zooming to Uncover Hidden Behavior**

Find all of the real zeros of $f(x) = x^4 - 1.4x^3 - 1.09x^2 - 0.214x - 0.012$.

SOLUTION

Solve Graphically

Because f is of degree 4, there are at most four zeros. The graph in Figure 2.9a suggests a single zero (multiplicity 1) around $x = 2$ and a triple zero (multiplicity 3) around $x = 0$. Closer inspection around $x = 0$ in Figure 2.9b reveals three separate zeros. Using the grapher, we find the four zeros to be $x \approx -0.3$, $x \approx -0.2$, $x \approx -0.1$, and $x \approx 2$.

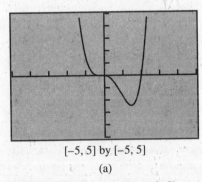
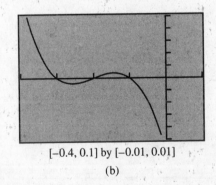

[−5, 5] by [−5, 5]   [−0.4, 0.1] by [−0.01, 0.01]

(a)   (b)

Figure 2.9 (Alternate Example 8)

Exercises for Alternate Example 8

In Exercises 19–24, find all of the real zeros of each function.

19. $f(x) = x^3 + 2.5x^2 + 1.06x + 0.12$ **20.** $f(x) = x^3 + 0.3x^2 - 7.18x - 8.16$ **21.** $f(x) = x^4 - 0.9x^3 + 26.99x^2 - 26.97x$

22. $f(x) = x^4 - 3x^3 - 0.01x^2 + 3.99x$ **23.** $f(x) = x^4 - 11.7x^3 + 45.62x^2 - 59.289x$

24. $f(x) = x^4 - 15.8x^3 + 93.59x^2 - 246.322x + 243.048$

2.4 Real Zeros of Polynomial Functions

Alternate Example 2 **Using the Remainder Theorem**

Find the remainder when $f(x) = 2x^2 - 4x + 2$ is divided by

(a) $x - 3$ (b) $x + 3$ (c) $x - 1$.

SOLUTION

Solve Numerically (by hand)

(a) We can find the remainder without doing long division. Using the Remainder Theorem with $k = 3$ we find that

$$r = f(3) = 2(3)^2 - 4(3) + 2 = 18 - 12 + 2 = 8$$

(b) $r = f(-3) = 2(-3)^2 - 4(-3) + 2 = 18 + 12 + 2 = 32$

(c) $r = f(1) = 2(1)^2 - 4(1) + 2 = 2 - 4 + 2 = 0$

Interpret

Because the remainder in part (c) is 0, $x - 1$ divides evenly into $f(x) = 2x^2 - 4x + 2$. So, $x - 1$ is a factor of $f(x) = 2x^2 - 4x + 2$, 1 is a solution of $2x^2 - 4x + 2 = 0$, and 1 is an x-intercept of the graph of $y = 2x^2 - 4x + 2$.

Support Numerically (using a grapher)

We can find the remainders of several division problems at once using the table feature of a grapher. (See Figure 2.10.)

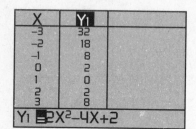

X	Y1	
-3	32	
-2	18	
-1	8	
0	2	
1	0	
2	2	
3	8	

Y1 ▍2X²-4X+2

Figure 2.10 Table for $f(x) = 2x^2 - 4x + 2$ showing the remainders obtained when $f(x)$ is divided by $x - k$, for $k = -4, -3, \ldots, 1, 2$.

Exercises for Alternate Example 2

In Exercises 1–6, use the Remainder Theorem to find the remainder when $f(x)$ is divided by the given linear expression.

1. $f(x) = 2x^2 + 4x + 3; x - 2$

2. $f(x) = -x^2 - 4x + 12; x - 1$

3. $f(x) = 3x^2 - 5x - 4; x + 3$

4. $f(x) = -2x^2 + 5x + 8; x + 2$

5. $f(x) = 2x^2 - 4x + 2; x - 5$

6. $f(x) = 3x^2 - 6x - 2; x - 0.5$

Alternate Example 5 **Finding the Rational Zeros**

Find the rational zeros of $f(x) = 2x^3 + x^2 - 2x - 1$.

SOLUTION

Because the leading coefficient is 2 and the constant coefficient is -1, the Rational Zeros Theorem gives us a list of several potential rational zeros of f. So we take an organized approach to our solution.

Potential Rational Zeros:

$$\frac{\text{Factors of } -1}{\text{Factors of } 2} : \frac{\pm 1}{\pm 1, \pm 2} : \pm 1, \pm \frac{1}{2}$$

Figure 2.11 suggests that, among our candidates, -1, 1 and possibly $-1/2$ are the most likely to be rational zeros. We use synthetic division because it tells us whether a number is a zero and, if so, how to factor the polynomial. To see whether 1 is a zero of f, we synthetically divide $f(x)$ by $x - 1$:

```
Zero of Divisor   1⌋   2    1   -2   -1    Dividend
                            2    3    1
                  ─────────────────────
                       2    3    1    0    Remainder
                         Quotient
```

[−5.1, 5.1] by [−4.7, 4.7]

Figure 2.11 The function $f(x) = 2x^3 + x^2 - 2x - 1$ has three real zeros.

So because the remainder is 0, $x - 1$ is a factor of $f(x)$ and 1 is a zero of f. By the Division Algorithm and factoring, we conclude

$$f(x) = 2x^3 + x^2 - 2x - 1$$
$$= (x - 1)(2x^2 + 3x + 1)$$
$$= (x - 1)(2x + 1)(x + 1)$$

So the rational zeros of f are -1, $-1/2$, and 1.

Exercises for Alternate Example 5

In Exercises 7–12, find the rational zeros of each function.

7. $f(x) = 6x^2 + x - 1$

8. $f(x) = 2x^3 + 5x^2 + 2x + 5$

9. $f(x) = 2x^4 - x^3 + x^2 - x - 1$

10. $f(x) = 3x^3 + x^2 - 6x - 2$

11. $f(x) = 3x^3 - 2x^2 - 9x + 6$

12. $f(x) = x^4 - 5x^2 + 6$

■

Alternate Example 6 **Establishing Bounds for Real Zeros**

Prove that all the real zeros of $f(x) = 2x^4 - x^3 - 7x^2 + 3x + 3$ must lie in the interval $[-5, 5]$.

SOLUTION

We must prove that 5 is an upper bound and -5 is a lower bound on the real zeros of f. The function f has a positive leading coefficient, so we employ the Upper and Lower Bound Tests, and use synthetic division:

$5\rfloor$	2	-1	-7	3	3	
		10	45	190	965	
	2	9	38	193	969	Last line
$-5\rfloor$	2	-1	-7	3	3	
		-10	55	-240	1185	
	2	-11	48	-237	1188	Last line

Because the last line in the first division scheme consists of all positive numbers, 5 is an upper bound. Because the last line in the second division consists of numbers of alternating signs, -5 is a lower bound. All of the real zeros of f must lie in the closed interval $[-5, 5]$.

Exercises for Alternate Example 6

In Exercises 13–18, use synthetic division to prove that all the real zeros of the given function must lie in the given interval.

13. $f(x) = x^2 - 4x + 3$; $[0, 5]$

14. $f(x) = x^2 + 2x - 3$; $[-4, 2]$

15. $f(x) = x^3 - 6x^2 + 11x - 6$; $[0, 7]$

16. $f(x) = x^3 - x^2 - 2x + 2$; $[-2, 2]$

17. $f(x) = x^4 - x^2 - 2$; $[-2, 2]$

18. $f(x) = x^4 - 2x^2 - 2$; $[-2, 2]$

■

Alternate Example 7 **Finding the Real Zeros of a Polynomial Function**

Find all of the real zeros of $f(x) = 2x^4 - x^3 - 7x^2 + 3x + 3$.

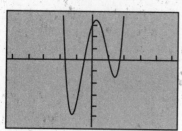

[−5.1, 5.1] by [−5.7, 3.7]

Figure 2.12 $f(x) = 2x^4 - x^3 - 7x^2 + 3x + 3$ has all of its real zeros in [−5, 5].

SOLUTION

From Alternate Example 6, we know that all of the real zeros of f must lie in the closed interval [−5, 5]. So in Figure 2.12 we set our Xmin and Xmax accordingly.

Next we use the Rational Zeros Theorem.

Potential Rational Zeros:

$$\frac{\text{Factors of } 3}{\text{Factors of } 2} : \frac{\pm 1, \pm 3}{\pm 1, \pm 2} : \pm 1, \pm \frac{1}{2}, \pm \frac{3}{2}, \pm 3$$

We compare the x-intercepts of the graph in Figure 2.12 and our list of candidates, and decide that 1 and −1/2 are the only potential rational zeros worth pursuing.

$$
\begin{array}{r|rrrrr}
1 & 2 & -1 & -7 & 3 & 3 \\
  &   & 2 & 1 & -6 & -3 \\
\hline
  & 2 & 1 & -6 & -3 & 0
\end{array}
$$

From this synthetic division we conclude

$$f(x) = 2x^4 - x^3 - 7x^2 + 3x + 3$$
$$= (x - 1)(2x^3 + x^2 - 6x - 3)$$

and now we divide the cubic factor $2x^3 + x^2 - 6x - 3$ by $x + 1/2$.

$$
\begin{array}{r|rrrr}
-1/2 & 2 & 1 & -6 & -3 \\
     &   & -1 & 0 & 3 \\
\hline
     & 2 & 0 & -6 & 0
\end{array}
$$

The second synthetic division allows us to complete the factoring of $f(x)$.

$$f(x) = (x - 1)(2x^3 + x^2 - 6x - 3)$$
$$= (x - 1)\left(x + \frac{1}{2}\right)(2x^2 - 6)$$
$$= 2(x - 1)\left(x + \frac{1}{2}\right)\left(x - \sqrt{3}\right)\left(x + \sqrt{3}\right)$$

So the zeros of f are the rational numbers 1 and $-\frac{1}{2}$ and the irrational numbers $-\sqrt{3}$ and $\sqrt{3}$.

Exercises for Alternate Example 7

In Exercises 19–24, find all of the real zeros of the given function.

19. $f(x) = 2x^3 - x^2 - 4x + 2$

20. $f(x) = 2x^4 - 3x^3 - 5x^2 + 9x - 3$

21. $f(x) = x^3 - 5x^2 + 8x - 4$

22. $f(x) = 2x^3 + x^2 - 15x + 7$

23. $f(x) = 3x^3 - 5x^2 - x + 2$

24. $f(x) = x^4 - 8x^2 + 7$

2.5 Complex Zeros and the Fundamental Theorem of Algebra

Alternate Example 2 **Finding a Polynomial from Given Zeros**

Write a polynomial function of minimum degree in standard form with real coefficients whose zeros include -2, 1, and $3 + i$.

SOLUTION

Because -2 and 1 are real zeros, $x + 2$ and $x - 1$ must be factors. Because the coefficients are real and $3 + i$ is a zero, $3 - i$ must also be a zero. Therefore, $x - (3 + i)$ and $x - (3 - i)$ must both be factors of $f(x)$. Thus,

$$f(x) = (x + 2)(x - 1)[x - (3 + i)][x - (3 - i)]$$
$$= (x^2 + x - 2)(x^2 - 6x + 10)$$
$$= x^4 - 5x^3 + 2x^2 + 22x - 20$$

is a polynomial of the type we seek. Any nonzero real number multiple of $f(x)$ will also be such a polynomial.

Exercises for Alternate Example 2

In Exercises 1–6, write a polynomial function of minimum degree in standard form with real coefficients whose zeros include the given numbers.

1. $2 - 3i$ **2.** 4 and $3i$ **3.** -1, 2, and $2 - i$ **4.** -3, 0, and $1 - 5i$ **5.** 1, 2, and $-2 + 4i$

6. i, $2i$, and $3i$

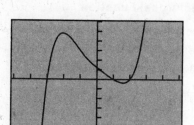

[−5.1, 5.1] by [−120, 120]

Figure 2.13 $f(x) = x^5 - 3x^3 + 6x^2 - 28x + 24$ has three real zeros.

Alternate Example 4 **Factoring a Polynomial with Complex Zeros**

Find all zeros of $f(x) = x^5 - 3x^3 + 6x^2 - 28x + 24$, and write $f(x)$ in its linear factorization.

SOLUTION

Figure 2.13 suggests that the real zeros of f are $x = -3$, $x = 1$, and $x = 2$.

Using synthetic division we can verify these zeros and show that $x^2 + 4$ is also a factor of f. So $x = 2i$ and $x = -2i$ are also zeros. Therefore,

$$f(x) = x^5 - 3x^3 + 6x^2 - 28x + 24$$
$$= (x + 3)(x - 1)(x - 2)(x^2 + 4)$$
$$= (x + 3)(x - 1)(x - 2)(x + 2i)(x - 2i).$$

Exercises for Alternate Example 4

In Exercises 7–12, find all zeros of f, and write $f(x)$ in its linear factorization.

7. $f(x) = x^4 - x^3 + x^2 - x$ **8.** $f(x) = x^4 - 16$ **9.** $f(x) = x^4 - x^3 + 3x^2 - 9x - 54$

10. $f(x) = x^5 - 9x^4 + 31x^3 - 49x^2 + 30x$ **11.** $f(x) = x^4 - 8x^3 + 32x^2 - 64x + 39$

12. $f(x) = x^5 - 6x^4 + 11x^3 + 10x^2 - 64x + 48$

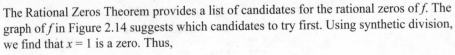

Alternate Example 6 **Factoring a Polynomial**

Write $f(x) = x^5 - x^4 - 2x^3 + 2x^2 - 3x + 3$ as a product of linear and irreducible quadratic factors, each with real coefficients.

SOLUTION

The Rational Zeros Theorem provides a list of candidates for the rational zeros of f. The graph of f in Figure 2.14 suggests which candidates to try first. Using synthetic division, we find that $x = 1$ is a zero. Thus,

$$\begin{aligned} f(x) &= (x-1)(x^4 - 2x^2 - 3) \\ &= (x-1)(x^2 - 3)(x^2 + 1) \\ &= (x-1)(x-\sqrt{3})(x+\sqrt{3})(x^2 + 1) \end{aligned}$$

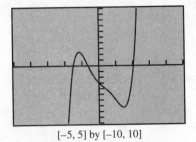

[-5, 5] by [-10, 10]

Figure 2.14 $f(x) = x^5 - x^4 - 2x^3 + 2x^2 - 3x + 3$ has three real zeros.

Because the zeros of $x^2 + 1$ are complex, any further factorization would introduce nonreal complex coefficients. We have taken the factorization of f as far as possible, subject to the condition that *each factor has real coefficients.*

Exercises for Alternate Example 6

In Exercises 13–18, write f as a product of linear and irreducible quadratic factors, each with real coefficients.

13. $f(x) = x^4 - x^2 - 6$

14. $f(x) = x^5 - 2x^4 + 2x^3 - 4x^2 - 15x + 30$

15. $f(x) = x^5 - x$

16. $f(x) = x^3 - x$

17. $f(x) = 2x^3 + x^2 - 14x - 7$

18. $f(x) = 2x^3 - 11x^2 + 20x + 13$

2.6 Graphs of Rational Functions

Alternate Example 2 **Transforming the Reciprocal Function**

Describe how the graph of the given function can be obtained by transforming the graph of the reciprocal function $f(x) = 1/x$. Identify the horizontal and vertical asymptotes and use limits to describe the corresponding behavior. Sketch the graph of the function.

(a) $g(x) = \dfrac{3}{x - 2}$

(b) $h(x) = \dfrac{2x + 1}{x + 2}$

SOLUTION

(a) $g(x) = \dfrac{3}{x - 2} = 3\left(\dfrac{1}{x - 2}\right) = 3f(x - 2)$

The graph of g is the graph of the reciprocal function shifted 2 units to the right and then stretched vertically by a factor of 3. So the lines $x = 2$ and $y = 0$ are vertical and horizontal asymptotes, respectively. Using limits we have $\lim\limits_{x \to \infty} g(x) = \lim\limits_{x \to -\infty} g(x) = 0$,

$\lim\limits_{x \to 2^+} g(x) = \infty$, and $\lim\limits_{x \to 2^-} g(x) = -\infty$. The graph is shown in Figure 2.15a.

(b) We begin with polynomial division:

$$\begin{array}{r} 2 \\ x+2 \overline{)\, 2x + 1} \\ \underline{2x + 4} \\ -3 \end{array}$$

So, $h(x) = \dfrac{2x + 1}{x + 2} = 2 - \dfrac{3}{x + 2} = -3f(x + 2) + 2$.

Thus the graph of h is the graph of the reciprocal function translated 2 units to the left, stretched by a factor of 3, followed by a reflection across the x-axis, and then translated 2 units up. (Note that the reflection must be executed before the vertical translation.) So, the lines $x = -2$ and $y = 2$ are the vertical and horizontal asymptotes, respectively. Using limits we have

$$\lim_{x \to \infty} h(x) = \lim_{x \to -\infty} h(x) = 2, \quad \lim_{x \to -2^+} h(x) = -\infty, \text{and} \quad \lim_{x \to -2^-} h(x) = \infty.$$

The graph is shown in Figure 2.15b.

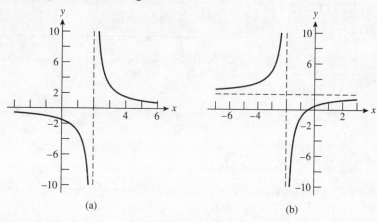

(a)   (b)

Figure 2.15 (a) The graphs of (a) $g(x) = \dfrac{3}{x-2}$ and (b) $h(x) = \dfrac{2x+1}{x+2}$ with asymptotes.

Exercises for Alternate Example 2

In Exercises 1–6, describe how the graph of the given function can be obtained by transforming the graph of the reciprocal function $f(x) = 1/x$. Identify the horizontal and vertical asymptotes and use limits to describe the corresponding behavior. Sketch the graph of the function.

1. $g(x) = \dfrac{2}{x+1}$ **2.** $h(x) = \dfrac{2x-1}{x+1}$ **3.** $g(x) = \dfrac{2}{x-2}$ **4.** $h(x) = \dfrac{3x+2}{x+2}$ **5.** $g(x) = \dfrac{1}{x-3}$

6. $h(x) = \dfrac{3x+2}{x-2}$

Alternate Example 4 Graphing a Rational Function

Find the asymptotes and intercepts of the function $f(x) = \dfrac{x^3}{x^2 - 4}$ and graph the function.

SOLUTION

The degree of the numerator is greater than the degree of the denominator, so the end behavior asymptote is the quotient polynomial. Using polynomial long division, we obtain

$$f(x) = \frac{x^3}{x^2 - 4} = x + \frac{4x}{x^2 - 4}.$$

So the quotient polynomial is $q(x) = x$, a slant asymptote. Factoring the denominator,

$$x^2 - 4 = (x + 2)(x - 2),$$

shows that the zeros of the denominator are $x = 2$ and $x = -2$. Consequently, $x = 2$ and $x = -2$ are the vertical asymptotes of f. The only zero of the numerator is 0, so $f(0) = 0$, and thus we see that the point $(0, 0)$ is the only x-intercept and the y-intercept of the graph of f.

The graph of f in Figure 2.16a passes through $(0, 0)$ and suggests the vertical asymptotes $x = 2$ and $x = -2$ and the slant asymptote $y = q(x) = x$. Figure 2.16b shows the graph of f with its asymptotes overlaid.

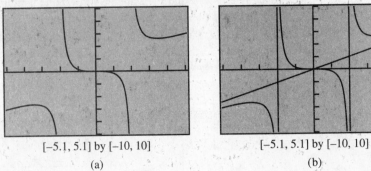

[-5.1, 5.1] by [-10, 10]
(a)

[-5.1, 5.1] by [-10, 10]
(b)

Figure 2.16 The graph of $f(x) = \dfrac{x^3}{x^2 - 4}$ (a) by itself and (b) with its asymptotes.

Exercises for Alternate Example 4

In Exercises 7–12, find the asymptotes and intercepts of each function and graph the function.

7. $f(x) = \dfrac{x^3}{x^2 - 1}$ **8.** $f(x) = \dfrac{2x^3}{x^2 - 4}$ **9.** $f(x) = \dfrac{x^3}{x^2 - 16}$ **10.** $f(x) = \dfrac{x^2 + 2x - 3}{x - 2}$

11. $f(x) = \dfrac{x^2 - 6x + 5}{x + 2}$ **12.** $f(x) = \dfrac{(x - 2)^2}{x + 4}$

> **Alternate Example 5** **Analyzing the Graph of a Rational Function**
>
> Find the intercepts, asymptotes, use limits to describe the behavior at the vertical asymptotes, and analyze and draw the graph of the rational function
>
> $$f(x) = \frac{x + 1}{x^2 - x - 6}.$$

SOLUTION

The numerator is 0 when $x = -1$, so the x-intercept is -1. Because $f(0) = -1/6$, the y-intercept is $-1/6$. The denominator factors as

$$x^2 - x - 6 = (x + 2)(x - 3),$$

so there are vertical asymptotes at $x = -2$ and $x = 3$. We know that the horizontal asymptote is $y = 0$. Figure 2.17 supports this information and allows us to conclude that

$$\lim_{x \to -2^-} f(x) = -\infty, \quad \lim_{x \to -2^+} f(x) = \infty, \quad \lim_{x \to 3^-} f(x) = -\infty, \text{ and } \lim_{x \to 3^+} f(x) = \infty.$$

Domain: $(-\infty, -2) \cup (-2, 3) \cup (3, \infty)$
Range: all reals
Continuity: all $x \neq -2, 3$
Decreasing on $(-\infty, -2) \cup (-2, 3) \cup (3, \infty)$
Not symmetric
Unbounded
No local extrema
Horizontal asymptotes: $y = 0$
Vertical asymptotes: $x = -2$ and $x = 3$
End behavior: $\lim\limits_{x \to \infty} f(x) = \lim\limits_{x \to -\infty} f(x) = 0$

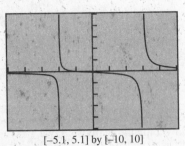

[-5.1, 5.1] by [-10, 10]

Figure 2.17 The graph of

$$f(x) = \frac{x + 1}{x^2 - x - 6}.$$

Exercises for Alternate Example 5

In Exercises 13–18, find the intercepts, asymptotes, use limits to describe the behavior at the vertical asymptotes, and analyze the graph of each rational function.

13. $f(x) = \dfrac{x}{x^2 - x - 6}$

14. $f(x) = \dfrac{x}{x^2 + 4x - 5}$

15. $f(x) = \dfrac{x + 1}{x^2 - 6x + 8}$

16. $f(x) = \dfrac{x}{x^2 + 5x - 6}$

17. $f(x) = \dfrac{x - 1}{x^2 + x - 6}$

18. $f(x) = \dfrac{x + 3}{x^2 + 7x + 10}$

Alternate Example 6 **Analyzing the Graph of a Rational Function**

Find the intercepts, analyze, and draw the graph of the rational function

$$f(x) = \frac{x^2 - 1}{x^2 - 9}$$

SOLUTION

The numerator factors as

$$x^2 - 1 = (x + 1)(x - 1),$$

so the x-intercepts are –1 and 1. The y-intercept is $f(0) = 1/9$. The denominator also factors as

$$x^2 - 9 = (x + 3)(x - 3),$$

so the vertical asymptotes are $x = -3$ and $x = 3$. We know that $y = 1$ (the leading coefficient of the numerator) is the horizontal asymptote. Figure 2.18 supports this information and allows us to conclude that

$$\lim_{x \to -3^-} f(x) = \infty, \quad \lim_{x \to -3^+} f(x) = -\infty, \quad \lim_{x \to 3^-} f(x) = -\infty, \text{ and } \lim_{x \to 3^+} f(x) = \infty.$$

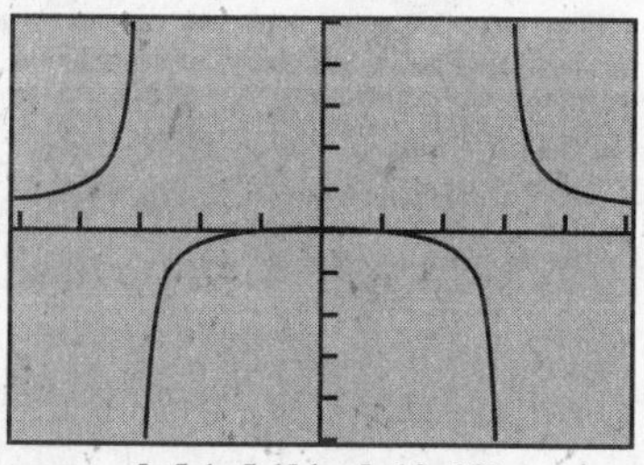

Domain: $(-\infty, -3) \cup (-3, 3) \cup (3, \infty)$
Range: $(-\infty, 1/9) \cup (1, \infty)$
Continuity: all $x \neq -3, 3$
Increasing on $(-\infty, -3)$ and $(-3, 0)$; decreasing on $(0, 3)$ and $(3, \infty)$
Symmetric with respect to the y-axis (an even function)
Unbounded
Local maximum of 1/9 at $x = 0$
Horizontal asymptotes: $y = 1$
Vertical asymptotes: $x = -3$ and $x = 3$
End behavior: $\lim_{x \to \infty} f(x) = \lim_{x \to -\infty} f(x) = 1$

[–5.1, 5.1] by [–10, 10]

Figure 2.18 The graph of
$f(x) = \dfrac{x^2 - 1}{x^2 - 9}$. It can be shown that
f takes on no value between 1/9, the
y-intercept, and 1, the horizontal
asymptote.

Exercises for Alternate Example 6

In Exercises 19–24, find the intercepts and analyze the graph of each rational function.

19. $f(x) = \dfrac{x^2 - 1}{x^2 - 16}$

20. $f(x) = \dfrac{x^2 - 1}{x^2 - 25}$

21. $f(x) = \dfrac{x^2 - 2}{x^2 - 9}$

22. $f(x) = \dfrac{2x^2 - 4}{x^2 - 25}$

23. $f(x) = \dfrac{3x^2 - 5}{x^2 - 16}$

24. $f(x) = \dfrac{2x^2 + 5}{x^2 - 9}$

2.7   Solving Equations in One Variable

Alternate Example 1 **Solving by Clearing Fractions**

Solve $2x - \dfrac{1}{x} = 1$.

SOLUTION
Solve Algebraically
The LCD is x.

$$2x - \frac{1}{x} = 1$$
$$2x^2 - 1 = x \qquad \text{Multiply by } x.$$
$$2x^2 - x - 1 = 0 \qquad \text{Subtract } x.$$
$$(2x + 1)(x - 1) = 0 \qquad \text{Factor.}$$
$$2x + 1 = 0 \quad \text{or} \quad x - 1 = 0 \qquad \text{Zero factor property}$$
$$x = -\frac{1}{2} \quad \text{or} \quad x = 1$$

Confirm Algebraically

For $x = -\dfrac{1}{2}$, $2x - \dfrac{1}{x} = 2\left(-\dfrac{1}{2}\right) - \dfrac{1}{-1/2} = 1$ and for $x = 1$, $2x - \dfrac{1}{x} = 2(1) - \dfrac{1}{1} = 1$.

Each value is a solution of the original equation.

Exercises for Alternate Example 1

In Exercises 1–6, solve each equation algebraically.

1. $\dfrac{2x-1}{x} = 5$
2. $2 - \dfrac{3}{x} = \dfrac{5}{x}$
3. $\dfrac{1}{x} - 4x = 3$
4. $\dfrac{2x+1}{x} = 3x$
5. $\dfrac{x+5}{x-3} = x$
6. $\dfrac{x}{x-3} = x$

Alternate Example 2 **Solving a Rational Equation**

Solve $2x - \dfrac{1}{x-3} = 0$.

SOLUTION
Solve Algebraically
The LCD is $x - 3$.

$$2x - \frac{1}{x-3} = 0$$
$$2x(x-3) - \frac{x-3}{x-3} = 0 \qquad \text{Multiply by } x-4.$$
$$2x^2 - 6x - 1 = 0 \qquad \text{Distributive property}$$
$$x = \frac{-(-6) \pm \sqrt{(-6)^2 - 4(2)(-1)}}{2(2)} \qquad \text{Quadratic formula}$$
$$x = \frac{6 \pm \sqrt{44}}{4} \qquad \text{Simplify.}$$
$$x = \frac{3 \pm \sqrt{11}}{2} \qquad \text{Simplify.}$$
$$x \approx -0.158, 3.158$$

Solve Graphically

The graph in Figure 2.19 suggests that the function $y = 2x - \dfrac{1}{x-3}$ has two zeros. We can use the graph to show that the zeros are about −0.158 and 3.158, agreeing with the values found algebraically.

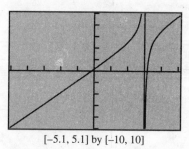

[−5.1, 5.1] by [−10, 10]

Figure 2.19 The graph of $f(x) = 2x - \dfrac{1}{x-3}$.

Exercises for Alternate Example 2

In Exercises 7–12, solve each equation algebraically. Support your answer graphically.

7. $\dfrac{x+3}{x+4} = x$ **8.** $\dfrac{1}{x} + \dfrac{1}{x+1} = 1$ **9.** $\dfrac{1}{x} - \dfrac{1}{x+1} = 1$ **10.** $\dfrac{2x-1}{3x} = x - 1$ **11.** $\dfrac{2x-1}{3x} = \dfrac{3x}{2x-1}$ **12.** $\dfrac{3x+1}{2x^2} = 2$ ■

Alternate Example 3 **Eliminating Extraneous Solutions**

Solve the equation

$$\frac{x}{x-3} - \frac{4}{x-1} = \frac{8}{x^2 - 4x + 3}.$$

SOLUTION

Solve Algebraically

The denominator of the right-hand side, $x^2 - 4x + 3$, factors into $(x-3)(x-1)$. So the least common denominator (LCD) of the equation is $(x-3)(x-1)$, and we multiply both sides of the equation by this LCD:

$$(x-3)(x-1)\left(\frac{x}{x-3} - \frac{4}{x-1} \right) = (x-3)(x-1)\left(\frac{8}{x^2-4x+3} \right)$$

$$x(x-1) - 4(x-3) = 8 \qquad \text{Distributive property}$$

$$x^2 - 5x + 4 = 0 \qquad \text{Distributive property}$$

$$(x-1)(x-4) = 0 \qquad \text{Factor.}$$

$$x = 1 \quad \text{or} \quad x = 4$$

Confirm Numerically

We replace x by 4 in the original equation:

$$\frac{4}{4-3} - \frac{4}{4-1} \overset{?}{=} \frac{8}{4^2 - 4(4) + 3}$$

$$4 - \frac{4}{3} \overset{?}{=} \frac{8}{3}$$

This equation is true, so $x = 4$ is a valid solution. The original equation is not defined for $x = 1$, so $x = 1$ is an extraneous solution.

Support Graphically

The graph of $f(x) = \dfrac{x}{x-3} - \dfrac{4}{x-1} - \dfrac{8}{x^2 - 4x + 3}$ in Figure 2.20 suggests that $x = 4$ is an x-intercept and $x = 1$ is not.

Interpret

Only $x = 4$ is a solution.

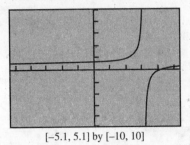

[−5.1, 5.1] by [−10, 10]

Figure 2.20 The graph of $f(x) = \dfrac{x}{x-3} - \dfrac{4}{x-1} - \dfrac{8}{x^2 - 4x + 3}$.

Exercises for Alternate Example 3

In Exercises 13–18, solve each equation algebraically. Support your answer graphically.

13. $\dfrac{3}{x-1} - \dfrac{2}{x-3} = \dfrac{5}{x^2 - 4x + 3}$

14. $\dfrac{1}{x+1} + \dfrac{3}{x-3} = \dfrac{2}{x^2 - 2x - 3}$

15. $\dfrac{4}{x-2} - \dfrac{1}{x-2} = \dfrac{3}{x^2 - 6x + 8}$

16. $x - \dfrac{2}{x-3} = \dfrac{1-x}{x-3}$

17. $\dfrac{2}{x} + \dfrac{5}{x-1} = \dfrac{8}{x^2 - x}$

18. $\dfrac{3}{x+1} = \dfrac{2}{x-2} + \dfrac{5}{x^2 - x - 2}$

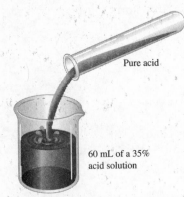

Figure 2.21 Mixing solutions (Alternate Example 5)

Acid Mixture

mL of acid

Intersection [135, 0.8]

Figure 2.22 (a) The graphs of $f(x) = \dfrac{x+21}{x+60}$ and $g(x) = 0.8$.

Alternate Example 5 **Calculating Acid Mixtures**

How much pure acid must be added to 60 mL of a 35% acid solution to produce a mixture that is 80% acid? (See Figure 2.21.)

SOLUTION

Model

Word Statement: $\dfrac{\text{mL of pure acid}}{\text{mL of mixture}} = $ concentration of acid

0.35×60 or $21 = $ mL of pure acid in 35% solution

$x = $ mL of acid needed

$x + 21 = $ mL of pure acid in resulting mixture

$x + 60 = $ mL of the resulting mixture

$\dfrac{x+21}{x+60} = $ concentration of acid

Solve Graphically

$\dfrac{x+21}{x+60} = 80$ Equation to be solved

Figure 2.22 shows the graphs of $f(x) = \dfrac{x+21}{x+60}$ and $g(x) = 0.8$. The point of intersection is $(135, 0.8)$.

Interpret

We need to add 135 mL of pure acid to the 35% acid solution to make a solution that is 80% acid.

Exercises for Alternate Example 5

In Exercises 19–24, how much pure acid must be added to the given volume of a 35% acid solution to produce a mixture that is the given percent acid?

19. given volume 100 mL; new solution: 75% acid

20. given volume 100 mL; new solution: 60% acid

21. given volume 200 mL; new solution: 90% acid

In Exercises 22–24, how much pure acid must be added to the given volume of a 25% acid solution to produce a mixture that is the given percent acid?

22. given volume 140 mL; new solution: 30% acid

23. given volume 120 mL; new solution: 50% acid

24. given volume 300 mL; new solution: 60% acid ∎

2.8 Solving Inequalities in One Variable

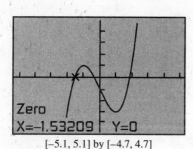

Figure 2.23 The graph of $f(x) = x^3 - 3x - 1$, with one of the three zeros highlighted

[–5.1, 5.1] by [–4.7, 4.7]

Alternate Example 3 **Solving A Polynomial Inequality Graphically**

Solve $x^3 \leq 3x + 1$ graphically.

SOLUTION

We first rewrite the inequality as $x^3 - 3x - 1 \leq 0$. Then we let $f(x) = x^3 - 3x - 1$ and find the real zeros of f graphically as shown in Figure 2.23. The three real zeros are approximately –1.53, –0.35, and 1.88. The solution consists of the x values for which the graph is on or below the x-axis. So the solution of $x^3 \leq 3x + 1$ is approximately

$(-\infty, -1.53] \cup [-0.35, 1.88]$.

The end points of these intervals are only accurate to two decimal places. We use square brackets because the zeros of the polynomial are solutions of the inequality, even though we only have approximations of their values.

Exercises for Alternate Example 3

In Exercises 1–6, solve each inequality graphically.

1. $x^2 \leq -x + 6$ **2.** $x^2 \geq 2x + 8$ **3.** $x^3 < -2x^2 + 3x$ **4.** $x^3 \geq 3x - 2$ **5.** $x^3 < 3x + 2$

6. $x^3 - 3x^2 - 4x + 12 \geq 0$ ∎

Alternate Example 5 **Creating a Sign Chart for a Rational Function**

Let $r(x) = \dfrac{2x + 3}{x(x - 3)}$. Determine the values of x that cause $r(x)$ to be **(a)** zero, **(b)** undefined. Then make a sign chart to determine the values of x that cause $r(x)$ to be **(c)** positive, **(d)** negative.

SOLUTION

(a) The real zeros of $r(x)$ are the real zeros of the numerator $2x + 3$. So $r(x)$ is zero if $x = -3/2$.

(b) $r(x)$ is undefined when the denominator $x(x - 3) = 0$. So $r(x)$ is undefined if $x = 0$ or $x = 3$.

These findings lead to the following chart, with three points of potential sign change:

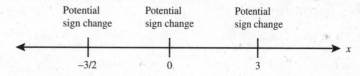

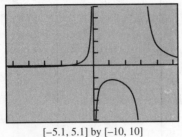

[−5.1, 5.1] by [−10, 10]

Figure 2.24 The graph of
$r(x) = \dfrac{2x+3}{x(x-3)}$.

Analyzing the factors of the numerator and denominator yields:

$$\xleftarrow{\qquad} \underset{\text{Negative}}{\overset{\frac{(-)}{(-)(-)}}{\quad}} \; \underset{-3/2}{0} \; \underset{\text{Positive}}{\overset{\frac{(+)}{(-)(-)}}{\quad}} \; \text{und.} \; \underset{0}{\overset{\frac{(+)}{(+)(-)}}{\quad}} \; \text{Negative} \; \underset{3}{\text{und.}} \; \underset{\text{Positive}}{\overset{\frac{(+)}{(+)(+)}}{\quad}} \xrightarrow{\quad} x$$

(c) So $r(x)$ is positive if $-3/2 < x < 0$ or $x > 3$, and the solution of $\dfrac{2x+3}{x(x-3)} > 0$ is $(-3/2, 0) \cup (3, \infty)$.

(d) Similarly $r(x)$ is negative if $x < -3/2$ or $0 < x < 3$, and the solution of $\dfrac{2x+3}{x(x-3)} < 0$ is $(-\infty, -3/2) \cup (0, 3)$.

Figure 2.24 supports our findings because the graph of r is above the x-axis for x in $(-3/2, 0) \cup (3, \infty)$ and is below the x-axis for x in $(-\infty, -3/2) \cup (0, 3)$.

Exercises for Alternate Example 5

In Exercises 7–12, determine the values of x that cause $r(x)$ to be **(a)** zero, **(b)** undefined. Then make a sign chart to determine the values of x that cause $r(x)$ to be **(c)** positive, **(d)** negative.

7. $r(x) = \dfrac{x}{x-2}$ **8.** $r(x) = \dfrac{x-1}{x-2}$ **9.** $r(x) = \dfrac{x-3}{x+1}$ **10.** $r(x) = \dfrac{x+1}{x^2-4}$ **11.** $r(x) = \dfrac{x+2}{x^2-3x-4}$

12. $r(x) = \dfrac{x^2}{x^2-x-2}$

Alternate Example 6   **Solving a Rational Inequality by Combining Fractions**

Solve $\dfrac{3}{x-3} - \dfrac{1}{x} > 0$.

SOLUTION

We combine the two fractions on the left-hand side of the inequality using the least common denominator $x(x-3)$:

$\dfrac{3}{x-3} - \dfrac{1}{x} > 0$	Original Inequality
$\dfrac{3x}{x(x-3)} - \dfrac{1(x-3)}{x(x-3)} > 0$	Use LCD to rewrite fractions.
$\dfrac{3x - x + 3}{x(x-3)} > 0$	Subtract fractions.
$\dfrac{2x+3}{x(x-3)} > 0$	Simplify.

This inequality matches Alternate Example 5c. So, the solution is $(-3/2, 0) \cup (3, \infty)$.

Exercises for Alternate Example 6

In Exercises 13–18, solve each inequality.

13. $\dfrac{3}{x-2} - 2 \le 0$ **14.** $\dfrac{x+1}{x-3} \ge 3$ **15.** $\dfrac{3}{x-1} - \dfrac{1}{x} < 0$ **16.** $\dfrac{2}{x-1} + \dfrac{1}{x+1} \ge 0$ **17.** $\dfrac{3}{2x+3} - \dfrac{1}{x} < 0$

18. $\dfrac{4}{x-3} - \dfrac{1}{2x} > 0$

Chapter 3

Exponential, Logistic, and Logarithmic Functions

3.1 Exponential and Logistic Functions

Alternate Example 2 **Computing Exponential Function Values for Rational Number Inputs**

For $f(x) = 3^x$, find the value of the function for the given value of x.

(a) $x = 3$ **(b)** $x = 0$ **(c)** $x = -2$ **(d)** $x = \dfrac{1}{4}$ **(e)** $x = -\dfrac{3}{2}$

SOLUTION

(a) $f(3) = 3^3 = 3 \cdot 3 \cdot 3 = 27$

(b) $f(0) = 3^0 = 1$

(c) $f(-2) = 3^{-2} = \dfrac{1}{3^2} = \dfrac{1}{9} = 0.111\ldots$

(d) $f\left(\dfrac{1}{4}\right) = 3^{1/4} = \sqrt[4]{3} = 1.31607\ldots$

(e) $f\left(-\dfrac{3}{2}\right) = 3^{-3/2} = \dfrac{1}{3^{3/2}} = \dfrac{1}{\sqrt{3^3}} = \dfrac{1}{\sqrt{27}} = 0.19245\ldots$

Exercises for Alternate Example 2

In Exercises 1–6, for $f(x) = 4^x$, find the value of the function for the given value of x.

1. $x = 2$ **2.** $x = 0$ **3.** $x = -3$ **4.** $x = \dfrac{1}{2}$ **5.** $x = -\dfrac{1}{2}$ **6.** $x = -\dfrac{3}{2}$

Alternate Example 3 **Finding an Exponential Function from Its Table of Values**

Determine formulas for the exponential functions g and h whose values are given below.

Values for Two Exponential Functions

x	$g(x)$	$h(x)$
-2	$3/4$	18
-1	$3/2$	6
0	3	2
1	6	$2/3$
2	12	$2/9$

For $g(x)$: $\times 2$ between successive values.
For $h(x)$: $\times \dfrac{1}{3}$ between successive values.

SOLUTION

Because g is exponential, $g(x) = ab^x$. Because $g(0) = 3$, the initial value a is 3. Because $g(1) = 3 \cdot b^1 = 6$, the base b is 2. So, $g(x) = 3 \cdot 2^x$.

Because h is exponential, $h(x) = ab^x$. Because $h(0) = 2$, the initial value a is 2. Because

$h(1) = 2 \cdot b^1 = \dfrac{2}{3}$, the base b is $\dfrac{1}{3}$. So, $h(x) = 2 \cdot \left(\dfrac{1}{3}\right)^x$.

Figure 3.1 shows the graphs of these functions pass through the points whose coordinates are given in the table.

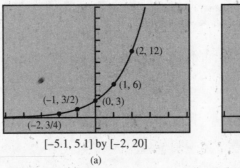
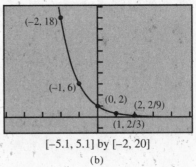

[−5.1, 5.1] by [−2, 20]	[−5.1, 5.1] by [−2, 20]
(a)	(b)

Figure 3.1 Graphs of (a) $g(x) = 3 \cdot 2^x$ and (b) $h(x) = 2 \cdot (1/3)^x$.

Exercises for Alternate Example 3

In Exercises 7–9, determine a formula for the exponential function g whose values are given in each table.

In Exercises 10–12, determine a formula for the exponential function h whose values are given in each table

7.

x	−2	−1	0	1	2
$g(x)$	$\dfrac{2}{25}$	$\dfrac{2}{5}$	2	10	50

10.

x	−2	−1	0	1	2
$h(x)$	48	12	3	$\dfrac{3}{4}$	$\dfrac{3}{16}$

8.

x	−2	−1	0	1	2
$g(x)$	$1\dfrac{3}{4}$	$3\dfrac{1}{2}$	7	14	28

11.

x	−2	−1	0	1	2
$h(x)$	100	20	4	0.8	0.16

9.

x	−2	−1	0	1	2
$g(x)$	$\dfrac{5}{9}$	$1\dfrac{2}{3}$	5	15	45

12.

x	−2	−1	0	1	2
$h(x)$	200	20	2	0.2	0.02

Alternate Example 4   **Transforming Exponential Functions**

Describe how to transform the graph of $f(x) = 3^x$ into the graph of the given function. Sketch the graphs by hand and support your answer with a grapher.

(a) $g(x) = 3^{x-2}$ **(b)** $h(x) = 3^{-x}$ **(c)** $k(x) = 2 \cdot 3^x$

SOLUTION

(a) The graph of $g(x) = 3^{x-2}$ is obtained by translating the graph of $f(x) = 3^x$ by two units to the right. (Figure 3.2a.).

(b) We can obtain the graph of $h(x) = 3^{-x}$ by reflecting the graph of $f(x) = 3^x$ across the y-axis (Figure 3.2b). Because $3^{-x} = (3^{-1})^x = (1/3)^x$, we can also think of h as an exponential function with an initial value of 1 and a base of 1/3.

(c) We can obtain the graph of $k(x) = 2 \cdot 3^x$ by vertically stretching the graph of $f(x) = 3^x$ by a factor of 2 (Figure 3-2c).

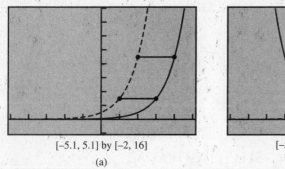
[-5.1, 5.1] by [-2, 16]
(a)

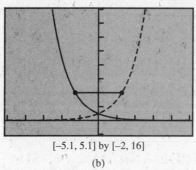
[-5.1, 5.1] by [-2, 16]
(b)

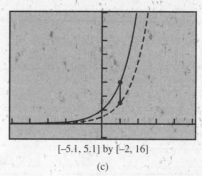
[-5.1, 5.1] by [-2, 16]
(c)

Figure 3.2 The graph of $f(x) = 3^x$ shown with (a) $g(x) = 3^{x-2}$, (b) $h(x) = 3^{-x}$, and (c) $k(x) = 2 \cdot 3^x$.

Exercises for Alternate Example 4

In Exercises 13–18, describe how to transform the graph of $f(x) = 5^x$ into the graph of the given function g. Sketch the graphs by hand and support your answer with a grapher.

13. $g(x) = 5^{x+3}$ **14.** $g(x) = 5^{x-1}$ **15.** $g(x) = -5^x$ **16.** $g(x) = 4 \cdot 5^x$ **17.** $k(x) = 7 \cdot 5^x$ **18.** $g(x) = 5^{x+5}$ ■

Alternate Example 6   Graphing Logistic Growth Functions

Graph the function. Find the y-intercept and the horizontal asymptotes.

(a) $f(x) = \dfrac{2}{1 + 2 \cdot 0.5^x}$ **(b)** $g(x) = \dfrac{15}{1 + 3 \cdot e^{-2x}}$

SOLUTION

(a) The graph of $f(x) = \dfrac{2}{1 + 2 \cdot 0.5^x}$ is shown in Figure 3.3a. The y-intercept is

$$f(0) = \frac{2}{1 + 2 \cdot 0.5^0} = \frac{2}{1 + 2} = \frac{2}{3}.$$

Because the limit to growth is 2, the horizontal asymptotes are $y = 0$ and $y = 2$.

(b) The graph of $g(x) = \dfrac{15}{1 + 3 \cdot e^{-2x}}$ is shown in Figure 3.3b. The y-intercept is

$$g(0) = \frac{15}{1 + 3 \cdot e^{-2(0)}} = \frac{15}{1 + 3} = \frac{15}{4}.$$

Because the limit to growth is 15, the horizontal asymptotes are $y = 0$ and $y = 15$.

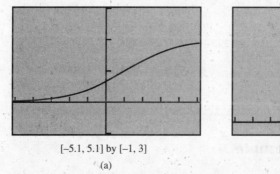
[-5.1, 5.1] by [-1, 3]
(a)

[-5.1, 5.1] by [-2, 20]
(b)

Figure 3.3 The graphs of (a) $f(x) = \dfrac{2}{1 + 2 \cdot 0.5^x}$ and (b) $g(x) = \dfrac{15}{1 + 3 \cdot e^{-2x}}$.

Exercises for Alternate Example 6

In Exercises 19–24, graph the function. Find the *y*-intercept and the horizontal asymptotes.

19. $f(x) = \dfrac{5}{1 + 3 \cdot 0.4^x}$ **20.** $g(x) = \dfrac{20}{1 + 2 \cdot e^{-2x}}$ **21.** $f(x) = \dfrac{1}{1 + 4 \cdot 0.2^x}$ **22.** $g(x) = \dfrac{12}{1 + 3 \cdot e^{-3x}}$

23. $f(x) = \dfrac{8}{1 + 3 \cdot 0.1^x}$ **24.** $g(x) = \dfrac{20}{1 + 4 \cdot e^{-x}}$

3.2 Exponential and Logistic Modeling

Alternate Example 3 Modeling Bacteria Growth

Suppose a culture of 200 bacteria is put into a petri dish and the culture doubles every hour. Predict when the number of bacteria will be 250,000.

SOLUTION

Model

$$400 = 200 \cdot 2 \qquad \text{Total bacteria after 1 hr}$$
$$800 = 200 \cdot 2^2 \qquad \text{Total bacteria after 2 hr}$$
$$1600 = 200 \cdot 2^3 \qquad \text{Total bacteria after 3 hr}$$
$$\vdots$$
$$P(t) = 200 \cdot 2^t \qquad \text{Total bacteria after } t \text{ hr}$$

So the function $P(t) = 200 \cdot 2^t$ represents the bacteria population t hr after it is placed in the Petri dish.

Solve Graphically

Figure 3.4 shows that the population function intersects $y = 250,000$ when $t \approx 10.2877$.

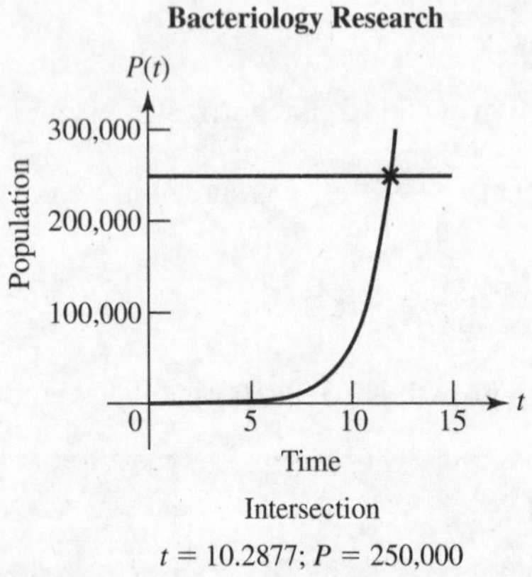

Bacteriology Research

Figure 3.4 Rapid growth of a bacteria population.

Interpret

The population of the bacteria in the petri dish will be 250,000 in about 10 hr and 18 min.

Exercises for Alternate Example 3

In Exercises 1–6, suppose a culture of bacteria is put into a petri dish and the culture doubles every hour. Predict when the number of bacteria will be the given number.

1. initial population: 300; target population: 240,000   **2.** initial population: 500; target population: 200,000

3. initial population: 240; target population: 240,000   **4.** initial population: 500; target population: 80,000

5. initial population: 100; target population: 30,000   **6.** initial population: 250; target population: 180,000

Alternate Example 4 **Modeling Radioactive Decay**

Suppose the half-life of a certain radioactive substance is 24 days and there are 10 g (grams) present initially. Find the time when there will be 2 g of the substance remaining.

Model

If t is the time in days, the number of half-lives will be $t/24$.

$$\frac{10}{2} = 10\left(\frac{1}{2}\right)^{24/24} \qquad \text{Grams after 24 days}$$

$$\frac{10}{4} = 10\left(\frac{1}{2}\right)^{48/24} \qquad \text{Grams after } 2(24) = 48 \text{ days}$$

$$\vdots$$

$$f(t) = 10\left(\frac{1}{2}\right)^{t/24} \qquad \text{Grams after } t \text{ days}$$

Thus the function $f(t) = 10 \cdot 0.5^{t/24}$ models the mass in grams of the radioactive substance at time t.

Solve Graphically

Figure 3.5 shows that the graph of $f(t) = 10 \cdot 0.5^{t/24}$ intersects $y = 2$ when $t \approx 55.7263$.

Interpret

There will be 2 g of the radioactive substance left after approximately 55.7 days, or about 55 days, 17 hr.

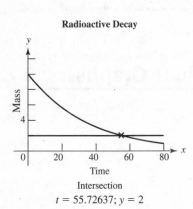

Radioactive Decay

Intersection
$t = 55.72637$; $y = 2$

Figure 3.5 Radioactive decay (Alternate Example 4).

Exercises for Alternate Example 4

In Exercises 7–12, the half-life of a certain radioactive substance is given and an initial amount is given. Find the time when there will be 15 g of the substance remaining.

7. half-life: 30 days; initial amount: 40 g

8. half-life: 10 days; initial amount: 30 g

9. half-life: 25 days; initial amount: 20 g

10. half-life: 60 days; initial amount: 60 g

11. half-life: 45 days; initial amount: 20 g

12. half-life: 40 days; initial amount: 40 g

Alternate Example 8 **Modeling a Rumor**

Jonesdale High School has 630 students. Josh, Mia, Tim, and Briana start a rumor, which spreads logistically so that $S(t) = 630/(1 + 29 \cdot e^{-0.8t})$ models the number of students who have heard the rumor by the end of t days, where $t = 0$ is the day the rumor begins to spread.

(a) How many students have heard the rumor by the end of Day 0?

(b) How long does it take for 500 students to hear the rumor?

SOLUTION

(a) $S(0) = \dfrac{630}{1 + 29 \cdot e^{-0.8(0)}} = \dfrac{630}{1 + 29} = 21$. So, 21 students have heard the rumor by the end of Day 0.

(b) We need to solve $\dfrac{630}{1 + 29 \cdot e^{-0.8(0)}} = 500$.

Figure 3.6 shows that the graph of $S(t) = 630/(1 + 29 \cdot e^{-0.8t})$ intersects $y = 500$ when $t \approx 5.90$. So toward the end of Day 6 the rumor has reached the ears of 500 students.

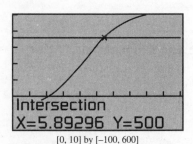

Intersection
X=5.89296 Y=500

[0, 10] by [–100, 600]

Figure 3.6 The spread of a rumor.

Exercises for Alternate Example 8

In Exercises 13–15, a rumor spreads logistically so that

$$S(t) = \frac{1500}{1 + 14 \cdot e^{-0.6t}}$$ models the number of people who have heard

the rumor by the end of t days, where $t = 0$ is the day the rumor begins to spread.

13. How many people have heard the rumor by the end of Day 0?

14. How long does it take for 400 people to hear the rumor?

15. How long does it take for 600 people to hear the rumor?

In Exercises 16–18, a rumor spreads logistically so that

$$S(t) = \frac{960}{1 + 19 \cdot e^{-0.7t}}$$ models the number of people who have heard

the rumor by the end of t days, where $t = 0$ is the day the rumor begins to spread.

16. How many people have heard the rumor by the end of Day 0?

17. How long does it take for 250 people to hear the rumor?

18. How long does it take for 480 people to hear the rumor?

3.3 Logarithmic Functions and Their Graphs

Alternate Example 1 **Evaluating Logarithms**

Evaluate each expression.

(a) $\log_3 27$

(b) $\log_2 \sqrt{2}$

(c) $\log_4 \frac{1}{16}$

(d) $\log_5 1$

(e) $\log_{10} 10$

SOLUTION

(a) $\log_3 27 = 3$ because $3^3 = 27$

(b) $\log_2 \sqrt{2} = \frac{1}{2}$ because $2^{1/2} = \sqrt{2}$

(c) $\log_4 \frac{1}{16} = -2$ because $4^{-2} = \frac{1}{16}$

(d) $\log_5 1 = 0$ because $5^0 = 1$

(e) $\log_{10} 10 = 1$ because $10^1 = 10$

Exercises for Alternate Example 1

In Exercises 1–8, evaluate each expression without using a calculator.

1. $\log_5 625$ **2.** $\log_3 \sqrt{3}$ **3.** $\log_5 \frac{1}{25}$ **4.** $\log_7 1$ **5.** $\log_3 3$ **6.** $\log_5 \sqrt{5}$

7. $\log_3 81$ **8.** $\log_{10} \sqrt[3]{10}$

Alternate Example 3 **Evaluating Logarithmic and Exponential Expressions – Base 10**

Evaluate each expression.

(a) $\log 1{,}000$

(b) $\log \sqrt[3]{10}$

(c) $\log \frac{1}{100}$

(d) $10^{\log 5}$

SOLUTION

(a) $\log 1,000 = 3$ because $10^3 = 1,000$

(b) $\log \sqrt[3]{10} = \log 10^{1/3} = \dfrac{1}{3}$

(c) $\log \dfrac{1}{100} = \log \dfrac{1}{10^2} = \log 10^{-2} = -2$

(d) $10^{\log 5} = 5$

Exercises for Alternate Example 3

In Exercises 9–16, evaluate each expression without using a calculator.

9. $\log 10,000$ **10.** $\log \sqrt[4]{10}$ **11.** $\log \dfrac{1}{10}$ **12.** $10^{\log 7}$ **13.** $\log 100,000$ **14.** $\log \sqrt[7]{10}$

15. $\log \dfrac{1}{10,000}$ **16.** $10^{\log 11}$

■

Alternate Example 6 **Evaluating Logarithmic and Exponential Expressions – Base *e***

Evaluate each expression.

(a) $\ln \sqrt[3]{e}$ (b) $\ln e^6$ (c) $e^{\ln 5}$

SOLUTION

(a) $\ln \sqrt[3]{e} = \log_e \sqrt[3]{e} = \dfrac{1}{3}$ because $e^{1/3} = \sqrt[3]{e}$

(b) $\ln e^6 = \log_e e^6 = 6$

(c) $e^{\ln 5} = 5$

Exercises for Alternate Example 6

In Exercises 17–26, evaluate each expression without using a calculator.

17. $\ln \sqrt[5]{e}$ **18.** $\ln e^{11}$ **19.** $e^{\ln e}$ **20.** $\ln \sqrt[7]{e}$ **21.** $\ln e^6$ **22.** $e^{\ln 5}$ **23.** $\ln e^{12}$

24. $\ln \sqrt[9]{e}$ **25.** $\ln e^8$ **26.** $\ln \sqrt[4]{e}$

■

Alternate Example 8 **Transforming Logarithmic Graphs**

Describe how to transform the graph of $y = \ln x$ or $y = \log x$ into the graph of the given function.

(a) $g(x) = \ln (x - 3)$

(b) $h(x) = \ln (2 - x)$

(c) $g(x) = 2 \log x$

(d) $h(x) = 2 + \log x$

SOLUTION

(a) The graph of $g(x) = \ln(x - 3)$ is obtained by translating the graph of $y = \ln x$ three units to the right. See Figure 3.7a.

(b) $h(x) = \ln(2 - x) = \ln[-(x - 2)]$. So we can obtain the graph of $h(x) = \ln(2 - x)$ from the graph of $y = \ln x$ by applying, in order, a reflection across the y-axis followed by a translation 2 units to the right. See Figure 3.7b.

(c) The graph of $g(x) = 2 \log x$ is obtained by vertically stretching the graph of $f(x) = \log x$ by a factor of 2. See Figure 3.7c.

(d) We can obtain the graph of $h(x) = 2 + \log x$ from the graph of $f(x) = \log x$ by a translation 2 units up. See Figure 3.7d.

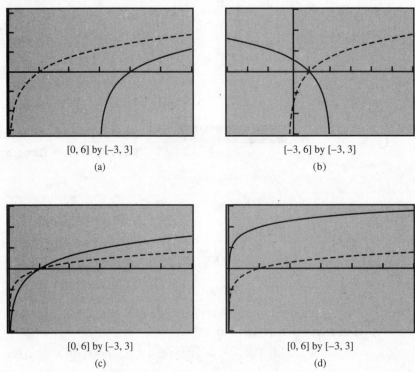

[0, 6] by [−3, 3]	[−3, 6] by [−3, 3]
(a)	(b)

[0, 6] by [−3, 3]	[0, 6] by [−3, 3]
(c)	(d)

Figure 3.7 Transforming $y = \ln x$ to obtain (a) $g(x) = \ln(x - 3)$ and (b) $h(x) = \ln(2 - x)$; and $y = \log x$ to obtain (c) $g(x) = 2 \log x$ and (d) $h(x) = 2 + \log x$.

Exercises for Alternate Example 8

In Exercises 27–34, describe how to transform the graph of $y = \ln x$ or $y = \log x$ into the graph of the given function.

27. $g(x) = \ln(x - 1)$ **28.** $h(x) = \ln(1 - x)$ **29.** $g(x) = 3 \log x$ **30.** $h(x) = 3 + \log x$

31. $g(x) = \ln(x + 1)$ **32.** $h(x) = \ln(4 - x)$ **33.** $g(x) = 3.5 \log x$ **34.** $h(x) = 1.5 + \log x$

3.4 Properties of Logarithmic Functions

Alternate Example 2 Expanding the Logarithm of a Product

Assuming x and y are positive, use properties of logarithms to write $\log(16x^2y^3)$ as a sum of logarithms or multiples of logarithms.

SOLUTION

$$\log(16x^2y^3) = \log 16 + \log x^2 + \log y^3 \qquad \text{Product rule}$$
$$= \log 2^4 + \log x^2 + \log y^3 \qquad 16 = 2^4$$
$$= 4 \log 2 + 2 \log x + 3 \log y \qquad \text{Power rule}$$

Exercises for Alternate Example 2

In Exercises 1–8, assuming x and y are positive, use properties of logarithms to write each expression as a sum of logarithms or multiples of logarithms.

1. $\log (25xy^3)$ **2.** $\log (100x^3y^4)$ **3.** $\log (144x^5y)$ **4.** $\log (8xy)$ **5.** $\log (64x^2y^3)$

6. $\log (81xy)$ **7.** $\log (x^2y^3)$ **8.** $\log (100xy)$

Alternate Example 3 **Expanding the Logarithm of a Quotient**

Assuming x is positive, use properties of logarithms to write $\ln \dfrac{\sqrt{x^2-3}}{x}$ as a sum or difference of logarithms or multiples of logarithms.

SOLUTION

$$\ln \frac{\sqrt{x^2-3}}{x} = \ln \frac{(x^2-3)^{1/2}}{x}$$

$$= \ln (x^2-3)^{1/2} - \ln x \qquad \text{Quotient rule}$$

$$= \frac{1}{2} \ln (x^2-3) - \ln x \qquad \text{Power rule}$$

Exercises for Alternate Example 3

In Exercises 9–16, assuming x is positive, use properties of logarithms to write each expression as a sum or difference of logarithms or multiples of logarithms.

9. $\ln \dfrac{x^2+1}{x}$ **10.** $\ln \dfrac{(x+3)^3}{x^2}$ **11.** $\ln \dfrac{\sqrt{x+1}}{x^2}$ **12.** $\ln \dfrac{\sqrt{x^3+1}}{x}$ **13.** $\ln \dfrac{(x-2)^2}{x^2}$

14. $\ln \dfrac{(x-2)^2}{(x+1)^2}$ **15.** $\ln \dfrac{\sqrt{x^2-100}}{x^6}$ **16.** $\ln \sqrt{\dfrac{x^2+1}{x^2-1}}$

Alternate Example 4 **Condensing a Logarithmic Expression**

Assuming x and y are positive, use properties of logarithms to write $\ln x^4 - 3 \ln (xy)$ as a single logarithm.

SOLUTION

$$\ln x^4 - 3 \ln (xy) = \ln x^4 - \ln (xy)^3 \qquad \text{Power rule}$$

$$= \ln x^4 - \ln (x^3y^3)$$

$$= \ln \frac{x^4}{x^3y^3} \qquad \text{Quotient rule}$$

$$= \ln \frac{x}{y^3}$$

Exercises for Alternate Example 4

In Exercises 17–24, assuming x and y are positive, use properties of logarithms to write each expression as a single logarithm.

17. $\ln x^2 - \ln (xy)$ **18.** $\ln x^2 - 4 \ln (xy)$ **19.** $\ln x^4 - 3 \ln (x^2y)$ **20.** $2 \ln x^4 - \ln (x^3y)$ **21.** $3 \ln x - 5 \ln (xy)$

22. $2 \ln x - 2 \ln (xy)$ **23.** $3 \ln x - \ln (xy^4)$ **24.** $3 \ln x^2 - 2 \ln (xy)$

Alternate Example 6 **Graphing Logarithmic Functions**

Describe how to transform the graph of $f(x) = \ln x$ into the graph of the given function. Sketch the graph by hand and support your answer with a grapher.

(a) $g(x) = \log_3 x$ **(b)** $h(x) = \log_{1/3} x$

SOLUTION

(a) $g(x) = \log_3 x = \ln x / \ln 3$, its graph is obtained by vertically shrinking the graph of $f(x) = \ln x$ by a factor of $1/\ln 3 \approx 0.91$. See Figure 3.8a.

(b) $h(x) = \log_{1/3} x = \dfrac{\ln x}{\ln 1/3} = \dfrac{\ln x}{\ln 1 - \ln 3} = \dfrac{\ln x}{-\ln 3} = -\dfrac{1}{\ln 3} \ln x$

So we can obtain the graph of h from the graph of $f(x) = \ln x$ by applying, in either order, a reflection across the x-axis and a vertical shrink by a factor of $1/\ln 3 \approx 0.91$. See Figure 3.8b.

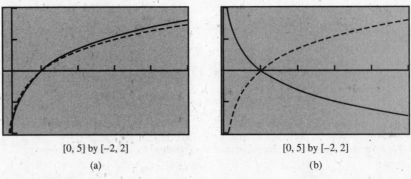

[0, 5] by [−2, 2]   [0, 5] by [−2, 2]

(a)   (b)

Figure 3.8 Transforming of $f(x) = \ln x$ to obtain (a) $g(x) = \log_3 x$ and (b) $h(x) = \log_{1/3} x$.

Exercises for Alternate Example 6

In Exercises 25–30, describe how to transform the graph of $f(x) = \ln x$ into the graph of the given function. Sketch the graph by hand and support your answer with a grapher.

25. $g(x) = \log_2 x$ **26.** $h(x) = \log_{1/2} x$ **27.** $g(x) = \log_4 x$ **28.** $h(x) = \log_{1/5} x$ **29.** $g(x) = \log_7 x$

30. $h(x) = \log_{1/6} x$

3.5 Equation Solving and Modeling

Alternate Example 1 **Solving an Exponential Equation Algebraically**

Solve $27(2/3)^{x/4} = 12$.

SOLUTION

$$27\left(\frac{2}{3}\right)^{x/4} = 12$$

$$\left(\frac{2}{3}\right)^{x/4} = \frac{4}{9} \qquad \text{Divide by 27.}$$

$$\left(\frac{2}{3}\right)^{x/4} = \left(\frac{2}{3}\right)^2 \qquad \frac{4}{9} = \left(\frac{2}{3}\right)^2$$

$$\frac{x}{4} = 2 \qquad \text{One - to - one property}$$

$$x = 8 \qquad \text{Multiply by 4.}$$

Exercises for Alternate Example 1

In Exercises 1–8, solve each equation.

1. $24\left(\dfrac{1}{2}\right)^{x/3} = 12$ **2.** $9\left(\dfrac{1}{3}\right)^{x/2} = 1$ **3.** $25\left(\dfrac{2}{5}\right)^{x/4} = 4$ **4.** $81\left(\dfrac{2}{3}\right)^{x/3} = 16$

5. $1000\left(\dfrac{1}{10}\right)^{x/2} = 1$ **6.** $64\left(\dfrac{1}{4}\right)^{x/3} = 1$ **7.** $64\left(\dfrac{3}{8}\right)^{x/4} = 9$ **8.** $125\left(\dfrac{3}{5}\right)^{x/3} = 27$

Alternate Example 3 **Solving a Logarithmic Equation**

Solve $\log x^4 = 4$.

SOLUTION
Method 1
Use the one-to-one property of logarithms.

$$\log x^4 = 4$$
$$\log x^4 = \log 10^4 \qquad\qquad y = \log 10^y$$
$$x^4 = 10^4 \qquad\qquad \text{One-to-one property}$$
$$x^4 = 10{,}000 \qquad\qquad 10^4 = 10{,}000$$
$$x = 10 \quad \text{or} \quad x = -10$$

Method 2
Change the equation from logarithmic to exponential form.

$$\log x^4 = 4$$
$$x^4 = 10^4 \qquad\qquad \text{Change to exponential form.}$$
$$x^4 = 10{,}000 \qquad\qquad 10^4 = 10{,}000$$
$$x = 10 \quad \text{or} \quad x = -10$$

Method 3 (Incorrect)
Use the power rule of logarithms.

$$\log x^4 = 4$$
$$4 \log x = 4 \qquad\qquad \text{Power rule}$$
$$\log x = 1 \qquad\qquad \text{Divide by 4.}$$
$$x = 10 \qquad\qquad \text{Change to exponential form.}$$

Support Graphically
Figure 3.9 shows that the graphs of $f(x) = \log x^4$ and $y = 4$ intersect when $x = -10$. From the symmetry of the graphs due to f being an even function, we can see that $x = 10$ is also a solution.

Interpret
Method 1 and 2 are correct. Method 3 fails because the domain of $\log x^2$ is all nonzero real numbers, but the domain of $\log x$ is only the positive real numbers. The correct solution includes both 10 and −10 because both of these x-values make the original equation true.

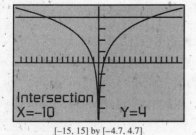

[−15, 15] by [−4.7, 4.7]

Figure 3.9 Graphs of $f(x) = \log x^4$ and $y = 4$.

Exercises for Alternate Example 3

In Exercises 9–16, solve each equation.

9. $\log x = 2$ **10.** $\log x = 3$ **11.** $\log 2x = 1$ **12.** $\log 4x = 3$ **13.** $\log x^2 = 8$ **14.** $\log x^2 = 3$

15. $\log 4x^2 = 2$ **16.** $\log 2x^2 = 2$

Alternate Example 7 Applying Newton's Law of Cooling

A hard-boiled egg at temperature 97°C is placed in 17°C water to cool. Three minutes later the temperature of the egg is 48°C. Use Newton's Law of Cooling to determine when the egg will be 20°C.

SOLUTION

Model

Because $T_0 = 97$ and $T_m = 17$, $T_0 - T_m = 80$ and $T(t) = T_m + (T_0 - T_m)e^{-kt} = 17 + 80e^{-kt}$.

To find the value of k we use the fact that $T = 48$ when $t = 3$.

$$48 = 17 + 80e^{-3k}$$

$$31 = 80e^{-3k}$$

$$\frac{31}{80} = e^{-3k} \qquad \text{Subtract 17, then divide by 80.}$$

$$\ln \frac{31}{80} = -3k \qquad \text{Change to logarithmic form.}$$

$$k = -\frac{\ln \dfrac{31}{80}}{3} \qquad \text{Divide by } -3.$$

$$k = 0.316\ldots$$

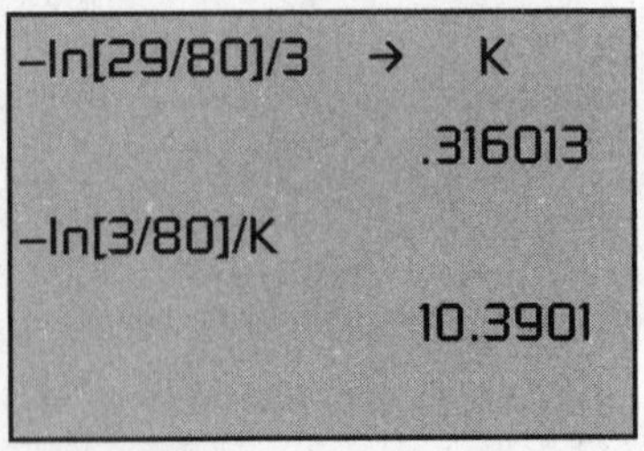

We save this k value because it is part of our model. (See Figure 3.10.)

```
-ln[29/80]/3   →   K
                    .316013
-ln[3/80]/K
                    10.3901
```

Figure 3.10 Saving and using the constant k.

Solve Algebraically

To find t when $T = 20°C$, we solve the equation:

$$20 = 17 + 80e^{-kt}$$

$$\frac{3}{80} = e^{-kt} \qquad \text{Subtract 17, then divide by 80.}$$

$$\ln \frac{3}{80} = -kt$$

$$t \approx -\frac{\ln \dfrac{3}{80}}{k} \approx 10.39 \qquad \text{See Figure 3.10.}$$

Interpret

The temperature of the egg will be 20°C after about 10.39 minutes (10 min 24 sec).

Exercises for Alternate Example 7

In Exercises 17–22, a hard-boiled egg at temperature 95°C is placed in 15°C water to cool. After the given number of minutes, the temperature of the egg is the given temperature. Use Newton's Law of Cooling to determine when the egg will be 20°C.

17. Two minutes later the temperature of the egg is 65°C.

18. Two minutes later the temperature of the egg is 55°C.

19. Three minutes later the temperature of the egg is 55°C.

20. Four minutes later the temperature of the egg is 55°C.

21. Five minutes later the temperature of the egg is 55°C.

22. Seven minutes later the temperature of the egg is 45°C.

Alternate Example 9 Selecting a Regression Model

Decide whether these data can be best represented by logarithmic, exponential, or power regression. Find the appropriate regression model.

x	1	2	3	4	5	6
y	5.3	7.2	9.6	12	17	23

SOLUTION

The shape of the data plot in Figure 3.11 suggests that the data could be modeled by an exponential or power function.

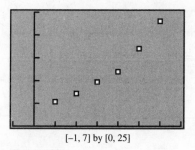

Figure 3.11 A scatter plot of the original data in Alternate Example 9

Figure 3.12a shows the $(x, \ln y)$ plot, and Figure 3.12b shows the $(\ln x, \ln y)$ plot. Of these two plots, the $(x, \ln y)$ plot appears to be more linear, so we find the exponential regression model for the original data.

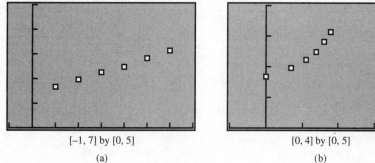

| [−1, 7] by [0, 5] | [0, 4] by [0, 5] |
| (a) | (b) |

Figure 3.12 (a) the $(x, \ln y)$ plot and (b) the $(\ln x, \ln y)$ plot (Alternate Example 9).

Figure 3.13 shows the scatter plot of the original (x, y) data with the graph of the exponential regression model $y = 3.9706(1.3360^x)$.

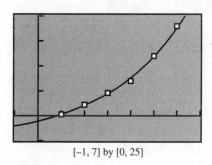

[−1, 7] by [0, 25]

Figure 3.13 An exponential regression model fits the data of Alternate Example 9.

Exercises for Alternate Example 9

In Exercises 23–28, tables of (x, y) data pairs are given. Determine whether a linear, logarithmic, exponential, or power regression equation is the best model for the data. Support your solution with a graph.

23.

x	1	2	3	4	5
y	7	11	24	48	96

24.

x	1	2	3	4	5
y	2	4.5	5	6	7

25.

x	1	2	3	4	5
y	3	9	16	23	36

26.

x	1	2	3	4	5
y	4	9	11	13	17

27.

x	1	2	3	4	5
y	3	2.3	1.9	1.6	1.4

28.

x	1	2	3	4	5
y	3	4.2	5	6	6.7

3.6 Mathematics of Finance

Alternate Example 3 **Finding the Time Period of an Investment**

Mazie has $600 to invest at 8% annual interest compounded monthly. How long will it take for her investment to grow to $2400?

SOLUTION

Let $P = 600$, $r = 0.08$, $k = 12$, and $A = 2400$ in the equation

$$A = P\left(1 + \frac{r}{k}\right)^{kt},$$

and solve for t.

Solve Graphically

For $2400 = 600\left(1 + \frac{0.08}{12}\right)^{12t}$, we let $f(t) = 600\left(1 + \frac{0.08}{12}\right)^{12t}$ and $y = 2400$,

and then find the point of intersection of the graphs. Figure 3.14 shows that this occurs at $t \approx 17.38$.

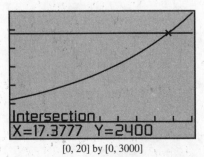

[0, 20] by [0, 3000]

Figure 3.14 Graph for Alternate Example 3.

Confirm Algebraically

$$2400 = 600\left(1 + \frac{0.08}{12}\right)^{12t}$$

$4 = 1.00667^{12t}$	Divide by 600.
$\ln 4 = \ln 1.00667^{12t}$	Apply ln to each side.
$\ln 4 = 12t \ln 1.00667$	Power rule
$t = \dfrac{\ln 4}{12 \ln 1.00667}$	Divide by 12 ln 1.00667.
$= 17.3777$	Calculate.

Interpret

So it will take Mazie 18 years for the value of the investment to reach (and slightly exceed) $2400.

Exercises for Alternate Example 3

In Exercises 1–6, an investment is made and interest is compounded monthly. How long will it take for the given investment to grow to the given goal?

1. principal: $1000, rate: 7%, target amount: $1500

2. principal: $1000, rate: 8%, target amount: $1800

3. principal: $1600, rate: 9%, target amount: $2400

4. principal: $1600, rate: 9%, target amount: $3200

5. principal: $500, rate: 6%, target amount: $1400

6. principal: $6000, rate: 8%, target amount: $10,000

Alternate Example 5 **Compounding Continuously**

Suppose Moesha invests $200 at 7% annual interest compounded continuously. Find the value of her investment at the end of each of the years 1, 2, . . . , 7.

SOLUTION

Substituting into the formula for continuous compounding, we obtain $A(t) = 200e^{0.07t}$. Figure 3.15 shows the values of $y_1 = A(x) = 200e^{0.07x}$ for $x = 1, 2, \ldots, 7$. For example the value of her investment is $ 264.63 at the end of 4 years, and $ 304.39 at the end of 6 years.

X	Y1	
1	214.50	
2	230.06	
3	246.74	
4	264.63	
5	283.81	
6	304.39	
7	326.46	

Y1=200e^[0.07X]

Figure 3.15 Table of values for Alternate Example 5.

Exercises for Alternate Example 5

In Exercises 7–12, interest is compounded continuously. Find the value of each investment at the end of each of the years 1, 2, . . . , 7.

7. principal: $800; rate: 8%
8. principal: $1200; rate: 6%
9. principal: $500; rate: 6%
10. principal: $1800; rate: 9%
11. principal: $900; rate: 7%
12. principal: $1500; rate: 6%

Alternate Example 8 **Calculating the Value of an Annuity**

At the end of each quarter year, Emile makes a $400 payment into the Smithville Financial Fund. If his investments earn 7.75% annual interest compounded quarterly, what will be the value of Emile's annuity in 25 years?

SOLUTION

Let $R = 400$, $i = 0.0775/4$, $n = 25(4) = 100$. Then,

$$FV = R\frac{(1 + i)^n - 1}{i}$$

$$FV = 400 \cdot \frac{(1 + 0.0775/4)^{100} - 1}{0.0775/4}$$

$$FV = 120{,}030$$

So the value of Emile's annuity in 25 years will be $120,030.

Exercises for Alternate Example 8

In Exercises 13–18, find the value of an annuity after investing the given payments for t years at the given interest rate, with payments made and interest credited k times per year.

13. $300 payment earning 6% annual interest, $t = 25$, $k = 4$.
14. $500 payment earning 7% annual interest, $t = 20$, $k = 4$.
15. $200 payment earning 3% annual interest, $t = 25$, $k = 12$.
16. $800 payment earning 6.5% annual interest, $t = 20$, $k = 2$.
17. $750 payment earning 5.5% annual interest, $t = 30$, $k = 6$.
18. $350 payment earning 5.25% annual interest, $t = 20$, $k = 4$.

Alternate Example 9 **Calculating Loan Payments**

Chantelle purchases a new car for $19,000. What are the monthly payments for a 3-year loan with a $3000 down payment if the annual interest rate (APR) is 3.1%?

SOLUTION

Model

The down payment is $3000, so the amount borrowed is $16,000. Since APR = 3.1%, $i = 0.031/12$ and the monthly payment is the solution to

$$16,000 = R\frac{1 - (1 + 0.031/12)^{-3(12)}}{0.031/12}$$

Solve Algebraically

$$R\left[1 - \left(1 + \frac{0.031}{12}\right)^{-3(12)}\right] = 16,000\left(\frac{0.031}{12}\right)$$

$$R = \frac{16,000(0.031/12)}{1 - (1 + 0.031/12)^{-36}}$$

$$= 466.00$$

Interpret

Chantelle will have to pay $466.00 per month for 35 months, and slightly less the last month.

Exercises for Alternate Example 9

In Exercises 19–24, what are the monthly payments for the given loan?

19. purchase price: $20,000; down payment: $2500; APR: 3%; time: 4 years

20. purchase price: $22,000; down payment: $4000; APR: 5%; time: 3 years

21. purchase price: $16,000; down payment: $3000; APR: 4.5%; time: 3 years

22. purchase price: $19,500; down payment: $4500; APR: 3.6%; time: 5 years

23. purchase price: $18,400; down payment: $3600; APR: 3.2%; time: 4 years

24. purchase price: $17,600; down payment: $4200; APR: 4.6%; time: 5 years

Chapter 4   Trigonometric Functions

4.1   Angles and Their Measures

Working with DMS Measure

(a) Convert 23.325° to DMS.

(b) Convert 30°30′18″ to degrees.

SOLUTION

(a) We need to convert the fractional part to minutes and seconds. First we convert 0.325° to minutes:

$$0.325\left(\frac{60'}{1°}\right) = 19.5'$$

Then we convert 0.5 minutes to seconds:

$$0.5'\left(\frac{60''}{1'}\right) = 30''$$

Putting it all together, we find that 23.325° = 23°19′30″.

(b) Each minute is $1/60^{\text{th}}$ of a degree, and each second is $1/3600^{\text{th}}$ of a degree. Therefore,

$$30°30'18'' = 30° + \left(\frac{30}{60}\right)° + \left(\frac{18}{3600}\right)° = 30.505°.$$

Exercises for Alternate Example 1

In Exercises 1–3, convert from decimal form to DMS.

1. 34.5°   **2.** 60.3°   **3.** 50.125°

In Exercises 4–6, convert from DMS to decimal form.

4. 10°30′45″   **5.** 24°12′18″   **6.** 29°36″ ■

Working with Radian Measure

(a) How many radians are in 45°?

(b) How many degrees are in $\pi/10$ radians?

(c) Find the length of an arc intercepted by a central angle of 1/8 radian in a circle of radius 24 inches.

(d) Find the radian measure of a central angle that intercepts an arc of length s in a circle of radius r.

SOLUTION

(a) Since π radians and 180° both measure a straight angle, we can use the conversion factor $(\pi \text{ radians})/(180°) = 1$ to convert radians to degrees:

$$45°\left(\frac{\pi \text{ radians}}{180°}\right) = \frac{45\pi}{180} \text{ radians} = \frac{\pi}{4} \text{ radians}.$$

(b) In this case, we use the conversion factor $(180°)/(\pi \text{ radians}) = 1$ to convert radians to degrees:

$$\left(\frac{\pi}{10}\text{ radians}\right)\left(\frac{180°}{\pi \text{ radians}}\right) = \frac{180°}{10} = 18°$$

(c) A central angle of 1 radian intercepts an arc of length 1 radius, which is 24 inches. Therefore, a central angle of 1/8 radian intercepts an arc of length 1/8 radius, which is 3 inches.

(d) We can solve this problem with ratios:

$$\frac{x \text{ radians}}{s \text{ units}} = \frac{1 \text{ radian}}{r \text{ units}}$$
$$xr = s$$
$$x = \frac{s}{r}$$

Exercises for Alternate Example 2

In Exercises 7–9, convert to radians.

7. 72°   **8.** 90°   **9.** 120°

In Exercises 10–12, convert to degrees.

10. $\pi/6$ radians   **11.** $\pi/5$ radians   **12.** $\pi/18$ radians

13. Find the length of an arc intercepted by a central angle of 1/6 radian in a circle of radius 30 inches.

14. Find the length of an arc intercepted by a central angle of 1/3 radian in a circle of radius 42 inches.

Alternate Example 3 **Perimeter of a Pizza Slice**

Find the perimeter of a 45° slice of a large (10 in. radius) pizza.

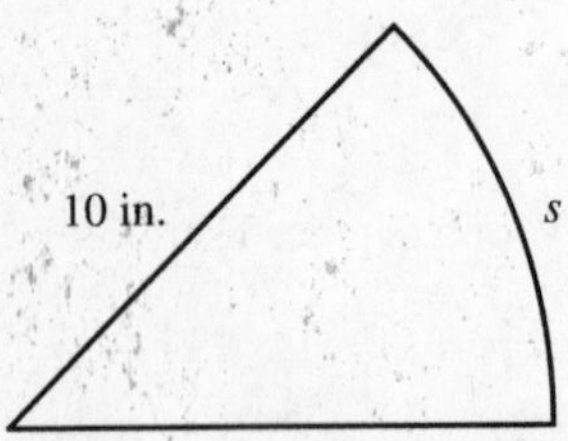

Figure 4.1 A 45° slice of a large pizza.

SOLUTION

The perimeter (Figure 4.1) is 10 in. + 10 in. + s in., where s is the arc length of the pizza's curved edge. By the arc length formula:

$$s = \frac{\pi(10)(45)}{180} = \frac{5\pi}{2} \approx 7.9$$

The perimeter is approximately 27.9 in.

Exercises for Alternate Example 3

In Exercises 15–20, find the perimeter of a slice of pizza given each central angle measure and pizza radius.

15. central angle: 60°; radius: 5 in.   **16.** central angle: 60°; radius: 6 in.   **17.** central angle: 60°; radius: 7 in.

18. central angle: 72°; radius: 5 in.   **19.** central angle: 72°; radius: 6 in.   **20.** central angle: 120°; radius: 4 in.

Alternate Example 5 **Using Angular Speed**

Denzel Murphy's truck has wheels 36 inches in diameter. If the wheels are rotating at 720 rpm (revolutions per minute), find the truck's speed in miles per hour.

SOLUTION

We convert revolutions per minute to miles per hour by a series of unit conversion factors. Note that the conversion factor 18 in./1 radian works for this example because the radius is 18 in.

$$\frac{720 \text{ rev}}{1 \text{ min}} \times \frac{60 \text{ min}}{1 \text{ hr}} \times \frac{2\pi \text{ radians}}{1 \text{ rev}} \times \frac{18 \text{ in.}}{1 \text{ radian}} \times \frac{1 \text{ ft}}{12 \text{ in.}} \times \frac{1 \text{ mi}}{5280 \text{ ft}} \approx 77.11 \frac{\text{mi}}{\text{hr}}$$

Exercises for Alternate Example 5

In Exercises 21–26, a truck has wheels 36 inches in diameter. If the wheels are rotating at the given number of revolutions per minute, find the truck's speed in miles per hour.

21. 600 rpm   **22.** 540 rpm   **23.** 620 rpm   **24.** 350 rpm   **25.** 480 rpm   **26.** 250 rpm

27. Clark's truck has wheels 30 inches in diameter. If the wheels are rotating at 600 rpm (revolutions per minute), find the truck's speed in miles per hour.

28. Liza Crane's car has wheels 28 inches in diameter. If the wheels are rotating at 700 rpm (revolutions per minute), find the car's speed in miles per hour.

4.2 Trigonometric Functions of Acute Angles

Alternate Example 3 **Using One Trigonometric Ratio to Find Them All**

Let θ be an acute angle such that $\cos \theta = 4/7$. Evaluate the other five trigonometric functions of θ.

SOLUTION

Sketch a triangle showing an acute angle θ. Label the adjacent side 4 and the hypotenuse 7. (See Figure 4.2.) Since $\cos \theta = 4/7$, this must be our angle. Now we need the other side of the triangle (labeled x in the figure).

From the Pythagorean theorem it follows that $x^2 + 4^2 = 7^2$, so $x = \sqrt{49 - 16} = \sqrt{33}$. Applying the definitions,

$$\sin \theta = \frac{opp}{hyp} = \frac{\sqrt{33}}{7} \approx 0.821 \qquad \csc \theta = \frac{hyp}{opp} = \frac{7}{\sqrt{33}} \approx 1.219$$

$$\cos \theta = \frac{adj}{hyp} = \frac{4}{7} \approx 0.571 \qquad \sec \theta = \frac{hyp}{adj} = \frac{7}{4} \approx 1.75$$

$$\tan \theta = \frac{opp}{adj} = \frac{\sqrt{33}}{4} \approx 1.436 \qquad \cot \theta = \frac{adj}{opp} = \frac{4}{\sqrt{33}} \approx 0.696$$

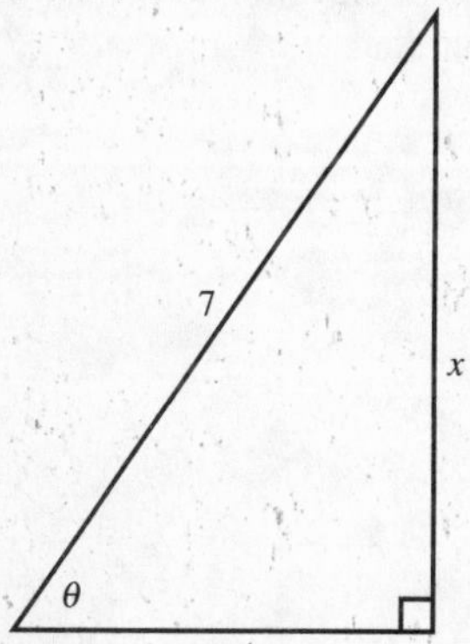

Figure 4.2 How to create an acute angle θ such that $\cos \theta = 4/7$.

Exercises for Alternate Example 3

In Exercises 1–6, let θ be an acute angle such that the given condition is met. Evaluate the other five trigonometric functions of θ.

1. $\tan \theta = 4/3$ **2.** $\cos \theta = 2/3$ **3.** $\sin \theta = 1/6$ **4.** $\sec \theta = 7/5$ **5.** $\sin \theta = 4/9$ **6.** $\csc \theta = 8/5$

Alternate Example 5 **Solving a Right Triangle**

A right triangle with hypotenuse 6 includes a 35° angle (Figure 4.3). Find the measures of the other two angles and the lengths of the other two sides.

SOLUTION

Since it is a right triangle, one of the other angles is 90°. That leaves $180° - 90° - 35° = 55°$ for the third angle.

Referring to the labels in Figure 4.3, we have

$$\sin 35° = \frac{a}{6} \qquad \cos 35° = \frac{b}{6}$$

$$a = 6 \sin 35° \qquad b = 6 \cos 35°$$

$$a \approx 3.44 \qquad b \approx 4.91$$

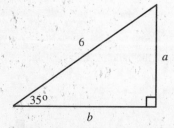

Figure 4.3 (Alternate Example 5)

Exercises for Alternate Example 5

In Exercises 7–12, solve each problem.

7. A right triangle with hypotenuse 10 includes a 65° angle. Find the measures of the other two angles and the lengths of the other two sides.

8. A right triangle with hypotenuse 20 includes a 42° angle. Find the measures of the other two angles and the lengths of the other two sides.

9. A right triangle with hypotenuse 8 includes a 38° angle. Find the measures of the other two angles and the lengths of the other two sides.

10. A right triangle with hypotenuse 25 includes a 70° angle. Find the measures of the other two angles and the lengths of the other two sides.

11. A right triangle with shorter leg of length 50 includes a 60° angle. Find the measures of the other two angles and the lengths of the other leg and the hypotenuse.

12. A right triangle with hypotenuse 45 includes a 25° angle. Find the measures of the other two angles and the lengths of the other two sides.

Alternate Example 6 **Finding the Height of a Building**

From a point 80 feet away from the base of a building, the angle of elevation to the top of the building is 72°. (See Figure 4.4.) Find the height h of the building.

72°

80 ft

Figure 4.4 (Alternate Example 6)

SOLUTION

We need a ratio that will relate an angle to its opposite and adjacent sides. The tangent function is the appropriate choice.

$$\tan 72° = \frac{h}{80}$$
$$h = 80 \tan 72°$$
$$h \approx 246$$

Exercises for Alternate Example 6

In Exercises 13–18, using the given distance from the base of the building and the angle of elevation to the top of the building, find the height of the building.

13. distance from base: 60 ft; angle of elevation: 70°

14. distance from base: 70 ft; angle of elevation: 75°

15. distance from base: 45 ft; angle of elevation: 68°

16. distance from base: 30 ft; angle of elevation: 62°

17. distance from base: 55 ft; angle of elevation: 69°

18. distance from base: 25 ft; angle of elevation: 80°

Trigonometry Extended: The Circular Functions

4.3

Evaluating Trig Functions Determined by a Point Not in QI

Let θ be any angle in standard position whose terminal side contains the point $(-2, 5)$. Find the six trigonometric functions of θ.

SOLUTION

The distance from $(-2, 5)$ to the origin is $\sqrt{29}$.

So

$$\sin \theta = \frac{5}{\sqrt{29}} \approx 0.928 \qquad \csc \theta = \frac{\sqrt{29}}{5} \approx 1.077$$

$$\cos \theta = \frac{-2}{\sqrt{29}} \approx -0.371 \qquad \sec \theta = \frac{\sqrt{29}}{-2} \approx -2.693$$

$$\tan \theta = \frac{5}{-2} = -2.5 \qquad \cot \theta = \frac{-2}{5} = -0.4$$

Exercises for Alternate Example 3

In Exercises 1–6, let θ be an angle whose terminal side contains the given point. Find the six trigonometric functions of θ.

1. $(3, 7)$ 2. $(-3, -3)$ 3. $(-5, 3)$ 4. $(-2, -5)$ 5. $(3, 2)$ 6. $(3, -5)$

Using One Trig Ratio to Find the Others

Find $\sin \theta$ and $\tan \theta$ by using the given information to construct a reference triangle.

(a) $\cos \theta = -\dfrac{4}{5}$ and $\tan \theta < 0$

(b) $\sec \theta = \dfrac{5}{3}$ and $\sin \theta > 0$

(c) $\cot \theta$ is undefined and $\sin \theta$ is positive

SOLUTION

(a) Since $\cos \theta$ is negative, the terminal side is either in QII or QIII. The added fact that $\tan \theta$ is negative means that the terminal side is in QII. We draw a reference triangle in QII with $r = 5$ and $x = 4$ (Figure 4.5); then we use the Pythagorean theorem to find that $y = \sqrt{5^2 - 4^2} = 3$. (Note that y is positive in QII.)

We then use the definitions to get

$$\sin \theta = \frac{3}{5} \text{ and } \tan \theta = -\frac{3}{4}.$$

(b) Since $\sec \theta$ is positive, the terminal side is either in QI or QIV. The added fact that $\sin \theta$ is positive means that the terminal side is in QI. We draw a reference triangle in QI with $r = 5$ and $x = 3$ (Figure 4.6); then we use the Pythagorean theorem to find that $y = \sqrt{5^2 - 3^2} = 4$. (Note that y is positive in QI.)

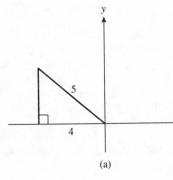

(a)

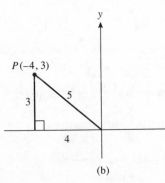

(b)

Figure 4.5

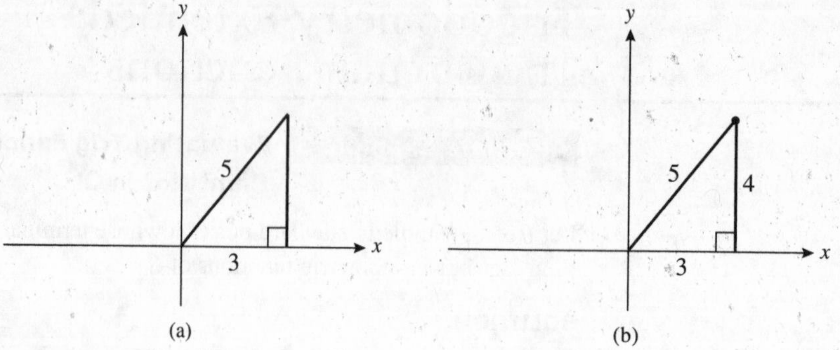

(a)   (b)

Figure 4.6

We then use the definitions to get

$$\sin \theta = \frac{4}{5} \text{ and } \tan \theta = \frac{4}{3}.$$

(c) Since cot θ is undefined, we conclude that $y = 0$ and that θ is a quadrantal angle on the x-axis. The added fact that sin θ is positive means that the terminal side lies along the positive x-axis. We choose the point $(1, 0)$ on the terminal side and use the definitions to get

$$\sin \theta = 1 \text{ and } \tan \theta = \frac{0}{1} = 0.$$

Exercises for Alternate Example 7

In Exercises 7–12, use the given information to find the values of the six trigonometric functions of θ without a calculator.

7. $\cos \theta = -\frac{3}{5}$ and $\tan \theta < 0$ **8.** $\cos \theta = \frac{3}{5}$ and $\tan \theta < 0$ **9.** $\csc \theta = 1$ and $\tan \theta$ is undefined

10. $\csc \theta = -1$ and $\tan \theta$ is undefined   **11.** $\tan \theta$ is undefined and $\csc \theta$ is positive

12. $\tan \theta$ is undefined and $\sin \theta$ is positive

Alternate Example 8   Using Periodicity

Find each of the following numbers without a calculator.

(a) $\sin\left(\dfrac{23{,}561\pi}{2}\right)$

(b) $\cos (134.5\pi) - \cos (142.5\pi)$

(c) $\tan\left(\dfrac{\pi}{4} - 11{,}111\pi\right)$

SOLUTION

(a) $\sin\left(\dfrac{23{,}561\pi}{2}\right) = \sin\left(\dfrac{\pi}{2} + \dfrac{23{,}560\pi}{2}\right) = \sin\left(\dfrac{\pi}{2} + 11{,}780\pi\right) = \sin\left(\dfrac{\pi}{2}\right) = 1$

Notice that $11{,}780\pi$ is just a multiple of 2π, so $\pi/2$ and $((\pi/2) + 11780\pi)$ wrap to the same point on the unit circle, namely $(0, 1)$.

(b) $\cos (134.5\pi) - \cos (142.5\pi) = \cos (134.5\pi) - \cos (134.5\pi + 8\pi) = 0$

Notice that 134.5π and $(134.5\pi + 8\pi)$ wrap to the same point on the unit circle, so the cosine of one is the same as the cosine of the other.

(c) Since the period of the tangent function is π rather than 2π, $11{,}111\pi$ is a large multiple of the period of the tangent function. Therefore,

$$\tan\left(\frac{\pi}{4} - 11{,}111\pi\right) = \tan\left(\frac{\pi}{4}\right) = 1.$$

Exercises for Alternate Example 8

In Exercises 13–18, find each of the following numbers without a calculator.

13. $\sin\left(\dfrac{4881\pi}{2}\right)$ **14.** $\cos\left(\dfrac{44{,}881\pi}{2}\right)$ **15.** $\tan\left(\dfrac{\pi}{4} - 3247\pi\right)$ **16.** $\tan\left(\dfrac{\pi}{4} - 88{,}888\pi\right)$

17. $\cos(371.9\pi) - \cos(379.9\pi)$ **18.** $\sin(894.5\pi) - \sin(902.5\pi)$

4.4 Graphs of Sine and Cosine: Sinusoids

Alternate Example 1   Vertical Stretch or Shrink and Amplitude

Find the amplitude of each function and use the language of transformations to describe how the graphs are related.

(a) $y_1 = \sin x$ **(b)** $y_2 = \dfrac{1}{3}\sin x$ **(c)** $y_3 = -2\sin x$

SOLUTION

Solve Algebraically

The amplitudes are **(a)** 1, **(b)** 1/3, and **(c)** $|-2| = 2$.

The graph of y_2 is a vertical shrink of the graph of y_1 by a factor of 1/3.

The graph of y_3 is a vertical stretch of the graph of y_1 by a factor of 2, and a reflection across the x-axis, performed in either order. (We do not call this a vertical stretch by a factor of –2, nor do we say that the amplitude is –2.)

Support Graphically

The graphs of the three functions are shown in Figure 4.7. You should be able to tell which is which quite easily by checking the amplitudes.

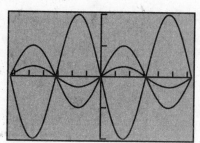

$[-2\pi, 2\pi]$ by $[-2, 2]$

Figure 4.7 Sinusoids (in this case, sine curves) of different amplitudes.

Exercises for Alternate Example 1

In Exercises 1–6, find the amplitude of each function and use the language of transformations to describe how the graphs are related.

1. $y_1 = \sin x$ and $y_2 = \dfrac{1}{2}\sin x$ **2.** $y_1 = \sin x$ and $y_2 = 3\sin x$ **3.** $y_1 = \cos x$ and $y_2 = \dfrac{1}{2}\cos x$

4. $y_1 = \sin x$ and $y_2 = -4\sin x$ **5.** $y_1 = \cos x$ and $y_2 = \dfrac{1}{4}\cos x$ **6.** $y_1 = \sin x$ and $y_2 = -\sin x$

Alternate Example 2   Horizontal Stretch or Shrink and Period

Find the period of each function and use the language of transformations to describe how the graphs are related.

(a) $y_1 = \sin x$ **(b)** $y_2 = -3\sin\left(\dfrac{x}{2}\right)$ **(c)** $y_3 = 2\sin(-2x)$

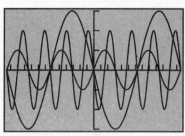

[−4π, 4π] by [−3, 3]

Figure 4.8 Sinusoids (in this case, sine curves) of different amplitudes and periods.

SOLUTION

Solve Algebraically

The periods are (a) 2π, (b) $2\pi/(1/2) = 4\pi$, and (c) $2\pi/|{-2}| = \pi$.

The graph of y_2 is a horizontal stretch of the graph of y_1 by a factor of 2, a vertical stretch by a factor of 3, and a reflection across the x-axis, performed in any order.

The graph of y_3 is a horizontal shrink of the graph of y_1 by a factor of 1/2, a vertical stretch by a factor of 2, and a reflection across the y-axis, performed in any order. (Note that we do not call this a horizontal shrink by a factor of $-1/2$, nor do we say that the period is $-\pi$.)

Support Graphically

The graphs of the three functions are shown in Figure 4.8. You should be able to tell which is which quite easily by checking the periods or the amplitudes.

Exercises for Alternate Example 2

In Exercises 7–12, find the period of each function and use the language of transformations to describe how the graphs are related.

7. $y_1 = \sin x$ and $y_2 = \dfrac{1}{2} \sin\left(\dfrac{x}{2}\right)$

8. $y_1 = \sin x$ and $y_2 = -\sin(3x)$

9. $y_1 = \cos x$ and $y_2 = 2\cos\left(\dfrac{x}{2}\right)$

10. $y_1 = \sin x$ and $y_2 = \sin(4x)$

11. $y_1 = \cos x$ and $y_2 = 3\cos\left(\dfrac{x}{3}\right)$

12. $y_1 = \cos x$ and $y_2 = -5\cos(2x)$

■

Alternate Example 3 **Finding the Frequency of a Sinusoid**

Find the frequency of the function $f(x) = 3\sin(3x/2)$ and interpret its meaning graphically.

Sketch the graph in the window $[-2\pi, 2\pi]$ by $[-3, 3]$.

SOLUTION

The frequency is $(3/2) \div 2\pi = 3/(4\pi)$. This is the reciprocal of the period, which is $4\pi/3$. The graphical interpretation is that the graph completes 1 full cycle per interval of length $4\pi/3$. (That, of course, is what having a period of $4\pi/3$ is all about.) The graph is shown in Figure 4.9.

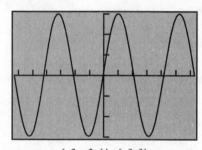

[−2π, 2π] by [−3, 3]

Figure 4.9 The graph of the function $f(x) = 3\sin(3x/2)$. It has frequency $3/(4\pi)$, so it completes 1 full cycle per interval of length $4\pi/3$.

Exercises for Alternate Example 3

In Exercises 13–18, find the frequency of each function and interpret its meaning graphically. Sketch the graph.

13. $f(x) = \sin(x/2)$

14. $f(x) = 3\sin(2x)$

15. $f(x) = 2\sin(2x/5)$

16. $f(x) = 2\sin(5x/2)$

17. $f(x) = 3\cos(3x/2)$

18. $f(x) = 3\cos(2x/3)$

■

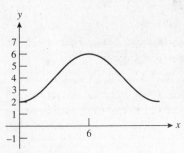

Figure 4.10 A sinusoid with specifications.

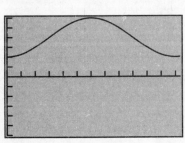

$[0, 4\pi]$ by $[-6, 6]$

Figure 4.11 The graph of the

function $y = -2 \cos\left(\dfrac{\pi x}{6}\right) + 4$.

Alternate Example 6   **Constructing a Sinusoid by Transformations**

Construct a sinusoid $y = f(x)$ that rises from a minimum value of $y = 2$ at $x = 0$ to a maximum of $y = 6$ at $x = 6$ (Figure 4.10).

SOLUTION
Solve Algebraically

The amplitude of this sinusoid is half the height of the graph: $(6 - 2)/2 = 2$. So $|a| = 2$. The period is 12 (since a full period goes from minimum to maximum and back down to the minimum). So we set $2\pi/|b| = 12$. Solving, we get $|b| = \pi/6$.

We need a sinusoid that takes on its minimum value at $x = 0$. We could shift the graph of sine or cosine horizontally, but it is easier to take the cosine curve (which assumes its maximum value at $x = 0$) and turn it upside down. This reflection can be obtained by letting $a = -2$ rather than 2.

So far we have:

$$y = -2 \cos\left(\pm\frac{\pi x}{6}\right)$$

$$= -2 \cos\left(\frac{\pi x}{6}\right) \qquad \text{(Since cos is an even function)}$$

Finally note that this function ranges from a minimum of -2 to a maximum of 2. We shift the graph vertically by 4 to obtain a function that ranges from a minimum of 2 to a maximum of 6, as required. Thus

$$y = -2 \cos\left(\frac{\pi x}{6}\right) + 4.$$

Support Graphically

We support our answer graphically by graphing the function (Figure 4.11).

Exercises for Alternate Example 6

In Exercises 19–24, construct a sinusoid $y = f(x)$ that rises from the given minimum value to the given maximum.

19. minimum: $y = 4$ at $x = 0$; maximum: $y = 8$ at $x = 6$

20. minimum: $y = 1$ at $x = 0$; maximum: $y = 5$ at $x = 4$

21. minimum: $y = 3$ at $x = 0$; maximum: $y = 9$ at $x = 5$

22. minimum: $y = 6$ at $x = 0$; maximum: $y = 8$ at $x = 2$

23. minimum: $y = 5$ at $x = 0$; maximum: $y = 11$ at $x = 6$

24. minimum: $y = 0$ at $x = 0$; maximum: $y = 8$ at $x = 5$

4.5 Graphs of Tangent, Cotangent, Secant, and Cosecant

Alternate Example 1   **Graphing a Tangent Function**

Describe the graph of $y = -\tan 3x$ in terms of a basic trigonometric function. Locate the vertical asymptotes and graph four periods of the function.

SOLUTION

The effect of the 3 is a horizontal shrink of the graph of $y = \tan x$ by a factor of 1/3, while the effect of the -1 is a reflection across the x-axis. Since the vertical asymptotes of $y = \tan x$ are all the odd multiples of $\pi/2$, the shrink factor causes the vertical asymptotes of $y = \tan 3x$ to be all odd multiples of $\pi/6$ (Figure 4.12a). The reflection across the x-axis (Figure 4.12b) does not change the asymptotes.

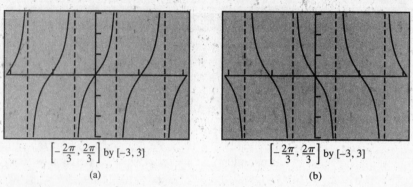

$$\left[-\frac{2\pi}{3}, \frac{2\pi}{3}\right] \text{ by } [-3, 3]$$

(a)

$$\left[-\frac{2\pi}{3}, \frac{2\pi}{3}\right] \text{ by } [-3, 3]$$

(b)

Figure 4.12 The graph of (a) $y = \tan 3x$ is reflected across the x-axis to produce the graph of (b) $y = -\tan 3x$.

Since the period of the function $y = \tan x$ is π, the period of the function $y = -\tan 3x$ is (thanks again to the shrink factor) $\pi/3$. Thus, any window of horizontal length $4\pi/3$ will show four periods of the graph. Figure 4.12b uses the window $[-2\pi/3, 2\pi/3]$ by $[-3, 3]$.

Exercises for Alternate Example 1

In Exercises 1–6, describe the graph of each function in terms of $y = \tan x$. Locate the vertical asymptotes and graph four periods of the function.

1. $y = -\tan x$ **2.** $y = \tan 2x$ **3.** $y = -\tan 2x$ **4.** $y = -\tan 4x$ **5.** $y = \tan 3x$ **6.** $y = \tan 4x$ ∎

Alternate Example 2   **Graphing a Cotangent Function**

Describe the graph of $f(x) = 2 \cot (x/2) - 1$ in terms of a basic trigonometric function. Locate the vertical asymptotes and graph four periods of the function.

SOLUTION

The graph is obtained from the graph of $y = \cot x$ by effecting a horizontal stretch by a factor of 2, a vertical stretch by a factor of 2, and a vertical translation down one unit. The horizontal stretch makes the period of the function 2π (twice the period of $y = \cot x$), and the asymptotes are at the even multiples of π. Figure 4.13 shows two periods of the graph of f.

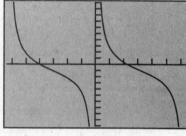

$[-2\pi, 2\pi]$ by $[-10, 10]$

Figure 4.13 Two periods of $f(x) = 2 \cot (x/2) - 1$.

Exercises for Alternate Example 2

In Exercises 7–12, describe the graph of each function in terms of $y = \cot x$. Locate the vertical asymptotes and graph four periods of the function.

7. $y = \cot x + 1$ **8.** $y = \cot x - 2$ **9.** $y = -\cot x$ **10.** $y = -2 \cot x$ **11.** $y = \cot (x/2) + 1$

12. $y = 2 \cot 2x + 1$ ∎

Alternate Example 3 **Solving a Trigonometric Equation Algebraically**

Find the value of x between $\pi/2$ and π that solves $\csc x = 2$.

SOLUTION

We construct a reference triangle in the second quadrant that has the appropriate ratio, *hyp/opp*, equal to 2. This is most easily accomplished by choosing a y-coordinate of 1 and a hypotenuse of 2. (Figure 4.14a) We recognize this as a 30°-60°-90° triangle that determines and angle of 150°, which converts to $5\pi/6$ radians.

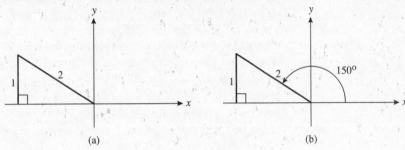

(a)   (b)

Figure 4.14 A reference triangle in the second quadrant (a) with *hyp/opp* = 2 determines an angle (b) of 150°, which converts to $5\pi/6$ radians.

Exercises for Alternate Example 3

In Exercises 13–18, find the value of x in the given interval. You should be able to find the solution without a calculator by using a reference triangle in the proper quadrant.

13. $\csc x = \dfrac{2}{\sqrt{3}}$; $0 < x < \pi/2$ **14.** $\csc x = \dfrac{2}{\sqrt{2}}$; $\pi/2 < x < \pi$ **15.** $\sec x = \dfrac{2}{\sqrt{2}}$; $3\pi/2 < x < 2\pi$

16. $\sec x = -\dfrac{2}{\sqrt{2}}$; $\pi/2 < x < \pi$ **17.** $\tan x = -1$; $\pi/2 < x < \pi$ **18.** $\csc x = -2$; $\pi < x < 2\pi$

4.6 Graphs of Composite Trigonometric Functions

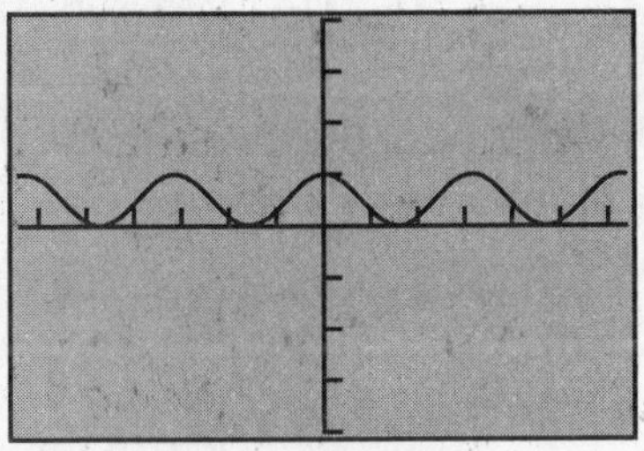

Alternate Example 2 **Verifying Periodicity Algebraically**

Verify algebraically that $f(x) = (\cos x)^2$ is periodic and determine its period graphically.

SOLUTION

We use the fact that the period of the basic cosine function is 2π, that is, $\cos(x + 2\pi) = \cos x$ for all x. It follows that

$$f(x + 2\pi) = (\cos(x + 2\pi))^2$$
$$= (\cos(x))^2 \qquad \text{By periodicity of cosine}$$
$$= f(x)$$

So $f(x)$ is also periodic, with some period that divides 2π. The graph in Figure 4.15 shows that the period is actually π.

$[-2\pi, 2\pi]$ by $[-4, 4]$

Figure 4.15 The graph of $f(x) = (\cos x)^2$. It exhibits periodic behavior over the interval $[-2\pi, 2\pi]$.

Exercises for Alternate Example 2

In Exercises 1–6, verify algebraically that the given function is periodic and determine its period graphically.

1. $f(x) = (\sin 3x)^2$ **2.** $f(x) = (\sin 2x)^2$ **3.** $f(x) = (\cos 2x)^2$ **4.** $f(x) = (\cos 4x)^2$ **5.** $f(x) = (-\cos x)^2$

6. $f(x) = (-\sin x)^2$

Alternate Example 4   Analyzing Nonnegative Periodic Functions

Find the domain, range, and period of each of the following functions. Sketch a graph showing four periods.

(a) $f(x) = |\cos x|$ **(b)** $g(x) = |\cot x|$

SOLUTION

(a) Whenever $\cos x$ is defined, so is $|\cos x|$. Therefore, the domain of f is the same as the domain of the cosine function, that is, all real numbers. Because $f(x) = |\cos x| \geq 0$ and the range of $\cos x$ is $[-1, 1]$, the range of f is $[0, 1]$. The period of f is only half the period of $y = \cos x$, for reasons that are apparent from viewing the graph. The negative sections of the cosine curve below the x-axis are reflected above the x-axis, where they become repetitions of the positive sections. The graph of $y = f(x)$ is shown in Figure 4.16.

(b) Whenever $\cot x$ is defined, so is $|\cot x|$. Therefore, the domain of g is the same as the domain of the cotangent function, that is, all real numbers except $n\pi$, $n = 0, \pm 1, \ldots$. Because $g(x) = |\cot x| \geq 0$ and the range of $\cot x$ is $(-\infty, \infty)$, the range of g is $[0, \infty)$. The period of g, like that of $y = \cot x$, is π. The graph of $y = g(x)$ is shown in Figure 4.17.

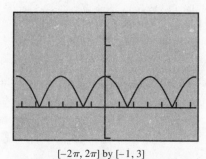

$[-2\pi, 2\pi]$ by $[-1, 3]$

Figure 4.16 $f(x) = |\cos x|$ has half the period as $y = \cos x$

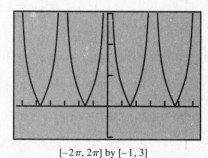

$[-2\pi, 2\pi]$ by $[-1, 3]$

Figure 4.17 $g(x) = |\cot x|$ has the same period as $y = \cot x$.

Exercises for Alternate Example 4

In Exercises 7–12, find the domain, range, and period of each of the following functions. Sketch a graph showing four periods.

7. $f(x) = |2 \cos x|^2$ **8.** $f(x) = (\sin x)^2$ **9.** $f(x) = (2 \cos x)^2$ **10.** $f(x) = -(0.5 \cos x)^2$ **11.** $f(x) = (\tan x)^2$
12. $f(x) = |2 \cot x|^2$

Alternate Example 7   Expressing the Sum of Sinusoids as a Sinusoid

Let $f(x) = 3 \sin x + 4 \cos x$. Since both $\sin x$ and $\cos x$ have period 2π, f is periodic and is a sinusoid.

(a) Find the period of f.

(b) Estimate the amplitude and phase shift graphically (to the nearest hundredth).

(c) Give a sinusoid in the form $a \sin (b(x - h))$ that approximates $f(x)$.

SOLUTION

(a) The period of f is the same as the period of $\sin x$ and $\cos x$, namely 2π.

Solve Graphically

(b) We will learn an algebraic way to find the amplitude and phase shift in the next chapter, but for now we will find this information graphically. Figure 4.18 suggests that f is indeed a sinusoid. That is, for some a and h,

$$3 \sin x + 4 \cos x = a \sin (x - h).$$

The *maximum value is* 5, so the amplitude of *f* is 5. The *x*-intercept closest to *x* = 0, rounded to the nearest hundredth, is –0.93, so the phase shift of the sine function is about –0.93. We conclude that

$$f(x) = a \sin (x - h) = 5 \sin (x + 0.93).$$

(c) We support our answer graphically by showing that the graphs of $y = 3 \sin x + 4 \cos x$ and $y = 5 \sin (x + 0.93)$ are virtually identical (Figure 4.19).

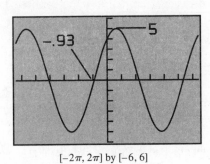

[–2π, 2π] by [–6, 6]

Figure 4.18 The sum of two sinusoids: $f(x) = 3 \sin x + 4 \cos x$.

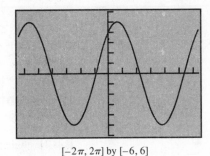

[–2π, 2π] by [–6, 6]

Figure 4.19 The graphs of $y = 3 \sin x + 4 \cos x$ and $y = 5 \sin (x + 0.93)$ are virtually identical.

Exercises for Alternate Example 7

In Exercises 13–18, **(a)** find the period of *f*, **(b)** estimate the amplitude and phase shift graphically (to the nearest hundredth), and **(c)** give a sinusoid $a \sin (x - h)$ that approximates $f(x)$.

13. $f(x) = \sin x + \cos x$ **14.** $f(x) = 4 \sin x + 3 \cos x$ **15.** $f(x) = 2 \sin x + \cos x$ **16.** $f(x) = 2 \sin x + 2 \cos x$

17. $f(x) = \sin x - \cos x$ **18.** $f(x) = 3 \sin x + \cos x$

4.7 Inverse Trigonometric Functions

Alternate Example 1 **Evaluating sin⁻¹ *x* without a Calculator**

Find the exact value of each expression without a calculator.

(a) $\sin^{-1}\left(-\dfrac{1}{2}\right)$ **(b)** $\sin^{-1}\left(\dfrac{\sqrt{3}}{2}\right)$ **(c)** $\sin^{-1}\left(-\dfrac{\pi}{2}\right)$

(d) $\sin^{-1}\left(\sin\left(\dfrac{\pi}{5}\right)\right)$ **(e)** $\sin^{-1}\left(\sin\left(\dfrac{3\pi}{4}\right)\right)$

SOLUTION

(a) Find the point on the right half of the unit circle whose *y*-coordinate is –1/2 and draw a reference triangle (Figure 4.20). We recognize this as one of our special ratios, and the angle in the interval [–π/2, π/2] whose sine is –1/2 is –π/6. Therefore

$$\sin^{-1}\left(-\frac{1}{2}\right) = -\frac{\pi}{6}.$$

(b) Find the point on the right half of the unit circle whose *y*-coordinate is $\sqrt{3}/2$ and draw a reference triangle (Figure 4.21). We recognize this as one of our special ratios,

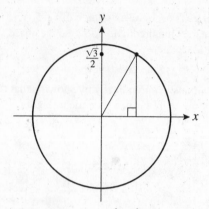

Figure 4.20 $\sin^{-1}\left(-\dfrac{1}{2}\right) = \dfrac{\pi}{6}$

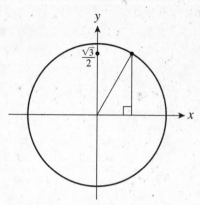

Figure 4.21 $\sin^{-1}\left(\dfrac{\sqrt{3}}{2}\right) = \dfrac{\pi}{3}$

and the angle in the interval $[-\pi/2, \pi/2]$ whose sine is $\sqrt{3}/2$ is $\pi/3$. Therefore

$$\sin^{-1}\left(\frac{\sqrt{3}}{2}\right) = \frac{\pi}{3}.$$

(c) $\sin^{-1}\left(-\dfrac{\pi}{2}\right)$ does not exist, because the domain of $\sin^{-1}$ is $[-1, 1]$ and $-\pi/2 < -1$.

(d) Draw an angle of $\pi/5$ in standard position and mark its y-coordinate on the y-axis (Figure 4.22). The angle in the interval $[-\pi/2, \pi/2]$ whose sine is this number is $\pi/5$. Therefore

$$\sin^{-1}\left(\sin\left(\frac{\pi}{5}\right)\right) = \frac{\pi}{5}.$$

(e) Draw an angle of $3\pi/4$ in standard position (notice that this angle is not in the interval $[-\pi/2, \pi/2]$) and mark its y-coordinate on the y-axis (Figure 4.23). The angle in the interval $[-\pi/2, \pi/2]$ whose sine is this number is $\pi - 3\pi/4$. Therefore,

$$\sin^{-1}\left(\sin\left(\frac{3\pi}{4}\right)\right) = \frac{\pi}{4}.$$

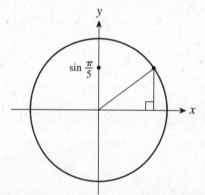

Figure 4.22 $\sin^{-1}\left(\sin\left(\dfrac{\pi}{5}\right)\right) = \dfrac{\pi}{5}$,

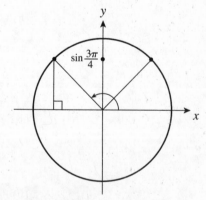

Figure 4.23 $\sin^{-1}\left(\sin\left(\dfrac{3\pi}{4}\right)\right) = \dfrac{\pi}{4}$

Exercises for Alternate Example 1

In Exercises 1–6, find the exact value of each expression without a calculator.

1. $\sin^{-1}\left(-\dfrac{\sqrt{2}}{2}\right)$　　**2.** $\sin^{-1}\left(-\dfrac{\sqrt{3}}{2}\right)$　　**3.** $\sin^{-1}\left(-\dfrac{\pi}{3}\right)$　　**4.** $\sin^{-1}\left(\sin\left(\dfrac{\pi}{11}\right)\right)$

5. $\sin^{-1}\left(\sin\left(\dfrac{2\pi}{3}\right)\right)$　　**6.** $\sin^{-1}\left(\sin\left(\dfrac{5\pi}{6}\right)\right)$

Alternate Example 2 **Evaluating $\sin^{-1} x$ with a Calculator**

Use a calculator in radian mode to evaluate these inverse sine values:

(a) $\sin^{-1}(-0.73)$

(b) $\sin^{-1}(\sin(3.45\pi))$

SOLUTION

(a) $\sin^{-1}(-0.73) = -0.8183219\ldots = -0.818$

(b) $\sin^{-1}(\sin(3.45\pi)) = -1.413716694\ldots \approx -1.414$

Although this is a calculator answer, we can use it to get an exact answer if we are alert enough to expect a multiple of π. Divide the answer by π:

$\text{Ans}/\pi = -0.45$

Therefore we can conclude that $\sin^{-1}(\sin(3.45\pi)) = -0.45\pi$.

You should also try to work Example 2b without a calculator. It is possible.

Exercises for Alternate Example 2

In Exercises 7–12, use a calculator in radian mode to evaluate these inverse sine values.

7. $\sin^{-1}(-0.51)$　　**8.** $\sin^{-1}(\sin(2.45\pi))$　　**9.** $\sin^{-1}(-0.89)$　　**10.** $\sin^{-1}(\sin(5.55\pi))$　　**11.** $\sin^{-1}(-0.21)$

12. $\sin^{-1}(\sin(4.78\pi))$

Alternate Example 3 **Evaluating Inverse Trig Functions without a Calculator**

Find the exact value of each expression without a calculator.

(a) $\cos^{-1}\left(\dfrac{\sqrt{3}}{2}\right)$　　　**(b)** $\tan^{-1} 1$　　**(c)** $\cos^{-1}(\cos(-1.5))$

SOLUTION

(a) Find the point on the top half of the unit circle whose x-coordinate is $\sqrt{3}/2$ and draw a reference triangle (Figure 4.24). We recognize this as one of our special ratios, and the angle in the interval $[0, \pi]$ whose cosine is $\sqrt{3}/2$ is $\pi/6$. Therefore

$$\cos^{-1}\left(\dfrac{\sqrt{3}}{2}\right) = \dfrac{\pi}{6}.$$

(b) Find the point on the right side of the unit circle whose y-coordinate equals the x-coordinate and draw a reference triangle (Figure 4.25). We recognize this as one of our special ratios, and the angle in the interval $(-\pi/2, \pi/2)$ whose tangent is 1 is $\pi/4$.

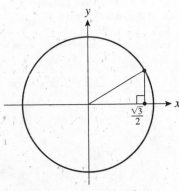

Figure 4.24　$\cos^{-1}\left(\dfrac{\sqrt{3}}{2}\right) = \dfrac{\pi}{6}$　　　$\tan^{-1} 1 = \dfrac{\pi}{4}$

(c) Draw an angle of −1.5 in standard position (notice that this angle is not in the interval $[0, \pi]$) and mark its x-coordinate on the x-axis (Figure 4.26). The angle in the interval $[0, \pi]$ whose cosine is this number is 1.5. Therefore

$$\cos^{-1}(\cos(-1.5)) = 1.5.$$

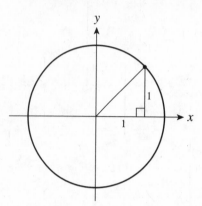

Figure 4.25 $\tan^{-1} 1 = \dfrac{\pi}{4}$

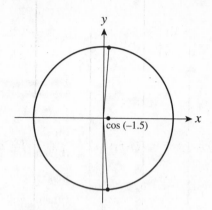

Figure 4.26 $\cos^{-1}(\cos(-1.5)) = 1.5$

Exercises for Alternate Example 3

In Exercises 13–18, find the exact value of each expression without a calculator.

13. $\cos^{-1}\left(-\dfrac{\sqrt{3}}{2}\right)$ **14.** $\cos^{-1}\left(-\dfrac{\sqrt{2}}{2}\right)$ **15.** $\tan^{-1}(-1)$ **16.** $\tan^{-1} 0$ **17.** $\cos^{-1}(\cos(-0.6))$

18. $\cos^{-1}(\cos(-1.8))$

4.8 Solving Problems with Trigonometry

Alternate Example 2 **Making Indirect Measurements**

From the top of the 120-ft tall Tallman Hall a man observes a car moving toward the building. If the angle of depression of the car changes from 25° to 45° during the period of observation, how far does the car travel?

SOLUTION

Solve Algebraically

Figure 4.27 models the situation. Notice that we have labeled the acute angles at the car's two positions as 25° and 45° (because the angle of elevation from the car equals the angle of depression from the building). Denote the distance the car moves as x. Denote its distance from the building at the second observation as d.

From the smaller right triangle we conclude:

$$\tan 45° = \frac{120}{d}$$

$$d = \frac{120}{\tan 45°}$$

Figure 4.27 A car approaches Tallman Hall.

From the larger right triangle we conclude:

$$\tan 25° = \frac{120}{x+d}$$

$$x + d = \frac{120}{\tan 25°}$$

$$x = \frac{120}{\tan 25°} - d$$

$$x = \frac{120}{\tan 25°} - \frac{120}{\tan 45°}$$

$$x \approx 137.3$$

Interpreting our answer, we find that the car travels about 137 feet.

Exercises for Alternate Example 2

In Exercises 1–6, from the top of a building h feet tall, a man observes a car moving toward the building. If the angle of depression of the car changes from a degrees to b degrees during the period of observation, how far does the car travel?

1. $h = 250$ ft; $a = 20°$; $b = 70°$ **2.** $h = 150$ ft; $a = 23°$; $b = 48°$ **3.** $h = 180$ ft; $a = 15°$; $b = 64°$

4. $h = 210$ ft; $a = 23°$; $b = 49°$ **5.** $h = 420$ ft; $a = 10°$; $b = 50°$ **6.** $h = 480$ ft; $a = 25°$; $b = 53°$

Alternate Example 4 **Using Trigonometry in Navigation**

A U.S. Coast Guard patrol boat leaves port and averages 35 knots (nautical mph) traveling for 2 hours on a course of 55° and then 3 hours on a course of 145°. What is the boat's bearing and distance from port?

SOLUTION

Figure 4.28 models the situation.

Solve Algebraically

In the diagram, line AB is a transversal that cuts a pair of parallel lines. Thus, $\beta = 55°$ because they are alternate interior angles. Angle α, as the supplement of a 145° angle is 35°. Consequently, $\angle ABC = 90°$ and AC is the hypotenuse of right $\triangle ABC$.

Use distance = rate × time to determine distances AB and BC.

$AB = (35 \text{ knots})(2 \text{ hours}) = 70$ nautical miles

$BC = (35 \text{ knots})(3 \text{ hours}) = 105$ nautical miles

Solve the right triangle fort AC and θ.

$$AC = \sqrt{70^2 + 105^2}$$

$$\approx 126.2$$

$$\theta = \tan^{-1}\left(\frac{105}{70}\right)$$

$$\theta \approx 56.3°$$

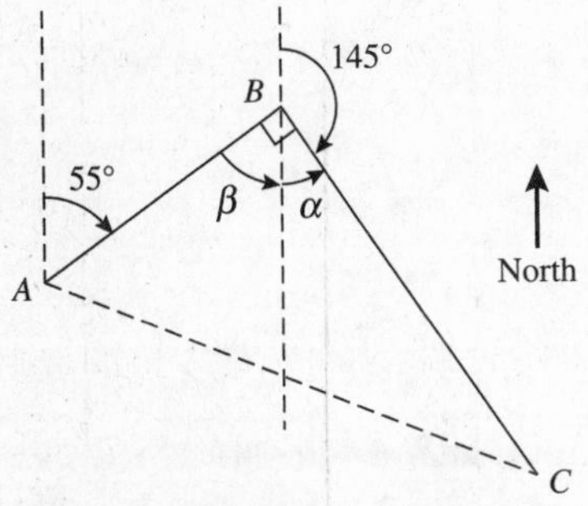

Figure 4.28 Path of travel for a Coast Guard boat that corners well at 35 knots.

Interpreting our answers, we find that the boat's bearing from port is $55° + \theta$, or approximately 111.3°. They are about 126 miles out.

Exercises for Alternate Example 4

In Exercises 7–10, a U.S. Coast Guard patrol boat leaves port and averages 35 knots (nautical mph) traveling for 1.5 hours on a course of $\alpha°$ and then 2.5 hours on a course of $\beta°$. What is the boat's bearing and distance from port?

7. $\alpha = 40°$; $\beta = 130°$ **8.** $\alpha = 52°$; $\beta = 142°$ **9.** $\alpha = 30°$; $\beta = 120°$ **10.** $\alpha = 58°$; $\beta = 148°$

In Exercises 11–14, A U.S. Coast Guard patrol boat leaves port and averages 30 knots (nautical mph) traveling for 3 hours on a course of $\alpha°$ and then 4 hours on a course of $\beta°$. What is the boat's bearing and distance from port?

11. $\alpha = 40°; \beta = 130°$ **12.** $\alpha = 42°; \beta = 132°$ **13.** $\alpha = 30°; \beta = 120°$ **14.** $\alpha = 58°; \beta = 148°$

Alternate Example 6   Calculating Harmonic Motion

A mass oscillating up and down on the bottom of a spring (assuming perfect elasticity and no friction or air resistance) can be modeled as harmonic motion. If the weight is displaced a maximum of 5 cm, find the modeling equation if it takes 3 seconds to complete one cycle. (See Figure 4.29.)

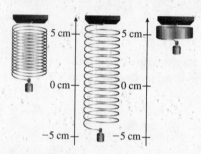

Figure 4.29 The mass and spring in Alternate Example 6.

SOLUTION

We have our choice between two equations $d = a \sin \omega t$ or $d = a \cos \omega t$. Assuming that the spring is at the origin of the coordinate system when $t = 0$, we choose the equation $d = a \sin \omega t$.

Because the maximum displacement is 5 cm, we conclude that the amplitude $a = 5$.

Because it takes 3 seconds to complete one cycle, we conclude that the period is 3 and the frequency is 1/3. Therefore,

$$\frac{\omega}{2\pi} = \frac{1}{3}$$
$$\omega = \frac{2\pi}{3}$$

Putting it all together, our modeling equation is $d = 5 \sin\left(\frac{2\pi}{3}t\right)$.

Exercises for Alternate Example 6

In Exercises 15–20, a mass is oscillating up and down on the bottom of a spring. If the weight is displaced a maximum of c cm, find the modeling equation if it takes t seconds to complete one cycle.

15. $c = 4$ and $t = 8$ **16.** $c = 7$ and $t = 6$ **17.** $c = 2$ and $t = 8$ **18.** $c = 1$ and $t = 1$ **19.** $c = 10$ and $t = 5$
20. $c = 16$ and $t = 10$

Chapter 5

Analytic Trigonometry

5.1 Fundamental Identities

Alternate Example 1   Using Identities

Find $\sin \theta$ and $\cos \theta$ if $\tan \theta = 3$ and $\cos \theta < 0$.

SOLUTION

We could solve this problem by the reference triangle techniques of Section 4.3 (see Alternate Example 7 of that section), but we will show an alternate solution here using identities.

First we note that $\sec^2 \theta = 1 + \tan^2 \theta = 1 + 3^2 = 10$, so $\sec \theta = \pm\sqrt{10}$.

Since $\sec \theta = \pm\sqrt{10}$, we have $\cos \theta = 1/\sec \theta = 1/\pm\sqrt{10}$.

But $\cos \theta < 0$, so $\cos \theta = -\sqrt{10}$.

Finally,

$$\tan \theta = 3$$

$$\frac{\sin \theta}{\cos \theta} = 3$$

$$\sin \theta = 3 \cos \theta = 3\left(-\frac{1}{\sqrt{10}}\right) = -\frac{3}{\sqrt{10}}$$

Therefore, $\sin \theta = -\dfrac{3}{\sqrt{10}}$ and $\cos \theta = -\dfrac{1}{\sqrt{10}}$.

Exercises for Alternate Example 1

In Exercises 1–6, find $\sin \theta$ and $\cos \theta$ without using a calculator. Use Pythagorean identities rather than reference triangles.

1. $\tan \theta = 1$ and $\sin \theta < 0$
2. $\tan \theta = 4$ and $\cos \theta > 0$
3. $\tan \theta = -3$ and $\cos \theta < 0$
4. $\tan \theta = -3$ and $\cos \theta > 0$
5. $\tan \theta = -2$ and $\sin \theta < 0$
6. $\tan \theta = 5$ and $\sin \theta > 0$

Alternate Example 3   Simplifying by Factoring and Using Identities

Simplify the expression $\cos^3 x + \cos x \sin^2 x$.

SOLUTION

Solve Algebraically

$$\cos^3 x + \cos x \sin^2 x = \cos x (\cos^2 x + \sin^2 x)$$

$$= \cos x (1) \qquad \text{Pythagorean identity}$$

$$= \cos x$$

Support Graphically

We recognize the graph of $y = \cos^3 x + \cos x \sin^2 x$ (Figure 5.1a) as the same as the graph of $y = \cos x$ (Figure 5.1b).

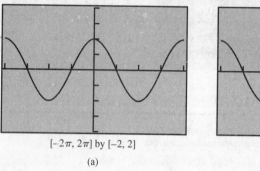

$[-2\pi, 2\pi]$ by $[-2, 2]$

(a)

$[-2\pi, 2\pi]$ by $[-2, 2]$

(b)

Figure 5.1 Graphical support of the identity $\cos^3 x + \cos x \sin^2 x = \cos x$.

Exercises for Alternate Example 3

In Exercises 7–12, simplify the expression by using basic identities.

7. $\cos^2 x + \cos^2 x \tan^2 x$

8. $\tan^2 x \cos^2 x - \sin^2 x \tan^2 x \cos^2 x$

9. $\sin^2 x \cot^2 x \sec x + \sin^2 x$

10. $2\sec^2 x - 2\sec^2 x \sin^2 x$

11. $\dfrac{\sin^2 x}{\sec x} + \sin^2 x \cos x$

12. $\tan^2 x \cos x \sec x - \cos^2 x \tan^2 x$

Alternate Example 6 **Solving a Trigonometric Equation**

Find all values of x in the interval $[0, 2\pi)$ that solve $\sin^3 x / \cos x = \tan x$.

SOLUTION

$$\frac{\sin^3 x}{\cos x} = \tan x$$

$$\frac{\sin^3 x}{\cos x} = \frac{\sin x}{\cos x}$$

$$\sin^3 x = \sin x \qquad \text{Multiply both sides by } \cos x.$$

$$\sin^3 x - \sin x = 0$$

$$(\sin x)(\sin^2 x - 1) = 0$$

$$(\sin x)(-\cos^2 x) = 0 \qquad \text{Pythagorean identity}$$

$$\sin x = 0 \text{ or } \cos x = 0$$

We reject the possibility that $\cos x = 0$ because that would make both sides of the original equation undefined.

The values in the interval $[0, 2\pi)$ that solve $\sin x = 0$ (and therefore $\sin^3 x / \cos x = \tan x$) are 0 and π.

Exercises for Alternate Example 6

In Exercises 13–18, find all values of x in the interval $[0, 2\pi)$ that solve the equation. You do not need a calculator.

13. $\dfrac{2\sin x \cos x}{\tan x} = \cos x$

14. $\tan x \sin x - \sin x = 0$

15. $4\sin^2 x = 3$

16. $4\cos^2 x = 1$

17. $\dfrac{\sin^3 x}{\cos x} = \tan x$

18. $\dfrac{\cos^3 x}{\sin x} = \cot x$

5.2 Proving Trigonometric Identities

Alternate Example 2 Proving an Identity

Prove the identity: $\sin x - \csc x = -\cos x \cot x$.

SOLUTION

We begin by deciding whether to start with the expression on the right or the left. It is usually best to start with the more complicated expression, as it is easier to proceed from the complex toward the simple than to go in the other direction. The expression on the left is slightly more complicated because it involves two terms.

$$\sin x - \csc x = \sin x - \frac{1}{\sin x}$$

$$= \frac{\sin^2 x}{\sin x} - \frac{1}{\sin x} \qquad \text{Setting up a common denominator}$$

$$= \frac{\sin^2 x - 1}{\sin x}$$

$$= \frac{-\cos^2 x}{\sin x} \qquad \text{Pythagorean identity}$$

$$= -\cos x \left(\frac{\cos x}{\sin x} \right)$$

$$= -\cos x \cot x \qquad \text{Basic identities}$$

(Remember that the "floaters" are not part of the proof.)

Exercises for Alternate Example 2

In Exercises 1–6, prove each identity.

1. $(1 - \cos x)^2 + 2\cos x = \sin^2 x + 2\cos^2 x$

2. $(\sec x \div \csc x) \cot x = 1$

3. $\dfrac{1 - \cos^2 x}{1 - \csc^2 x} = -\dfrac{\sin^4 x}{\cos^2 x}$

4. $\cos x + \sec x = \dfrac{2 - \sin^2 x}{\cos x}$

5. $1 + \tan^2 x = \dfrac{1}{1 - \sin^2 x}$

6. $\dfrac{\sin^2 x \csc^2 x}{\csc^2 x - \cot^2 x} = 1$

■

Alternate Example 4 Setting Up a Difference of Squares

Prove the identity: $\dfrac{\sin t}{1 + \cos t} = \dfrac{1 - \cos t}{\sin t}$.

SOLUTION

The left-hand expression is slightly more complicated, because we can handle extra terms in the numerator more easily than in the denominator. So we begin with the left.

$$\frac{\sin t}{1 + \cos t} = \frac{\sin t}{1 + \cos t} \cdot \frac{1 - \cos t}{1 - \cos t} \qquad \text{Setting up a difference of squares}$$

$$= \frac{\sin t (1 - \cos t)}{1 - \cos^2 t}$$

$$= \frac{\sin t(1 - \cos t)}{\sin^2 t} \qquad \text{Pythagorean identity}$$

$$= \frac{1 - \cos t}{\sin t}$$

Exercises for Alternate Example 4

In Exercises 7–12, prove each identity.

7. $\dfrac{1}{1 - \sin t} = \dfrac{1 + \sin t}{\cos^2 t}$

8. $\dfrac{1}{1 - \cos t} = \dfrac{1 + \cos t}{\sin^2 t}$

9. $\dfrac{1 + \cos t}{1 - \cos t} = \dfrac{(1 + \cos t)^2}{\cos^2 t}$

10. $\dfrac{1 - \sin t}{1 + \sin t} = \dfrac{(1 - \sin t)^2}{\cos^2 t}$

11. $\dfrac{\cos t - \sin t}{\cos t + \sin t} = \dfrac{(\cos t - \sin t)^2}{1 - 2\sin^2 t}$

12. $\dfrac{1}{\sec t + \tan t} = \sec t - \tan t$

■

Alternate Example 5   Working from Both Sides

Prove the identity: $\dfrac{\tan^2 x}{1 + \sec x} = \dfrac{\sin^2 x}{\cos x + \cos^2 x}$.

SOLUTION

Both sides are fairly complicated, but the left-hand side looks like it needs more work. We start on the left.

$$\frac{\tan^2 x}{1 + \sec x} = \frac{\sec^2 x - 1}{1 + \sec x} \qquad \text{Pythagorean identity}$$

$$= \frac{(\sec x - 1)(1 + \sec x)}{1 + \sec x} \qquad \text{Factor the numerator}$$

$$= \sec x - 1$$

At this point it is not clear how we can get from this expression to the one on the right-hand side of our identity. However, we now have reason to believe that the right hand side must simplify to $\sec x - 1$, so we try simplifying the right-hand side.

$$\frac{\sin^2 x}{\cos x + \cos^2 x} = \frac{1 - \cos^2 x}{\cos x(1 + \cos x)} \qquad \text{Pythagorean identity}$$

$$= \frac{(1 - \cos x)(1 + \cos x)}{\cos x(1 + \cos x)} \qquad \text{Factor}$$

$$= \frac{1 - \cos x}{\cos x}$$

$$= \frac{1}{\cos x} - 1 \qquad \text{Distribute the product}$$

$$= \sec x - 1$$

Now we can reconstruct the proof by going through $\sec x - 1$ as an intermediate step.

$$\frac{\tan^2 x}{1 + \sec x} = \frac{\sec^2 x - 1}{1 + \sec x} = \frac{(\sec x - 1)(1 + \sec x)}{1 + \sec x}$$

$$= \sec x - 1$$

$$= \frac{1}{\cos x} - 1$$

$$= \frac{1 - \cos x}{\cos x}$$

$$= \frac{(1 - \cos x)(1 + \cos x)}{\cos x(1 + \cos x)}$$

$$= \frac{1 - \cos^2 x}{\cos x(1 + \cos x)}$$

$$= \frac{\sin^2 x}{\cos x + \cos^2 x}$$

Exercises for Alternate Example 5

In Exercises 13–18, prove each identity.

13. $\dfrac{\cos^2 x}{1 + \sin x} = \dfrac{1 - \sin x \cos x - \sin x + \cos}{1 + \cos x}$

14. $\dfrac{\cos^2 x - \sin^2 x}{\cos x + \sin x} = \dfrac{\csc x - \sec x}{\sec x \csc x}$

15. $\dfrac{\sec^2 x - \tan^2 x}{\sec x + \tan x} = \dfrac{1 - \sin x}{\cos x}$

16. $\dfrac{\csc^2 x - \cot^2 x}{\csc x + \cot x} = \dfrac{1 - \cos x}{\sin x}$

17. $\dfrac{1}{\cos x - \sin x} + \dfrac{1}{\cos x + \sin x} = \dfrac{2}{\sec x(1 - 2\sin^2 x)}$

18. $\dfrac{1}{\sec x - 1} + \dfrac{1}{\sec x + 1} = \dfrac{2}{\cos x \tan^2 x}$

5.3 Sum and Difference Identities

Alternate Example 1 **Using the Cosine-of-a-Difference Identity**

Find the exact value of $\cos 105°$ without using a calculator.

SOLUTION

The trick is to write $\cos 105°$ as $\cos (135° - 30°)$; then we can use our knowledge of the special angles.

$$\cos 105° = \cos (135° - 30°)$$

$$= \cos 135° \cos 30° + \sin 135° \sin 30° \qquad \text{Cosine difference identity}$$

$$= \left(-\frac{\sqrt{2}}{2} \right)\left(\frac{\sqrt{3}}{2} \right) + \left(\frac{\sqrt{2}}{2} \right)\left(\frac{1}{2} \right)$$

$$= \frac{-\sqrt{6} + \sqrt{2}}{4}$$

Exercises for Alternate Example 1

In Exercises 1–6, find the exact value of the cosine without using a calculator by using the cosine difference identity.

1. $\cos 15°$ **2.** $\cos 300°$ **3.** $\cos 75°$ **4.** $\cos (-15°)$ **5.** $\cos (-75°)$ **6.** $\cos 255°$

Alternate Example 3 **Using the Sum/Difference Formulas**

Write each of the following expressions as the sine or cosine of an angle.

(a) $\sin 25° \cos 5° + \cos 25° \sin 5°$

(b) $\cos \dfrac{\pi}{4} \cos \dfrac{\pi}{5} + \sin \dfrac{\pi}{4} \sin \dfrac{\pi}{5}$

(c) $\sin 3x \sin 5x - \cos 3x \cos 5x$

SOLUTION

The key in each case is recognizing which formula applies. (Indeed, the real purpose of such exercises is to help you remember the formulas.)

(a) $\sin 25° \cos 5° + \cos 25° \sin 5° \qquad \text{Recognizing sine of sum formula}$

$$= \sin (25° + 5°) = \sin 30°$$

(b) $\cos \dfrac{\pi}{4} \cos \dfrac{\pi}{5} + \sin \dfrac{\pi}{4} \sin \dfrac{\pi}{5}$

$$= \cos \left(\dfrac{\pi}{4} - \dfrac{\pi}{5} \right) = \cos \dfrac{\pi}{20}$$

Recognizing cosine of difference formula

(c) $\sin 3x \sin 5x - \cos 3x \cos 5x$

$$= -(\cos 3x \cos 5x - \sin 3x \sin 5x)$$

Recognizing *opposite* of cos sum formula

$$= -\cos (3x + 5x)$$

Applying formula

$$= -\cos 8x$$

Exercises for Alternate Example 3

In Exercises 7–14, write each of the following expressions as the sine or cosine of an angle.

7. $\sin 55° \cos 3° + \cos 55° \sin 3°$

8. $\sin 75° \cos 34° - \cos 75° \sin 34°$

9. $\cos \dfrac{\pi}{3} \cos \dfrac{\pi}{6} + \sin \dfrac{\pi}{3} \sin \dfrac{\pi}{6}$

10. $\sin \dfrac{\pi}{4} \cos \dfrac{\pi}{5} - \cos \dfrac{\pi}{4} \sin \dfrac{\pi}{5}$

11. $\sin 4x \sin 7x - \cos 4x \cos 7x$

12. $\cos 4x \cos 7x + \sin 4x \sin 7x$

13. $\sin 42° \cos 13° + \cos 42° \sin 13°$

14. $\cos \dfrac{\pi}{3} \cos \dfrac{\pi}{2} + \sin \dfrac{\pi}{3} \sin \dfrac{\pi}{2}$

Alternate Example 4 **Proving Reduction Formulas**

Prove the reduction formulas.

(a) $\sin (\pi - x) = \sin x$

(b) $\cos \left(x - \dfrac{3\pi}{2} \right) = -\sin x$

SOLUTION

(a) $\sin (\pi - x) = \sin \pi \cos x - \sin x \cos \pi$

$$= \sin x \cdot (0) - \sin x \cdot (-1)$$

$$= \sin x$$

(b) $\cos \left(x - \dfrac{3\pi}{2} \right) = \cos x \cos \dfrac{3\pi}{2} + \sin x \sin \dfrac{3\pi}{2}$

$$= \cos x \cdot (0) + \sin x \cdot (-1)$$

$$= -\sin x$$

Exercises for Alternate Example 4

In Exercises 15–20, prove the identity.

15. $\cos \left(x - \dfrac{3\pi}{2} \right) = -\sin x$

16. $\sin \left(x - \dfrac{3\pi}{2} \right) = \cos x$

17. $\sin (2\pi - x) = -\sin x$

18. $\sin \left(\dfrac{3\pi}{2} - x \right) = -\cos x$

19. $\sin (x + \pi) = -\sin x$

20. $\cos (x + \pi) = -\cos x$

5.4 Multiple-Angle Formulas

Alternate Example 1 **Proving a Double-Angle Identity**

Prove the identity: $\cos 2u = \cos^2 u - \sin^2 u$.

SOLUTION

$$\cos 2u = \cos(u + u)$$
$$= \cos u \cos u - \sin u \sin u \qquad \text{Cosine of a sum } (v = u)$$
$$= \cos^2 u - \sin^2 u$$

Exercises for Alternate Example 1

In Exercises 1–6, prove the identity.

1. $\cos 4x = \cos^2 2x - \sin^2 2x$ **2.** $\sin 4x = 2 \sin 2x \cos 2x$ **3.** $\cos 6x = \cos^2 3x - \sin^2 3x$

4. $\sin 6x = 2 \sin 3x \cos 3x$ **5.** $\cos nx = \cos^2\left(\dfrac{n}{2}x\right) - \sin^2\left(\dfrac{n}{2}x\right)$ **6.** $\sin nx = 2\cos\left(\dfrac{n}{2}x\right)\sin\left(\dfrac{n}{2}x\right)$

Alternate Example 4 **Using a Double-Angle Identity**

Solve algebraically in the interval $[0, 2\pi)$: $\cos 2x = \sin x$.

SOLUTION

$$\cos 2x = \sin x$$
$$\cos^2 x - \sin^2 x = \sin x$$
$$1 - \sin^2 x - \sin^2 x = \sin x$$
$$2 \sin^2 x + \sin x - 1 = 0$$
$$(2 \sin x - 1)(\sin x + 1) = 0$$
$$\sin x = \frac{1}{2} \qquad \text{or} \qquad \sin x = -1$$

[0, 2π] by [−2, 2]

Figure 5.2 The function $y = \cos 2x - \sin x$ for $0 \le x < 2\pi$. The scale on the *x*-axis shows intervals of length $\pi/6$. This graph supports the solution found algebraically in Alternate Example 4.

The two solutions of $\sin x = \dfrac{1}{2}$ are $x = \dfrac{\pi}{6}$ and $\dfrac{5\pi}{6}$. The solution of $\sin x = -1$ is $x = \dfrac{3\pi}{6}$.

Therefore, the solutions of $\cos 2x = \sin x$ are

$$\frac{\pi}{6}, \frac{5\pi}{6}, \text{ and } \frac{3\pi}{6}.$$

We can support this result graphically by verifying the three x-intercepts of the function $y = \cos 2x - \sin x$ in the interval $[0, 2\pi)$ (Figure 5.2).

Exercises for Alternate Example 4

In Exercises 7–12, solve algebraically for exact solutions in the interval $[0, 2\pi)$.

7. $\cos 2x - \cos x = 0$ **8.** $\sin 2x - 2 \sin x = 0$ **9.** $\cos 2x - \sin x = 0$ **10.** $\cos 2x = \cos 4x$

11. $\cos x = -3 \sin (x/2) + 2$ **12.** $\cos x = \sin 0.5x$

 **Using Half-Angle Identities**

Solve $\cos^2 x = (\sin (x/2))^2$.

SOLUTION

The graph of $y = \cos^2 x - (\sin (x/2))^2$ in Figure 5.3 suggests that this function is periodic with period 2π and that the equation $\cos^2 x = (\sin (x/2))^2$ has three solutions in $[0, 2\pi)$.

Solve Algebraically

$$\cos^2 x = \sin^2\left(\frac{x}{2}\right)$$

$$\cos^2 x = \frac{1 - \cos x}{2}$$

$$2\cos^2 x = 1 - \cos x$$

$$2\cos^2 x + \cos x - 1 = 0$$

$$(2\cos x - 1)(\cos x + 1) = 0$$

$$\cos x = \frac{1}{2} \qquad \text{or} \qquad \cos x = -1$$

$$x = \frac{\pi}{3}, \frac{5\pi}{3} \qquad \text{or} \qquad x = \pi$$

The rest of the solutions are obtained by periodicity:

$$x = \frac{\pi}{3} + 2\pi n, \qquad x = 2\pi n, \qquad x = \frac{5\pi}{3} + 2\pi n, \qquad n = 0, \pm 1, \pm 2, \ldots$$

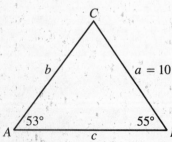

$[-2\pi, 2\pi]$ by $[-2, 2]$

Figure 5.3 The graph of $y = \cos^2 x - (\sin (x/2))^2$ suggests that $\cos^2 x = (\sin (x/2))^2$ has three solutions in $[0, 2\pi)$.

Exercises for Alternate Example 5

In Exercises 13–18, use the half-angle identities to find the general solution for each equation.

13. $\sin^2\left(\frac{x}{2}\right) = \frac{1}{4}$

14. $\cos^2\left(\frac{x}{2}\right) = \frac{3}{4}$

15. $\sin\left(\frac{x}{2}\right) = \cos x$

16. $\cos\left(\frac{x}{2}\right) = \sin x$

17. $\cos^2\left(\frac{x}{2}\right) = \sin^2\left(\frac{x}{2}\right)$

18. $2\sin\left(\frac{x}{2}\right)\cos\left(\frac{x}{2}\right) = 1$

5.5 The Law of Sines

Alternate Example 1 **Solving a Triangle Given Two Angles and a Side**

Solve $\triangle ABC$ given that $\angle A = 53°$, $\angle B = 55°$, and $a = 10$. (See Figure 5.4.)

SOLUTION

First we note that $\angle C = 180° - 53° - 55° = 72°$.

Then we apply the Law of Sines:

$$\frac{\sin A}{a} = \frac{\sin B}{b} \qquad\qquad \frac{\sin A}{a} = \frac{\sin C}{c}$$

$$\frac{\sin 53°}{10} = \frac{\sin 55°}{b} \qquad\qquad \frac{\sin 53°}{10} = \frac{\sin 72°}{c}$$

$$b = \frac{10 \sin 55°}{\sin 53°} \qquad\qquad c = \frac{10 \sin 72°}{\sin 53°}$$

$$b \approx 10.257 \qquad\qquad c \approx 11.909$$

Figure 5.4 A triangle determined by AAS.

The six parts of the triangle are

$$\angle A = 53° \qquad a = 10$$
$$\angle B = 55° \qquad b \approx 10.257$$
$$\angle C = 72° \qquad c \approx 11.909$$

Exercises for Alternate Example 1

In Exercises 1–6, solve $\triangle ABC$ given the measures of two angles and the length of one side.

1. $\angle A = 40°$, $\angle B = 70°$, and $a = 10$

2. $\angle A = 45°$, $\angle B = 70°$, and $a = 12$

3. $\angle A = 100°$, $\angle B = 20°$, and $a = 18$

4. $\angle A = 33°$, $\angle B = 71°$, and $a = 30$

5. $\angle A = 10°$, $\angle B = 21°$, and $a = 8$

6. $\angle A = 68°$, $B = 15°$, and $a = 34$

Alternate Example 2 **Solving a Triangle Given Two Sides and an Angle**

Solve $\triangle ABC$ given that $a = 10$, $b = 12$, and $\angle A = 27°$. (See Figure 5.5.)

SOLUTION

By drawing a reasonable sketch (Figure 5.5), we can assure ourselves that this is not the ambiguous case. In fact, this is the case in which $BC \geq AB$ and therefore the solution is unique.)

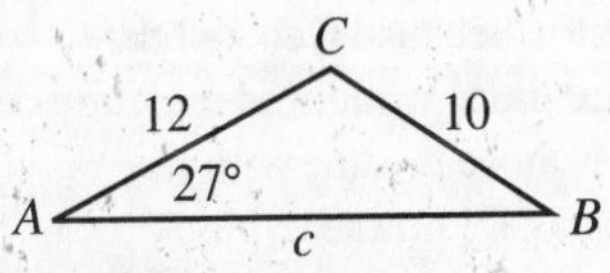

Begin by solving for the acute angle B, using the Law of Sines:

$$\frac{\sin A}{a} = \frac{\sin B}{b} \qquad \text{Law of Sines}$$

$$\frac{\sin 27°}{10} = \frac{\sin B}{12}$$

$$\sin B = \frac{12 \sin 27°}{10}$$

$$B = \sin^{-1}\left(\frac{12 \sin 27°}{10} \right)$$

$$B = 33° \qquad \text{Round to match accuracy of given angle.}$$

Then we find the obtuse angle C by subtraction:

$$C = 180° - 27° - 33° = 120°$$

Finally find side c:

$$\frac{\sin A}{a} = \frac{\sin C}{c}$$

$$\frac{\sin 27°}{10} = \frac{\sin 120°}{c}$$

$$c = \frac{10 \sin 120°}{\sin 27°}$$

$$c \approx 19.1$$

The six parts of the triangle are:

$$\angle A = 27° \qquad a = 10$$
$$\angle B = 33° \qquad b = 12$$
$$\angle C = 120° \qquad c \approx 19.1$$

Figure 5.5 A triangle determined by SSA.

Exercises for Alternate Example 2

In Exercises 7–12, solve $\triangle ABC$ given the lengths of two sides and the measure of one angle.

7. $a = 10$, $b = 10$, and $\angle A = 41°$

8. $a = 9$, $b = 11$, and $\angle A = 35°$

9. $a = 15$, $b = 12$, and $\angle A = 23°$

10. $a = 18$, $b = 18$, and $\angle A = 32°$

11. $a = 20$, $b = 17$, and $\angle A = 54°$

12. $a = 19$, $b = 18$, and $\angle A = 50°$

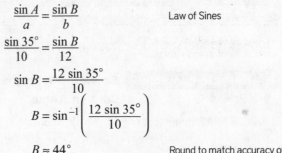

Figure 5.6 Two triangles determined by the same SSA values. (Alternate Example 3)

Alternate Example 3 **Handling the Ambiguous Case**

Solve $\triangle ABC$ given that $a = 10$, $b = 12$, and $\angle A = 35°$.

SOLUTION

By drawing a reasonable sketch (Figure 5.6), we see that two triangles are possible with the given information. We keep this in mind as we proceed.

We begin by using the Law of Sines to find angle B.

$$\frac{\sin A}{a} = \frac{\sin B}{b} \qquad \text{Law of Sines}$$

$$\frac{\sin 35°}{10} = \frac{\sin B}{12}$$

$$\sin B = \frac{12 \sin 35°}{10}$$

$$B = \sin^{-1}\left(\frac{12 \sin 35°}{10}\right)$$

$$B \approx 44° \qquad \text{Round to match accuracy of given angle.}$$

Notice that the calculator gave us one value for B, not two. That is because we used the *function* $\sin^{-1}$, which cannot give two output values for the same input value. Indeed, the function $\sin^{-1}$ will *never give an obtuse angle*, which is why we chose to start with the acute angle in Alternate Example 3. In this case, the calculator has found the angle B shown in Figure 5.6a.

Then we find the obtuse angle C by subtraction:

$$C = 180° - 35° - 44° = 101°$$

Finally find side c:

$$\frac{\sin A}{a} = \frac{\sin C}{c}$$

$$\frac{\sin 35°}{10} = \frac{\sin 101°}{c}$$

$$c = \frac{10 \sin 101°}{\sin 35°}$$

$$c \approx 17.1$$

So under the assumption that angle B is acute (See Figure 5.6a), the six parts of the triangle are:

$$\angle A = 35° \qquad a = 10$$
$$\angle B \approx 44° \qquad b = 12$$
$$\angle C \approx 101° \qquad c \approx 17.1$$

If angle B is obtuse, then we can see from Figure 5.6b that it has measure $180° - 44° = 136°$.

By subtraction, the acute angle $C = 180° - 35° - 136° = 9°$. We then recompute c:

$$c = \frac{10 \sin 9°}{\sin 35°} \approx 2.7 \qquad \text{Substitute 9° for 101° in earlier computation.}$$

So, under the assumption that angle B is obtuse (see Figure 5.6b), the six parts of the triangle are:

$$\angle A = 35° \qquad a = 10$$
$$\angle B \approx 136° \qquad b = 12$$
$$\angle C \approx 9° \qquad c \approx 2.7$$

Exercises for Alternate Example 3

In Exercises 13–18, solve $\triangle ABC$ given the lengths of two sides and the measure of one angle.

13. $a = 9$, $b = 11$, and $\angle A = 34°$ **14.** $a = 15$, $b = 24$, and $\angle A = 20°$ **15.** $a = 13$, $b = 19$, and $\angle A = 31°$

16. $a = 25$, $b = 30$, and $\angle A = 48°$ **17.** $a = 8$, $b = 12$, and $\angle A = 27°$ **18.** $a = 13$, $b = 16$, and $\angle A = 29°$ ■

5.6 The Law of Cosines

Alternate Example 1   Solving a Triangle (SAS)

Solve $\triangle ABC$ given that $a = 10$, $b = 4$, and $C = 32°$. (See Figure 5.7.)

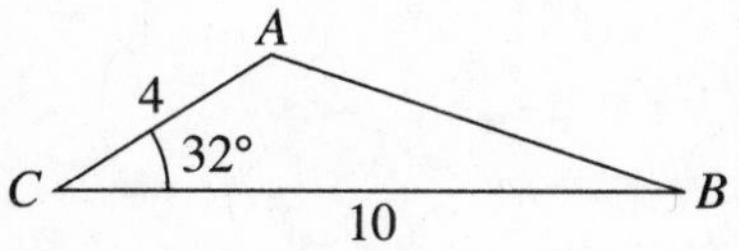

Figure 5.7 A triangle with two sides and an included angle known.

SOLUTION

$$c^2 = a^2 + b^2 - 2ab \cos C$$
$$= 10^2 + 4^2 - 2(10)(4) \cos 32°$$
$$= 48.1562\ldots$$
$$c = \sqrt{48.1562\ldots} \approx 6.9$$

We could now use either the Law of Cosines or the Law of Sines to find one of the two unknown angles. As a general rule, it is better to use the Law of Cosines to find angles, since the arccosine function will distinguish obtuse angles from acute angles.

$$a^2 = b^2 + c^2 - 2bc \cos A$$
$$10^2 = 4^2 + (6.93946)^2 - 2(4)(6.93946) \cos A$$
$$\cos A = \frac{4^2 + (6.93946)^2 - 10^2}{2(4)(6.93946)}$$
$$A = \cos^{-1}\left(\frac{4^2 + (6.93946)^2 - 10^2}{2(4)(6.93946)} \right)$$
$$\approx 130.2°$$
$$B = 180° - 130.2° - 32°$$
$$B \approx 17.8°$$

The six parts of the triangle are:

$$A \approx 130.2° \qquad a \approx 10$$
$$B \approx 17.8° \qquad b \approx 4$$
$$C \approx 32° \qquad c \approx 6.9$$

Exercises for Alternate Example 1

In Exercises 1–6, solve $\triangle ABC$ given the values of a, b, and C.

1. $a = 15$, $b = 7$, and $C = 40°$ **2.** $a = 20$, $b = 20$, and $C = 35°$ **3.** $a = 6$, $b = 12$, and $C = 70°$

4. $a = 1$, $b = 1$, and $C = 60°$ **5.** $a = 15$, $b = 20$, and $C = 50°$ **6.** $a = 35$, $b = 30$, and $C = 80°$ ■

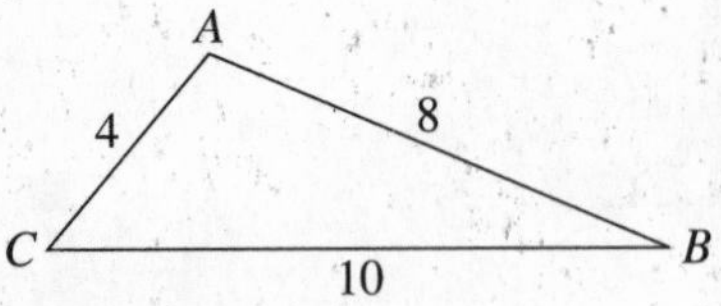

Alternate Example 2 Solving a Triangle (SSS)

Solve $\triangle ABC$ given that $a = 10$, $b = 4$, and $c = 8$. (See Figure 5.8.)

SOLUTION

We can use the Law of Cosines to find two of the angles. The third angle can be found by subtraction from 180°.

Figure 5.8 A triangle with three sides known.

$$a^2 = b^2 + c^2 - 2bc \cos A$$
$$10^2 = 4^2 + 8^2 - 2(4)(8) \cos A$$
$$64 \cos A = -20$$
$$A = \cos^{-1}\left(-\frac{20}{64}\right)$$
$$A \approx 108.2°$$

$$b^2 = a^2 + c^2 - 2ac \cos B$$
$$4^2 = 10^2 + 8^2 - 2(10)(8) \cos B$$
$$160 \cos B = 148$$
$$A = \cos^{-1}\left(\frac{148}{160}\right)$$
$$A \approx 22.3°$$

Then $C = 180° - 108.2° - 22.3° \approx 49.5°$.

Exercises for Alternate Example 2

In Exercises 7–12, solve $\triangle ABC$ given the lengths of the three sides.

7. $a = 10$, $b = 15$, and $c = 11$ 8. $a = 24$, $b = 20$, and $c = 16$ 9. $a = 7$, $b = 14$, and $c = 15$

10. $a = 3$, $b = 4$, and $c = 5$ 11. $a = 9$, $b = 9$, and $c = 16$ 12. $a = 50$, $b = 100$, and $c = 120$

Alternate Example 4 Using Heron's Formula

Find the area of a triangle with sides 13, 17, and 20.

SOLUTION

First we compute the semiperimeter: $s = (13 + 17 + 20)/2 = 25$.

Then we use Heron's Formula

$$\text{Area} = \sqrt{25(25 - 13)(25 - 17)(25 - 20)}$$
$$= \sqrt{25(12)(8)(5)}$$
$$= \sqrt{12{,}000}$$
$$= 20\sqrt{30}$$

The approximate area is 110 square units.

Exercises for Alternate Example 4

In Exercises 13–18, use Heron's Formula to find the area of a triangle with the given sides.

13. 15, 15, and 25   14. 24, 30, and 45   15. 18, 18, and 30   16. 3, 4, and 5   17. 13, 13, and 13

18. 100, 60, and 50

Chapter 6 Applications of Trigonometry

6.1 Vectors in the Plane

Alternate Example 3 **Performing Vector Operations**

Let $\mathbf{u} = \langle 3, 2 \rangle$ and $\mathbf{v} = \langle 5, -1 \rangle$. Find the component form of the following vectors:

(a) $2\mathbf{v}$ **(b)** $\mathbf{u} + \mathbf{v}$ **(c)** $(-2)\mathbf{u} + 3\mathbf{v}$

SOLUTION

Recall that the product of a scalar k and a vector $\mathbf{v} = \langle v_1, v_2 \rangle$ is

$$k\mathbf{v} = k\langle v_1, v_2 \rangle = \langle kv_1, kv_2 \rangle.$$

The sum of the vectors $\mathbf{u} = \langle u_1, u_2 \rangle$ and $\mathbf{v} = \langle v_1, v_2 \rangle$ is

$$\mathbf{u} + \mathbf{v} = \langle u_1, u_2 \rangle + \langle v_1, v_2 \rangle = \langle u_1 + v_1, u_2 + v_2 \rangle.$$

(a) $2\mathbf{v} = 2\langle 5, -1 \rangle = \langle 10, -2 \rangle$

(b) $\mathbf{u} + \mathbf{v} = \langle 3, 2 \rangle + \langle 5, -1 \rangle = \langle 3 + 5, 2 + (-1) \rangle = \langle 8, 1 \rangle$

(c) $(-2)\mathbf{u} + 3\mathbf{v} = (-2)\langle 3, 2 \rangle + 3\langle 5, -1 \rangle$
$$= \langle -6, -4 \rangle + \langle 15, -3 \rangle$$
$$= \langle -6 + 15, -4 - 3 \rangle = \langle 9, -7 \rangle$$

Exercises for Alternate Example 3

In Exercises 1–4, let $\mathbf{u} = \langle 2, -1 \rangle$, $\mathbf{v} = \langle 1, 3 \rangle$, and $\mathbf{w} = \langle -1, -1 \rangle$. Find the component form of the vector.

1. $-\mathbf{u} + \mathbf{v}$ **2.** $\mathbf{u} + (-1)\mathbf{w}$

3. $-\mathbf{u} + \mathbf{w}$ **4.** $3\mathbf{u} + 2\mathbf{v}$

In Exercises 5–8, let $\mathbf{u} = \langle -2, -2 \rangle$, and $\mathbf{w} = \langle 1, 4 \rangle$. Find the component form of the vector.

5. $\mathbf{u} + 2\mathbf{w}$ **6.** $(-1)\mathbf{u} + \mathbf{w}$

7. $2\mathbf{u} + 3\mathbf{w}$ **8.** $(-1)\mathbf{u} - 2\mathbf{w}$

Alternate Example 4 **Finding a Unit Vector**

Find a unit vector in the direction of $\mathbf{v} = \langle 5, -1 \rangle$.

SOLUTION

First, find the magnitude of the vector $\mathbf{v}$.

$$|\mathbf{v}| = |\langle 5, -1 \rangle| = \sqrt{5^2 + (-1)^2} = \sqrt{26}$$

Then multiply the vector by the reciprocal of its magnitude to find the unit vector.

$$\frac{\mathbf{v}}{|\mathbf{v}|} = \frac{1}{\sqrt{26}}\langle 5, -1 \rangle = \left\langle \frac{5}{\sqrt{26}}, \frac{-1}{\sqrt{26}} \right\rangle$$

Verify that this is a unit vector by finding its magnitude.

$$\left| \left\langle \frac{5}{\sqrt{26}}, \frac{-1}{\sqrt{26}} \right\rangle \right| = \sqrt{\left(\frac{5}{\sqrt{26}} \right)^2 + \left(\frac{-1}{\sqrt{26}} \right)^2} = \sqrt{\frac{25}{26} + \frac{1}{26}} = \sqrt{\frac{26}{26}} = 1$$

Exercises for Alternate Example 4

In Exercises 9–16, find the unit vector in the direction of the given vector.

9. $\mathbf{u} = \langle 2, -1 \rangle$ 10. $\mathbf{u} = \langle 2, 3 \rangle$ 11. $\mathbf{v} = 3\mathbf{i} + 2\mathbf{j}$ 12. $\mathbf{w} = -\mathbf{i} - \mathbf{j}$ 13. $\mathbf{u} = \langle 8, 6 \rangle$

14. $\mathbf{v} = \langle -1, -3 \rangle$ 15. $\mathbf{v} = \langle -4, 3 \rangle$ 16. $\mathbf{v} = 12\mathbf{i} - 3\mathbf{j}$

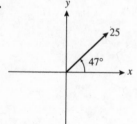

Figure 6.1 The direction angle of **u** is 290°.

Alternate Example 5 **Finding the Components of a Vector**

Find the components of a vector **u** with direction angle 290° and magnitude 4 (Figure 6.1).

SOLUTION

If a and b are the horizontal components, respectively, of the vector **u**, then

$$\mathbf{u} = \langle a, b \rangle = \langle 4\cos 290°, 4\sin 290° \rangle .$$

Use a calculator to compute cos 290° and sin 290°. So, $a = 4 \cos 290° \approx 0.342$ and $b = 4 \sin 290° \approx -0.940$.

The component form of the vector is $\mathbf{u} = \langle 0.342, -0.940 \rangle$.

Exercises for Alternate Example 5

In Exercises 17–22, find the components of a given vector.

17. **u** with direction angle 105° and magnitude 8   18. **v** with direction angle 75° and magnitude 7.5

19. **w** with direction angle 208° and magnitude 4   20. **v** with direction angle 120° and magnitude 16

21.

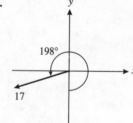

22.

Alternate Example 6 **Finding the Direction Angle of a Vector**

Find the magnitude and direction angle of each vector:

(a) $\mathbf{u} = \langle 1, 4 \rangle$ (b) $\mathbf{v} = \langle -5, 4 \rangle$ (c) $\mathbf{w} = \langle 3, -2 \rangle$

SOLUTION

Plot each vector in a coordinate system as in Figure 6.2. Note that the sign of the sine and the cosine of the direction angle are the same as the sign of the *y*- and the *x*-components of the vector, respectively.

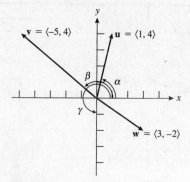

Figure 6.2 The three vectors of Alternate Example 4.

(a) If α is the direction angle of **u**, then $\mathbf{u} = \langle 1, 4 \rangle = \langle |\mathbf{u}|\cos\alpha, |\mathbf{u}|\sin\alpha \rangle$.

First find the magnitude of the vector: $|\mathbf{u}| = \sqrt{1^2 + 4^2} = \sqrt{17}$.

Then write the equation for one of the components of the vector:

$$1 = |\mathbf{u}|\cos\alpha \qquad \text{Horizontal component of } \mathbf{u}$$

$$1 = \sqrt{17}\cos\alpha \qquad \text{Substitute } |\mathbf{u}|$$

$$\frac{1}{\sqrt{17}} = \cos\alpha$$

$$\alpha = \cos^{-1}\left(\frac{1}{\sqrt{17}}\right) \approx 75.96° \qquad \alpha \text{ is in Quadrant I}$$

(b) If β is the direction angle of **v**, then $\mathbf{v} = \langle -5, 4 \rangle = \langle |\mathbf{v}|\cos\beta, |\mathbf{v}|\sin\beta \rangle$.

First find the magnitude of the vector: $|\mathbf{v}| = \sqrt{(-5)^2 + 4^2} = \sqrt{41}$.

Then write the equation for one of the components of the vector:

$$-5 = |\mathbf{v}|\cos\beta \qquad \text{Horizontal component of } \mathbf{v}$$

$$-5 = \sqrt{41}\cos\beta \qquad \text{Substitute } |\mathbf{v}|$$

$$\frac{-5}{\sqrt{41}} = \cos\beta$$

$$\beta = \cos^{-1}\left(\frac{-5}{\sqrt{41}}\right) \approx 141.34° \qquad \beta \text{ is in Quadrant II}$$

(c) If γ is the direction angle of **w**, then $\mathbf{w} = \langle 3, -2 \rangle = \langle |\mathbf{w}|\cos\gamma, |\mathbf{w}|\sin\gamma \rangle$.

First find the magnitude of the vector: $|\mathbf{w}| = \sqrt{3^2 + (-2)^2} = \sqrt{13}$.

Then write the equation for one of the components of the vector:

$$3 = |\mathbf{v}|\cos\gamma \qquad \text{Horizontal component of } \mathbf{w}$$

$$3 = \sqrt{41}\cos\gamma \qquad \text{Substitute } |\mathbf{w}|$$

$$\frac{3}{\sqrt{13}} = \cos\gamma$$

$$\gamma = 360° - \cos^{-1}\left(\frac{3}{\sqrt{13}}\right) \approx 326.31° \qquad \gamma \text{ is in Quadrant IV}$$

Exercises for Alternate Example 6

In Exercises 23–30, find the magnitude and the direction angle of the given vector.

23. $\mathbf{u} = \langle 2, -1 \rangle$ **24.** $\mathbf{v} = \langle 4, 3 \rangle$ **25.** $\mathbf{w} = \langle -2, -7 \rangle$ **26.** $\mathbf{u} = \langle -1, -1 \rangle$ **27.** $\mathbf{v} = \langle 0.5, -2 \rangle$

28. $\mathbf{w} = \langle -2.5, 5 \rangle$ **29.** $\mathbf{u} = \langle 6, -4 \rangle$ **30.** $\mathbf{v} = \langle 1.5, 2.5 \rangle$

6.2 Dot Product of Vectors

Alternate Example 1 **Finding Dot Products**

Find the dot product of the given vectors.

(a) $\mathbf{u} = \langle 3, 1 \rangle$, $\mathbf{v} = \langle 2, -5 \rangle$

(b) $\mathbf{u} = \langle 5, 4 \rangle$, $\mathbf{v} = \langle -4, 5 \rangle$

(c) $\mathbf{u} = (-2\mathbf{i} + 2\mathbf{j})$, $\mathbf{v} = (3\mathbf{i} - 5\mathbf{j})$

SOLUTION

(a) $\langle 3, 1 \rangle \cdot \langle 2, -5 \rangle = 3(2) + 1(-5) = 6 - 5 = 1$

(b) $\langle 5, 4 \rangle \cdot \langle -4, 5 \rangle = 5(-4) + 4(5) = -20 + 20 = 0$

(c) You can rewrite each vector in the component form or multiply the components directly.

$$(-2\mathbf{i} + 2\mathbf{j}) \cdot (3\mathbf{i} - 5\mathbf{j}) = \langle -2, 2 \rangle \cdot \langle 3, -5 \rangle = -2(3) + 2(-5) = -6 - 10 = -16$$

Exercises for Alternate Example 1

In Exercises 1–8, find the dot products of vectors **u** and **v**.

1. $\mathbf{u} = \langle 2, 5 \rangle$, $\mathbf{v} = \langle 4, 1 \rangle$

2. $\mathbf{u} = \langle 3, 2 \rangle$, $\mathbf{v} = \langle 2, 1 \rangle$

3. $\mathbf{u} = \langle 6, 2 \rangle$, $\mathbf{v} = \langle -1, 3 \rangle$

4. $\mathbf{u} = \langle 4, 3 \rangle$, $\mathbf{v} = \langle -3, -1 \rangle$

5. $\mathbf{u} = \langle 7.5, 2.5 \rangle$, $\mathbf{v} = \langle 2, 3 \rangle$

6. $\mathbf{u} = \langle 1.2, 1 \rangle$, $\mathbf{v} = \langle 0.5, 1 \rangle$

7. $\mathbf{u} = (3\mathbf{i} + 4\mathbf{j})$, $\mathbf{v} = (-2\mathbf{i} + 2\mathbf{j})$

8. $\mathbf{u} = (\mathbf{i} - 6\mathbf{j})$, $\mathbf{v} = (3\mathbf{i} + 2\mathbf{j})$

Alternate Example 3 **Finding the Angle Between Vectors**

Find the angle between the vectors $\mathbf{u} = \langle 2, 3 \rangle$ and $\mathbf{v} = \langle 4, -1 \rangle$ (Figure 6.3).

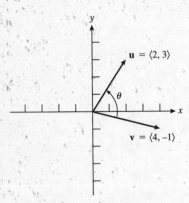

Figure 6.3 Find angle θ between vectors **u** and **v**.

SOLUTION

First, find the dot products of vector **u** and **v** and their magnitudes.

$$\mathbf{u} \cdot \mathbf{v} = \langle 2, 3 \rangle \cdot \langle 4, -1 \rangle = 5$$
$$|\mathbf{u}| = |\langle 2, 3 \rangle| = \sqrt{2^2 + 3^2} = \sqrt{13}$$
$$|\mathbf{v}| = |\langle 4, -1 \rangle| = \sqrt{4^2 + (-1)^2} = \sqrt{17}$$

Then use the Angle Between Two Vectors Theorem.

$$\cos \theta = \frac{\mathbf{u} \cdot \mathbf{v}}{|\mathbf{u}||\mathbf{v}|} = \frac{5}{\sqrt{13}\sqrt{17}}$$

So,

$$\theta = \cos^{-1}\left(\frac{5}{\sqrt{13}\sqrt{17}}\right) \approx 70.3°.$$

Exercises for Alternate Example 3

In Exercises 9–16, find the angle between the two given vectors.

9. $\mathbf{u} = \langle 2, 5 \rangle, \mathbf{v} = \langle 3, 1 \rangle$

10. $\mathbf{u} = \langle 3, 2 \rangle, \mathbf{v} = \langle 2, 1 \rangle$

11. $\mathbf{u} = \langle 6, 2 \rangle, \mathbf{v} = \langle -1, 3 \rangle$

12. $\mathbf{u} = \langle 4, 3 \rangle, \mathbf{v} = \langle -3, -1 \rangle$

13. $\mathbf{u} = \langle -1, 3 \rangle, \mathbf{v} = \langle -1, 5 \rangle$

14. $\mathbf{u} = \langle 5, 3 \rangle, \mathbf{v} = \langle -1, -3 \rangle$

15. $\mathbf{u} = \mathbf{i} + \sqrt{2}\mathbf{j}, \mathbf{v} = -\sqrt{2}\mathbf{i} - 4\mathbf{j}$

16. $\mathbf{u} = 2\mathbf{i} + 2\mathbf{j}, \mathbf{v} = -\mathbf{i}$

Alternate Example 5 **Decomposing a Vector into Perpendicular Components**

(a) Find the projection of vector $\mathbf{u} = \langle 4, 3 \rangle$ onto vector $\mathbf{v} = \langle 6, 2 \rangle$.

(b) Write $\mathbf{u}$ as a sum of two orthogonal vectors, one of which is $\text{proj}_{\mathbf{v}}\mathbf{u}$.

SOLUTION

(a) $\mathbf{u}_1 = \text{proj}_{\mathbf{v}}\mathbf{u} = \left(\dfrac{\mathbf{u} \cdot \mathbf{v}}{|\mathbf{v}|^2} \right) \mathbf{v} = \dfrac{30}{40} \langle 6, 2 \rangle = \left\langle \dfrac{9}{2}, \dfrac{3}{2} \right\rangle$

(b) Write $\mathbf{u} = \mathbf{u}_1 + \mathbf{u}_2$, where $\mathbf{u}_1 = \text{proj}_{\mathbf{v}}\mathbf{u}$. Then $\mathbf{u}_2 = \mathbf{u} - \mathbf{u}_1$.

$$\mathbf{u}_2 = \mathbf{u} - \mathbf{u}_1 = \langle 4, 3 \rangle - \left\langle \dfrac{9}{2}, \dfrac{3}{2} \right\rangle = \left\langle -\dfrac{1}{2}, \dfrac{3}{2} \right\rangle$$

Figure 6.4 shows these vectors.

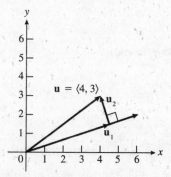

Figure 6.4 The vectors $\mathbf{u} = \langle 4, 3 \rangle$, $\mathbf{v} = \langle 6, 2 \rangle$, $\mathbf{u}_1 = \text{proj}_{\mathbf{v}}\mathbf{u}$, and $\mathbf{u}_2 = \mathbf{u} - \mathbf{u}_1$.

Exercises for Alternate Example 5

In Exercises 17–24, find the projection of vector $\mathbf{u}$ onto vector $\mathbf{v}$. Then write $\mathbf{u}$ as a sum of two orthogonal vectors, one of which is $\text{proj}_{\mathbf{v}}\mathbf{u}$.

17. $\mathbf{u} = \langle -6, 3 \rangle, \mathbf{v} = \langle 8, 2 \rangle$

18. $\mathbf{u} = \langle 5, 3 \rangle, \mathbf{v} = \langle 8, -3 \rangle$

19. $\mathbf{u} = \langle -2, 7 \rangle, \mathbf{v} = \langle 8, 3 \rangle$

20. $\mathbf{u} = \langle 3, 4 \rangle, \mathbf{v} = \langle -4, 8 \rangle$

21. $\mathbf{u} = \langle -3, 2 \rangle, \mathbf{v} = \langle -5, 0 \rangle$

22. $\mathbf{u} = \langle 7, -3 \rangle, \mathbf{v} = \langle 6, 6 \rangle$

23. $\mathbf{u} = \langle -1, -3 \rangle, \mathbf{v} = \langle -1, 3 \rangle$

24. $\mathbf{u} = \langle -4, -3 \rangle, \mathbf{v} = \langle -6, -2 \rangle$

6.3   Parametric Equations and Motion

Alternate Example 2   **Eliminating the Parameter**

Eliminate the parameter and identify the graph of the parametric curve

$$x = 3t - 2, \quad y = 1 - 2t, \quad -\infty < t < \infty.$$

SOLUTION

Solve one equation for t.

$$x = 3t - 2$$

$$3t = x + 2$$

$$t = \frac{1}{3}(x + 2)$$

Then substitute the expression for t into the other equation.

$$y = 1 - 2t$$

$$y = 1 - 2\left(\frac{1}{3}\right)(x + 2)$$

$$y = 1 - \frac{2}{3}x - \frac{4}{3}$$

$$y = \frac{2}{3}x - \frac{1}{3}$$

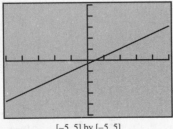

[−5, 5] by [−5, 5]

Figure 6.5 The graph of
$x = 3t - 2, \ y = 1 - 2t, \ -\infty < t < \infty$
or of $y = \frac{2}{3}x - \frac{1}{3}$.

The graph of the equation $y = \frac{2}{3}x - \frac{1}{3}$ is a line with slope $\frac{2}{3}$ and y-intercept $-\frac{1}{3}$ (Figure 6.5).

Exercises for Alternate Example 2

In Exercises 1–8, eliminate the parameter and identify the graph of the parametric curve.

1. $x = 2 - t, \ y = 2t + 1$
2. $x = 3t + 2, \ y = 4 + t$
3. $x = 2t - 3, \ y = 2 - 2t$
4. $x = 2.5 - 0.5t, \ y = 1 - 2t$
5. $x = t, \ y = 1 - t^2$
6. $x = t - 2, \ y = 1 + 2t^2$
7. $x = 2t - 1, \ y = 2t^2 + 1$
8. $x = t^2, \ y = 2 - t$

Alternate Example 4   **Eliminating the Parameter**

Eliminate the parameter and identify the graph of the parametric curve

$$x = 3 \cos t, \quad y = 3 \sin t, \quad 0 \le t \le 2\pi.$$

SOLUTION

The graph of the parametric equations (Figure 6.6) appears to be a circle of radius 3 centered at the origin. You can use the trigonometric identity $\cos^2 t + \sin^2 t = 1$.

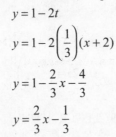

[−6, 6] by [−4, 4]

Figure 6.6 The graph of
$x = 3 \cos t, \ y = 3 \sin t,$
$0 \le t \le 2\pi$ is a circle.

$$x^2 + y^2 = (3 \cos t)^2 + (3 \sin t)^2$$

$$= 9 \cos^2 t + 9 \sin^2 t$$

$$= 9(\cos^2 t + \sin^2 t)$$

$$= 9(1) \qquad \qquad \cos^2 t + \sin^2 t = 1$$

$$= 9 = 3^2$$

The graph of $x^2 + y^2 = 9$ is a circle of radius 3 centered at the origin. The parameter t is the central angle, so the condition $0 \le t \le 2\pi$ causes the grapher to trace the circle complete circle starting at $(1, 0)$. Changing the endpoints of the parameter interval moves the starting and final points for the grapher. If the length of the parameter interval is greater than 2π then the grapher traces a part of the circle more than once. If the length of the parameter interval is less than 2π then the grapher traces a portion of the circle.

Exercises for Alternate Example 4

In Exercises 9–14, eliminate the parameter and identify the graph of the parametric curve.

9. $x = 6 \cos t$, $y = 6 \sin t$, $0 \le t \le 2\pi$

10. $x = 2.5 \cos t$, $y = 2.5 \sin t$, $0 \le t \le 2\pi$

11. $x = 2 \sin t$, $y = 2 \cos t$, $-\pi \le t \le \pi$

12. $x = 4 \cos t$, $y = 4 \sin t$, $-\dfrac{\pi}{2} \le t \le \dfrac{\pi}{2}$

13. $x = 3 \cos t$, $y = 3\text{s in } t$, $-2\pi \le t \le 2\pi$

14. $x = 5 \cos t$, $y = 5 \sin t$, $0 \le t \le \dfrac{2\pi}{3}$

Alternate Example 6 **Finding Parametric Equations for a Line Segment**

Find the parametrization of the line segment with endpoints $A = (-3, 1)$ and $B = (3, 3)$.

SOLUTION

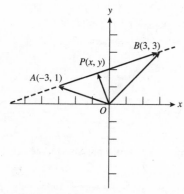

Figure 6.7 As parameter t changes, point P moves from A to B.

First, identify the parametrization of the line that passes through A and B. Consider a point $P(x, y)$ as it moves along the line AB (Figure 6.7). You can write the vector $\overrightarrow{OP}$ as the sum of two other vectors.

$$\overrightarrow{OP} = \overrightarrow{OA} + \overrightarrow{AP}$$

Vector $\overrightarrow{AP}$ is parallel to the vector $\overrightarrow{AB}$, so it can be written as a multiple of $\overrightarrow{AB}$ with parameter t taking on all possible real values.

$$\overrightarrow{AP} = t\,\overrightarrow{AB}$$

Now we can rewrite the original expression for $\overrightarrow{OP}$.

$$\overrightarrow{OP} = \overrightarrow{OA} + t\,\overrightarrow{AB}$$
$$\langle x, y \rangle = \langle -3, 1 \rangle + t\langle 3 - (-3), 3 - 1 \rangle$$
$$\langle x, y \rangle = \langle -3, 1 \rangle + t\langle 6, 2 \rangle$$
$$\langle x, y \rangle = \langle -3 + 6t, 1 + 2t \rangle$$

Two vectors are equal if their components are equal, so the vector equation represents a pair of parametric equations $x = -3 + 6t$ and $y = 1 + 2t$.

When does point P coincide with the endpoints A and B? The vector equation suggests the answer. $\overrightarrow{OP} = \overrightarrow{OA}$ when $t = 0$ and $\overrightarrow{OP} = \overrightarrow{OB} = \overrightarrow{OA} + \overrightarrow{AB}$ when $t = 1$.

So the complete parametric form for the interval with endpoints $A = (3, 1)$ and $B = (3, 3)$ is $x = -3 + 6t$, $y = 1 + 2t$, $0 \le t \le 1$.

Exercises for Alternate Example 6

In Exercises 15–20, find the parametrization of the line segment with endpoints A and B.

15. $A = (2, 1), B = (6, 1)$

16. $A = (1, -1), B = (-2, 7)$

17. $A = (12, -2), B = (4, 3)$

18. $A = (1, 1), B = (3, 3)$

19. $A = (5, 7), B = (-4, -1)$

20. $A = (0, 1), B = (1, 0)$

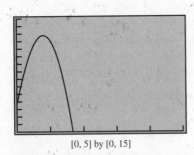

[0, 5] by [0, 15]

Figure 6.8 The graph of height as a function of time.

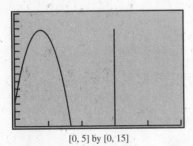

[0, 5] by [0, 15]

Figure 6.9 Simultaneous graphing of $x_1 = t$, $y_1 = -16t^2 + 25t + 3$ (height against time) and $x_2 = 3$, $y = -16t^2 + 25t + 3$ (the actual path of the toy rocket).

| **Alternate Example 8** | **Simulating Projectile Motion** |

A toy rocket is launched straight up from a cart. The toy's initial position is 3 ft above ground and its initial velocity is 25 ft/sec. Graph the toy's height against time, give the height of the toy above the ground, and simulate the toy's motion for each length of time.

(a) 0.5 sec   **(b)** 0.75 sec   **(c)** 1 sec   **(d)** 1.25 sec

SOLUTION

The equation that models the vertical position t seconds after launch is

$$y = -16t^2 + 25t + 3.$$

A graph of the rocket's height against time can be found using the parametric equations

$$x_1 = t, \quad y_1 = -16t^2 + 25t + 3.$$

(See Figure 6.8.)

To simulate the rocket's flight straight up and its fall to the ground, use the parametric equations

$$x_2 = 3, \quad y = -16t^2 + 25t + 3.$$

Note that we chose $x_2 = 3$ so that the two graphs would not intersect.

Figure 6.9 shows the two graphs in simultaneous graphing mode for $0 \leq t \leq 2$. We can use the TRACE feature to find the height above the ground. At $t = 0.5$ sec, the rocket is 11.5 ft above the ground; at $t = 0.75$ sec, the rocket is 12.75 ft above the ground; at $t = 1$ sec, the rocket is 12 ft above the ground; and at $t = 1.25$ sec, the rocket is 9.25 ft above the ground.

Exercises for Alternate Example 8

In Exercises 21–23, use parametric equations to model the motion of the object. Find the maximum height of the object by tracing the graph.

21. A gymnast throws a club straight up in the air, then catches it. The initial speed of the club is 25 ft/sec and the gymnast releases the club 4.5 ft above ground.

22. A referee throws a basketball up in the air for tip-off. The referee releases the ball 6 ft above the floor and gives it an initial speed of 14 ft/sec.

23. A cat tosses a toy up in the air. The toy's initial position is 1 ft above ground and its initial speed is 9 ft/sec.

In Exercises 24–26, use parametric equations to model the motion of the object. Use TRACE to find long it takes for the object to hit the ground.

24. A hovering helicopter drops off a supplies package for a geology expedition from the altitude of 100 ft.

25. A small package is dropped from a plane flying at the altitude of 200 ft.

26. A rock falls off a cliff 500 ft high.

6.4 Polar Coordinates

| **Alternate Example 3** | **Converting from Polar to Rectangular Coordinates** |

Find the rectangular coordinates of the points with the given polar coordinates.

(a) $P(4, 2\pi/3)$ **(b)** $Q(5, -75°)$

SOLUTION

(a) For $P(4, 2\pi/3)$, $r = 4$ and $\theta = 2\pi/3$:

$$x = r\cos\theta \qquad\qquad y = r\sin\theta$$

$$x = 4\cos\frac{2\pi}{3} \qquad \text{and} \qquad y = 4\sin\frac{2\pi}{3}$$

$$x = 4\left(-\frac{1}{2}\right) = -2 \qquad\qquad y = 4\left(\frac{\sqrt{3}}{2}\right) \approx 3.46$$

The rectangular coordinates for P are $(-2, \sqrt{3}/2) \approx (-2, 3.46)$ (Figure 6.10a).

(b) For $Q(5, -75°)$, $r = 5$ and $\theta = -75°$:

$$x = r\cos\theta \qquad\qquad y = r\sin\theta$$

$$x = 5\cos(-75°) \approx 1.29 \qquad \text{and} \qquad y = 5\sin(-75°) \approx -4.83$$

The rectangular coordinates for Q are $(1.29, -4.83)$ (Figure 6.10b).

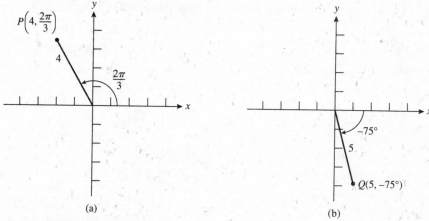

Figure 6.10 The points P and Q in Alternate Example 3.

Exercises for Alternate Example 3

In Exercises 1–8, find the rectangular coordinates of a point with the given polar coordinates.

1. $P(3, 3\pi/4)$ **2.** $Q(2, 7\pi/6)$ **3.** $R(3, -5\pi/6)$ **4.** $S(5, -5\pi/8)$ **5.** $T(2, -215°)$ **6.** $U(4, -100°)$

7. $V(6, 80°)$ **8.** $W(12, 300°)$

■

Alternate Example 4 **Converting from Rectangular to Polar Coordinates**

Find two polar coordinate pairs for the points with the given rectangular coordinates.

(a) $P(1, 2)$ **(b)** $Q(-4, 4)$

SOLUTION

(a) For $P(1, 2)$, $x = 1$ and $y = 2$:

$$r^2 = x^2 + y^2 \qquad\qquad \tan\theta = \frac{y}{x}$$

$$r^2 = 1^2 + 2^2 \qquad\qquad \tan\theta = \frac{2}{1} = 2$$

$$r = \pm\sqrt{5} \qquad\qquad \theta = \tan^{-1}2 + n\pi \approx 1.11 + n\pi$$

We can use the values 0 and 1 for n. Then the angles are $\theta = 1.11$ and $\theta = 1.11 + \pi \approx 4.25$. The first angle gives us the positive value of r. For the second angle, the P is on the ray opposite the terminal side of the angle 1.11 radians. So the value of r for this angle is negative. So two polar coordinate pairs of point P are

$$\left(\sqrt{5},\, 1.11\right) \text{ and } \left(-\sqrt{5},\, 4.25\right) \text{(Figure 6.11a).}$$

(b) For $Q(4, -4)$, $x = 4$ and $y = -4$:

$$r^2 = x^2 + y^2 \qquad\qquad \tan\theta = \frac{y}{x}$$

$$r^2 = 4^2 + (-4)^2 \qquad\quad \tan\theta = \frac{-4}{4} = -1$$

$$r = \pm\sqrt{32} = \pm 4\sqrt{2} \qquad \theta = \tan^{-1}(-1) + n\pi \approx -\frac{\pi}{4} + n\pi$$

We can use the angles $-\pi/4$ and $-\pi/4 + \pi = 3\pi/4$. Because Q is on the terminal side of $-\pi/4$, the value of r corresponding to this angle is positive. Because Q is on the ray opposite the terminal side of $3\pi/4$, the value of r corresponding to this angle is negative. So two polar coordinate pairs of point Q are

$$\left(4\sqrt{2},\, -\frac{\pi}{4}\right) \text{ and } \left(-4\sqrt{2},\, \frac{3\pi}{4}\right) \text{(Figure 6.11b).}$$

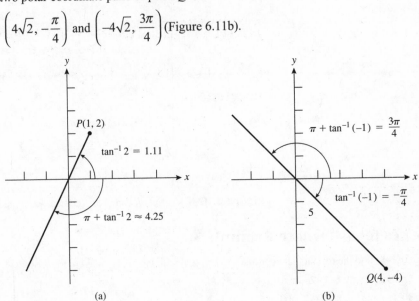

Figure 6.11 Infinitely many polar coordinate pairs correspond to each pair of rectangular coordinates.

Exercises for Alternate Example 4

In Exercises 9–14, for the point with the given rectangular coordinates, find all polar coordinates that satisfy $-\pi \leq \theta \leq \pi$.

9. $P(-3, 3)$ **10.** $Q(2, 5)$ **11.** $R(3, -5)$ **12.** $S(3, -1)$ **13.** $T(-2, -2)$ **14.** $U(-4, 10)$

15. $V(-5, 12)$ **16.** $W\left(\sqrt{2},\, \sqrt{2}\right)$

Alternate Example 5   **Converting from Polar Form to Rectangular Form**

Convert $r = 4\cos\theta$ to rectangular form and identify the graph. Support your answer with a polar graphing utility.

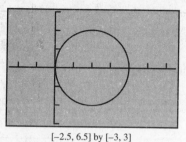

[−2.5, 6.5] by [−3, 3]

Figure 6.12 The graph of the circle $r = 4 \cos \theta$
$((x-2)^2 + y^2 = 2^2)$.

SOLUTION

$$r = 4\cos\theta$$
$$r^2 = 4r\cos\theta \qquad \text{Multiply by } r$$
$$x^2 + y^2 = 4x \qquad r\cos\theta = x \text{ and } r^2 = x^2 + y^2$$
$$(x^2 - 4x + 4) + y^2 = 4 \qquad \text{Subtract } 4x \text{ and then complete the square.}$$
$$(x-2)^2 + y^2 = 2^2 \qquad \text{Perfect square trinomial}$$

The graph is a circle with center $(2, 0)$ and radius 2 (Figure 6.12).

Exercises for Alternate Example 5

In Exercises 15–20, convert the polar equation to rectangular form and identify the graph. Support your answer with a polar graphing utility.

17. $r = 4\csc\theta$ **18.** $r = 2\cos\theta + \sin\theta$ **19.** $r = -6\sin\theta$ **20.** $r = -3\sec\theta$

21. $r = 3\sin\theta - 5\cos\theta$ **22.** $r\sec\theta = -12$

Alternate Example 6   **Converting from Rectangular Form to Polar Form**

Convert $(x-2)^2 + (y+2)^2 = 8$ to polar form. Then graph the polar equation.

SOLUTION

$$(x-2)^2 + (y+2)^2 = 8$$
$$x^2 - 4x + 4 + y^2 + 4y + 4 = 8$$
$$x^2 + y^2 - 4x + 4y = 0$$

Then substitute r^2 for $x^2 + y^2$, $r\cos\theta$ for x, and $r\sin\theta$ for y.

$$r^2 - 4r\cos\theta + 4r\sin\theta = 0$$
$$r(r - 4\cos\theta + 4\sin\theta) = 0$$
$$r = 0 \quad \text{or} \quad r - 4\cos\theta + 4\sin\theta = 0$$

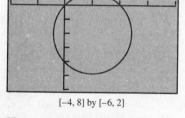

[−4, 8] by [−6, 2]

Figure 6.13 The graph of the circle $r = 4 \cos \theta - 4 \sin \theta$.

The graph of $r = 0$ is a single point that is on the graph of the second equation. So the polar form is

$$r = 4\cos\theta - 4\sin\theta.$$

The graph of $r = 4\cos\theta - 4\sin\theta$ for $0 \leq \theta \leq 2\pi$ is shown in Figure 6.13 and appears to be a circle with center at

$(2, -2)$ and radius $\sqrt{8}$ or $2\sqrt{2}$.

Exercises for Alternate Example 6

In Exercises 23–28, convert the rectangular equation to polar form. Then graph the polar equation.

23. $x^2 + (y-4)^2 = 16$ **24.** $(x+1)^2 + (y-3)^2 = 10$ **25.** $(x-3)^2 + (y-4)^2 = 25$ **26.** $y = 3$ **27.** $x + 5y = 3$

28. $3y - 2x = 3$

6.5   Graphs of Polar Equations

Alternate Example 1   **Testing for Symmetry**

Use the polar symmetry tests to determine if the graph of $r = -3\cos 2\theta$ is symmetric about the x-axis, y-axis, or the origin.

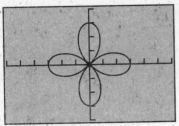

[−6, 6] by [−4, 4]

Figure 6.14 The graph of $r = -3\cos 2\theta$ is symmetric about the x-axis, about the y-axis, and about the origin.

SOLUTION

Figure 6.14 suggests that the graph of $r = -3\cos 2\theta$ is symmetric about the x-axis, about the y-axis, and about the origin.

To test for each type of symmetry, replace (r, θ) with a corresponding pair:

about x-axis	$(r, -\theta)$ or $(-r, \pi - \theta)$
about the y-axis	$(-r, -\theta)$ or $(r, \pi - \theta)$
about the origin	$(-r, \theta)$ or $(r, \theta + \pi)$

To test if $r = -3\cos 2\theta$ is symmetric about the x-axis, replace (r, θ) by $(r, -\theta)$.

$r = -3\cos 2\theta$

$r = -3\cos 2(-\theta)$

$r = -3\cos(-2\theta)$ cos θ is an even function of θ

$r = -3\cos 2\theta$ (same as original)

Because the equation $r = -3\cos 2\theta$ is equivalent to $r = -3\cos 2(-\theta)$, the graph is symmetric about the x-axis.

To test if $r = -3\cos 2\theta$ is symmetric about the y-axis, replace (r, θ) by $(-r, -\theta)$.

$r = -3\cos 2\theta$

$-r = -3\cos 2(-\theta)$

$-r = -3\cos(-2\theta)$

$-r = -3\cos 2\theta$

$r = 3\cos 2\theta \neq -3\cos 2\theta$

Since the graph seems to be symmetric about the y-axis, try the other substitution:

$r = -3\cos 2\theta$

$r = -3\cos 2(\pi - \theta)$

$r = -3\cos(2\pi - 2\theta)$ Since the period of cos θ is 2π, cos(2π − 2θ) = cos(−2θ)

$r = -3\cos(-2\theta)$

$r = -3\cos 2\theta$ Same as original

The graph of $r = -3\cos 2\theta$ is symmetric about the y-axis.

To test if $r = -3\cos 2\theta$ is symmetric about the origin, replace (r, θ) by $(-r, \theta)$.

$r = -3\cos 2\theta$

$-r = -3\cos 2\theta$

$r = 3\cos 2\theta \neq -3\cos 2\theta$

Since the graph seems to be symmetric about the origin, try the other substitution:

$r = -3\cos 2\theta$

$r = -3\cos 2(\theta + \pi)$ Since the period of cos θ is 2π, cos (2π − 2θ) = cos (−2θ)

$r = -3\cos(2\theta + 2\pi)$

$r = -3\cos 2\theta$ Same as original

So, the graph of $r = -3\cos 2\theta$ is symmetric about the x-axis, y-axis, and the origin.

Exercises for Alternate Example 1

In Exercises 1–6, use the polar symmetry tests to determine if the graph is symmetric about the x-axis, y-axis , or the origin.

1. $r = 3 - 3\cos\theta$ **2.** $r = 2\cos 5\theta$ **3.** $r = 1 + 2\sin\theta$ **4.** $r = 5\sin 2\theta$ **5.** $r = \dfrac{3}{1 - \sin\theta}$

6. $r = \dfrac{-10}{5 + \cos\theta}$

Alternate Example 3 Finding Maximum *r*-Values

Identify the points on the graph of $r = 4\cos 3\theta$ for $0 \le \theta \le 2\pi$ (Figure 6.15) that give maximum *r*-values.

SOLUTION

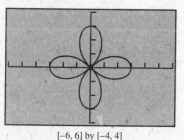

To identify the maximum points, sketch the graph of $y = 4\cos 3x$ for $0 \le x \le 2\pi$ in rectangular coordinates. (Figure 6.16a) Then sketch the graph of $y = |4\cos 3x|$ for $0 \le x \le 2\pi$. This graph (Figure 6.16b) shows the distance of each point on the curve from the origin. Note that the values $\theta = 0$ and $\theta = 2\pi$ correspond to the same point, so there are six extreme values of *y* (and, therefore, of *r*). The maximum value of 4 occurs at the following polar coordinate pairs:

$$(4, 0), (-4, \pi/3), (4, 2\pi/3), (-4, \pi), (4, 4\pi/3), (-4, 5\pi/3).$$

This may appear odd because the graph in polar coordinates appears to show only three extreme values of *r*. But the coordinate pairs $(4, 0)$ and $(-4, \pi)$ correspond to the same point on the graph. Pairs $(-4, \pi/3)$ and $(4, 4\pi/3)$ also correspond to a single point, as do pairs $(4, 2\pi/3)$ and $(-4, 5\pi/3)$. In each case, the maximum *r*-value is 4.

Figure 6.15 The graph of $r = 4\cos 3\theta$ appears to have three maximum points, but there are actually six of them for $0 \le \theta \le 2\pi$ because the grapher traces the curve twice.

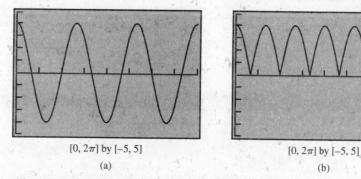

[0, 2π] by [−5, 5]
(a)

[0, 2π] by [−5, 5]
(b)

Figure 6.16 The graphs of (a) $y = 4\cos 3x$ and (b) $y = |4\cos 3x|$.

Exercises for Alternate Example 3

In Exercises 7–12, identify the points for $0 \le \theta \le 2\pi$ where maximum *r*-values occur on the graph of the polar equation. State the maximum *r*-value.

7. $r = 5\cos 4\theta$ **8.** $r = -2\sin 5\theta$ **9.** $r = 3\sin 4\theta$ **10.** $r = 1 + 3\cos\theta$ **11.** $r = 2 - 2\cos\theta$

12. $r = 4 + 3\sin\theta$

Alternate Example 4 Analyzing a Rose Curve

Analyze the graph of the rose curve $r = 3\cos 2\theta$.

SOLUTION

The graph of $r = 3\cos 2\theta$ is a 4-petal rose curve (Figure 6.17). It appears to be symmetric about the *x*-axis, about the *y*-axis, and about the origin. It is easy to verify that the curve does possess each of the symmetries by using the substitutions in Alternate Example 1. For example, to verify the *y*-axis symmetry of the graph, we can substitute $(r, \pi - \theta)$ for (r, θ), as follows:

$$r = 3\cos 2\theta$$
$$r = 3\cos 2(\pi - \theta)$$
$$r = 3\cos (2\pi - 2\theta)$$
$$r = 3\cos (-2\theta)$$
$$r = 3\cos 2\theta$$

Figure 6.17 The graph of the rose curve $r = 3\cos 2\theta$.

The new polar equation is identical to the original equation, so the graph is symmetric about the *y*-axis. The proofs that the graph is symmetric about the *x*-axis and about the origin are similar.

We can use the graph $y = |3 \cos 2x|$ to find the maximum *r*-values and their location. (Figure 6.18). The maximum *r*-value is 3.

We summarize the analysis as follows:

Domain: All reals.

Range [−3, 3]

Continuous

Symmetric about the *x*-axis, the *y*-axis, and the origin.

Bounded

Maximum *r*-value: 3

No asymptotes

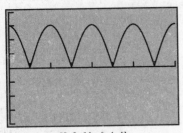

[0, 2π] by [−4, 4]

Figure 6.18 Use the graph of $y = |3 \cos 2x|$ to find the maximum *r*-value.

Exercises for Alternate Example 4

In Exercises 13–18, analyze the graph of the polar curve.

13. $r = 4 \sin 3\theta$ **14.** $r = 5 \cos 2\theta$ **15.** $r = -4 \sin 4\theta$ **16.** $r = -3 \cos 5\theta$ **17.** $r = -2 \sin 6\theta$

18. $r = 5 \cos 3\theta$

Alternate Example 6 **Analyzing a Limaçon Curve**

Analyze the graph of the limaçon curve $r = 1 + 2 \sin \theta$.

SOLUTION

Figure 6.19 shows that the curve is symmetric about the *y*-axis, but not about the *x*-axis or the origin. We can easily verify this observation through substitution, as in Alternate Example 1.

The graph has an inner loop. The maximum *r*-value is 3, and the maximum *r*-value in the inner loop is 1.

Domain: All reals.

Range [−1, 3]

Continuous

Symmetric about the *y*-axis.

Bounded

Maximum *r*-value: 3

No asymptotes

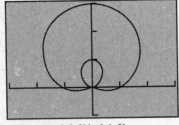

[−3, 3] by [−1, 3]

Figure 6.19 The graph of the limaçon curve $r = 1 + 2 \sin \theta$.

Exercises for Alternate Example 6

In Exercises 19–24, analyze the graph of the polar curve.

19. $r = 1 + 2 \cos \theta$ **20.** $r = 2 + 4 \sin \theta$ **21.** $r = 5 - 4 \cos \theta$ **22.** $r = 6 - \sin \theta$ **23.** $r = 3 + 6 \cos \theta$

24. $r = 3 + 3 \sin \theta$

6.6 De Moivre's Theorem and nth Roots

Alternate Example 2 **Finding Trigonometric Forms**

Find the trigonometric form with $0 \le \theta < 2\pi$ for the complex number.

(a) $2\sqrt{3} + 2i$ **(b)** $5 - 5i$ **(c)** $-2 - 5i$

SOLUTION

(a) For $2\sqrt{3} + 2i$,

$$r = \left|2\sqrt{3} + 2i\right| = \sqrt{(2\sqrt{3})^2 + 2^2} = 4.$$

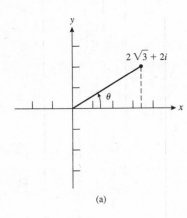

(a)

We can find the tangent of the reference angle θ by finding the ratio of the imaginary part to the real part ($\tan\theta = \dfrac{b}{a}$):

$$\tan\theta = \frac{2}{2\sqrt{3}} = \frac{1}{\sqrt{3}}.$$

Since the terminal side of θ is in the first quadrant,

$$\theta = \tan^{-1}\frac{1}{\sqrt{3}} = \frac{\pi}{6}.$$

Therefore,

$$2\sqrt{3} + 2i = 4\left(\cos\frac{\pi}{6} + i\sin\frac{\pi}{6}\right) \text{ (Figure 6.20a)}.$$

(b) For $5 - 5i$,

$$r = \left|5 - 5i\right| = \sqrt{5^2 + (-5)^2} = 5\sqrt{2}.$$

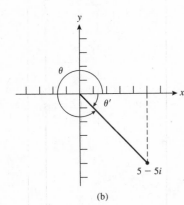

(b)

Now find the tangent of the reference angle θ' for θ by finding the absolute value of the ratio of the imaginary part to the real part:

$$\tan\theta' = \left|\frac{-5}{5}\right| = 1.$$

Since the terminal side of θ is in the fourth quadrant,

$$\theta = 2\pi - \theta' = 2\pi - \tan^{-1}1 = 2\pi - \frac{\pi}{4} = \frac{7\pi}{4}$$

Therefore,

$$5 - 5i = 5\sqrt{2}\left(\cos\left(-\frac{\pi}{4}\right) + i\sin\left(-\frac{\pi}{4}\right)\right) \text{ (Figure 6.20b)}.$$

(c) For $-2 - 5i$,

$$r = \left|-2 - 5i\right| = \sqrt{(-2)^2 + (-5)^2} = \sqrt{29}.$$

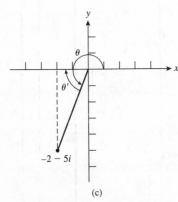

(c)

Figure 6.20 We can find the argument θ by comparing it to the reference angle θ'.

We can find the tangent of the reference angle θ' for θ by finding the absolute value of the ratio of the imaginary part to the real part:

$$\tan\theta' = \frac{5}{2} = 2.5$$

$$\theta' = \tan^{-1}2.5 \approx 1.190$$

Since the terminal side of θ is in the third quadrant,

$$\theta = \pi + \theta' \approx 4.33 \, .$$

Therefore,

$$-2 - 5i = \sqrt{29}(\cos 4.33 + \sin 4.33) \text{ (Figure 6.20c)}.$$

Exercises for Alternate Example 2

In Exercises 1–6, find the trigonometric form of the complex number where the argument satisfies $0 \le \theta < 2\pi$.

1. $3 + 3i$　　**2.** $-6i$　　**3.** $\dfrac{1}{\sqrt{3}} - \dfrac{1}{3}i$　　**4.** $-4 + 8\sqrt{3}i$　　**5.** $-2 - 7i$　　**6.** $-7 + 12i$

Alternate Example 3　Multiplying Complex Numbers

Express the product of z_1 and z_2 in standard form:

$$z_1 = 12\sqrt{3}\left(\cos\frac{\pi}{6} + i\sin\frac{\pi}{6}\right), \quad z_2 = \sqrt{6}\left(\cos\frac{-\pi}{4} + i\sin\frac{-\pi}{4}\right).$$

SOLUTION

$$z_1 \cdot z_2 = 12\sqrt{3}\left(\cos\frac{\pi}{6} + i\sin\frac{\pi}{6}\right) \cdot \sqrt{6}\left(\cos\frac{-\pi}{4} + i\sin\frac{-\pi}{4}\right)$$

$$= 12\sqrt{3} \cdot \sqrt{6}\left[\cos\left(\frac{\pi}{6} + \frac{-\pi}{4}\right) + i\sin\left(\frac{\pi}{6} + \frac{-\pi}{4}\right)\right]$$

$$= 36\sqrt{2}\left(\cos\frac{-\pi}{12} + i\sin\frac{-\pi}{12}\right)$$

$$= 49.18 - 13.18i$$

Exercises for Alternate Example 3

In Exercises 7–12, find the product of z_1 and z_2. Leave the answer in trigonometric form.

7. $z_1 = 3(\cos 17° + i\sin 17°)$
　　$z_2 = 11(\cos 63° + i\sin 63°)$

8. $z_1 = \sqrt{5}(\cos 35° + i\sin 35°)$
　　$z_2 = \dfrac{1}{2}(\cos 55° + i\sin 55°)$

9. $z_1 = 2(\cos 173° + i\sin 173°)$
　　$z_2 = 7(\cos(-31°) + i\sin(-31°))$

10. $z_1 = \sqrt{2}\left(\cos\dfrac{3\pi}{4} + i\sin\dfrac{3\pi}{4}\right)$
　　$z_2 = \sqrt{3}\left(\cos\dfrac{2\pi}{3} + i\sin\dfrac{2\pi}{3}\right)$

11. $z_1 = 2\left(\cos\dfrac{5\pi}{6} + i\sin\dfrac{5\pi}{6}\right)$
　　$z_2 = 5\left(\cos\left(-\dfrac{\pi}{3}\right) + i\sin\left(-\dfrac{\pi}{3}\right)\right)$

12. $z_1 = \dfrac{\sqrt{3}}{3}\left(\cos\dfrac{\pi}{4} + i\sin\dfrac{\pi}{4}\right)$
　　$z_2 = \dfrac{\sqrt{3}}{2}\left(\cos\dfrac{4\pi}{3} + i\sin\dfrac{4\pi}{3}\right)$

Alternate Example 4　Dividing Complex Numbers

Express the quotient z_1/z_2 in standard form:

$$z_1 = 4\left(\cos\frac{\pi}{3} + i\sin\frac{\pi}{3}\right), \quad z_2 = \sqrt{2}\left(\cos\frac{5\pi}{4} + i\sin\frac{5\pi}{4}\right).$$

SOLUTION

$$\frac{z_1}{z_2} = \frac{4\left(\cos\frac{\pi}{3} + i\sin\frac{\pi}{3}\right)}{\sqrt{2}\left(\cos\frac{5\pi}{4} + i\sin\frac{5\pi}{4}\right)}$$

$$= \frac{4}{\sqrt{2}}\left(\cos\left(\frac{\pi}{3} - \frac{5\pi}{4}\right) + i\sin\left(\frac{\pi}{3} - \frac{5\pi}{4}\right)\right)$$

$$= 2\sqrt{2}\left(\cos\left(-\frac{11\pi}{12}\right) + i\sin\left(-\frac{11\pi}{12}\right)\right)$$

Exercises for Alternate Example 4

In Exercises 13–18, find the trigonometric form of the quotient.

13. $\dfrac{3(\cos 45° + i\sin 45°)}{2(\cos 25° + i\sin 25°)}$

14. $\dfrac{7(\cos 240° + i\sin 240°)}{28(\cos 115° + i\sin 115°)}$

15. $\dfrac{8(\cos 125° + i\sin 125°)}{\sqrt{2}(\cos 325° + i\sin 325°)}$

16. $\dfrac{6(\cos 3\pi + i\sin 3\pi)}{2(\cos 4\pi + i\sin 4\pi)}$

17. $\dfrac{3(\cos(2\pi/3) + i\sin(2\pi/3))}{2(\cos(\pi/2) + i\sin(\pi/2))}$

18. $\dfrac{\cos(\pi/3) + i\sin(\pi/3)}{\cos(\pi/6) + i\sin(\pi/6)}$

Alternate Example 7 Finding Fourth Roots

Find the four fourth roots of $2(\cos(2\pi/3) + i\sin(2\pi/3))$

SOLUTION

The fourth roots of z are the complex numbers

$$\sqrt[4]{2}\left(\cos\frac{2\pi/3 + 2\pi k}{4} + i\sin\frac{2\pi/3 + 2\pi k}{4}\right),$$

for $k = 0, 1, 2, 3$. Taking into account that $\left(2\pi/3 + 2\pi k\right)/4 = \pi/6 + \pi k/2$, the four complex numbers are

$$z_1 = \sqrt[4]{2}\left(\cos\left(\frac{\pi}{6} + \frac{0}{2}\right) + i\sin\left(\frac{\pi}{6} + \frac{0}{2}\right)\right) = \sqrt[4]{2}\left(\cos\frac{\pi}{6} + i\sin\frac{\pi}{6}\right)$$

$$z_2 = \sqrt[4]{2}\left(\cos\left(\frac{\pi}{6} + \frac{\pi}{2}\right) + i\sin\left(\frac{\pi}{6} + \frac{\pi}{2}\right)\right) = \sqrt[4]{2}\left(\cos\frac{2\pi}{3} + i\sin\frac{2\pi}{3}\right)$$

$$z_3 = \sqrt[4]{2}\left(\cos\left(\frac{\pi}{6} + \frac{2\pi}{2}\right) + i\sin\left(\frac{\pi}{6} + \frac{2\pi}{2}\right)\right) = \sqrt[4]{2}\left(\cos\frac{7\pi}{6} + i\sin\frac{7\pi}{6}\right)$$

$$z_4 = \sqrt[4]{2}\left(\cos\left(\frac{\pi}{6} + \frac{3\pi}{2}\right) + i\sin\left(\frac{\pi}{6} + \frac{3\pi}{2}\right)\right) = \sqrt[4]{2}\left(\cos\frac{5\pi}{3} + i\sin\frac{5\pi}{3}\right)$$

Exercises for Alternate Example 7

In Exercises 19–24, find the nth root of the complex number for the specified value of n. Note that in Exercises 21–24, you will have to write the complex number in trigonometric form first.

19. $\cos 3\pi + i\sin 3\pi,\ n = 6$

20. $16\left(\cos\frac{\pi}{2} + i\sin\frac{\pi}{2}\right),\ n = 4$

21. $-1,\ n = 5$

22. $i,\ n = 4$

23. $\dfrac{\sqrt{3}}{2} - \dfrac{1}{2}i,\ n = 3$

24. $4 + 4i,\ n = 6$

Chapter 7 Systems and Matrices

7.1 Solving Systems of Two Equations

Alternate Example 1 Using the Substitution Method

Solve the system by substitution.

$$x + 3y = 8$$
$$3x - 2y = 13$$

SOLUTION

Solve Algebraically

Solving the first equation for x yields $x = 8 - 3y$. Then substitute the expression for x into the second equation.

$3x - 2y = 13$	Second equation
$3(8 - 3y) - 2y = 13$	Replace x by $8 - 3y$.
$24 - 9y - 2y = 13$	Distributive property
$-11y = -11$	Collect like terms.
$y = 1$	Divide by -11.
$x = 5$	Use $x = 8 - 3y$.

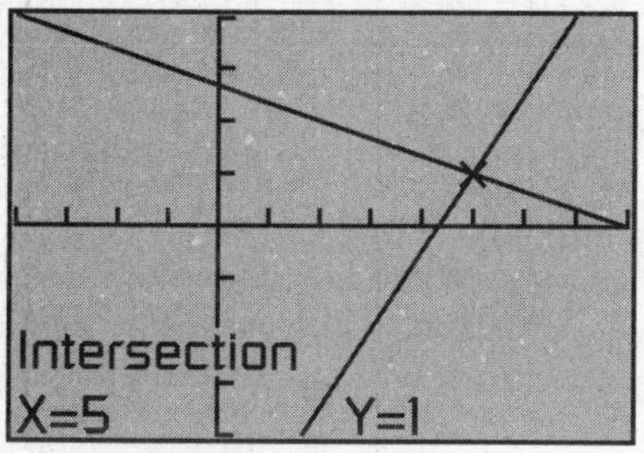

Support Graphically

The graph of each equation is a line. Figure 7.1 shows that the two lines intersect at a single point (5, 1).

Interpret

The solution of the system is $x = 5$, $y = 1$ or the ordered pair (5, 1).

Intersection
X=5   Y=1

[–4, 8] by [–4, 4]

Figure 7.1 The two lines $x + 3y = 8$ and $3x - 2y = 13$ intersect at the point (5, 1).

Exercises for Alternate Example 1

In Exercises 1–6, solve the system by substitution.

1. $2x + y = -1$
   $4x + 3y = 1$

2. $x + 2y = 0$
   $2x + 5y = -1$

3. $-3x + y = -5$
   $2x + 3y = 6$

4. $x - y = 2$
   $3x - 5y = 0$

5. $4x + y = 11$
   $3x - 3y = -3$

6. $-x + 5y = 8$
   $4x + 2y = -10$

Alternate Example 2 Solving a Nonlinear System by Substitution

Find the dimensions of a rectangular room that has perimeter 60 ft and the area 180 ft^2.

SOLUTION

Model

Let x and and y be the lengths of two adjacent sides of the room. Then

$$2x + 2y = 60 \qquad \text{Perimeter is 60.}$$

$$xy = 180 \qquad \text{Area is 180.}$$

Solve Algebraically

Solving the first equation for y yields $y = 30 - x$. Then substitute the expression for y into the second equation.

$xy = 180$	Second equation
$x(30 - x) = 180$	Replace y by $30 - x$.
$30x - x^2 = 180$	Distributive property
$x^2 - 30x + 180 = 0$	
$x = \dfrac{30 \pm \sqrt{(-30)^2 - 4(180)}}{2}$	Quadratic formula
$x = 21.708\ldots$ or $x = 8.291\ldots$	Evaluate.
$y = 8.291\ldots$ or $y = 21.708\ldots$	Use $y = 30 - x$.

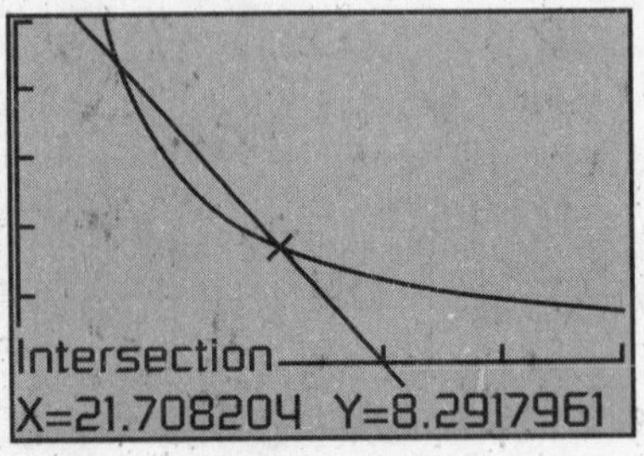

Support Graphically

Figure 7.2 shows that the graphs of $y = 30 - x$ and $y = 180/x$ have two points of intersection.

Interpret

The two ordered pairs $(21.708\ldots, 8.291\ldots)$ and $(8.291\ldots, 21.708\ldots)$ represent the same rectangle whose dimensions are approximately 22 ft by 8 ft.

Intersection
X=21.708204 Y=8.2917961

[0, 50] by [10, 25]

Figure 7.2 We can assume that $x \geq 0$ and $y \geq 0$, because x and y are lengths.

Exercises for Alternate Example 2

In Exercises 7–12, solve the system by substitution.

7. $2x + y = 20$
    $100 + 2y^2 = 60x$

8. $y = x^2$
    $2x + y = 15$

9. $x + y = 9$
    $x^2 + 2y = 9x$

10. $x - y = 2$
    $2x^2 - 3y = 10x$

11. $x^2 + y^2 = 100$
    $4x + 2y = 40$

12. $2x + y = 5$
    $x^2 - y^2 = 3y$

Alternate Example 4 **Using the Elimination Method**

Solve the system by elimination.

$$2x - 8y = 24$$

$$3x + 5y = 2$$

SOLUTION

Solve Algebraically

Multiply the first equation by 5 and the second by 8 to obtain

$$10x - 40y = 120$$

$$24x + 40y = 16.$$

Then add the two equations to eliminate the variable y.

$$34x = 136$$

Next divide by 34 to solve for x.

$$x = 4$$

Finally, substitute $x = 4$ into either original equation to determine that $y = -2$. For example,

$$2x - 8y = 24$$
$$2(4) - 8y = 24$$
$$8 - 8y = 24$$
$$-8y = 16$$
$$y = -2$$

The solution of the original system is $(4, -2)$.

Exercises for Alternate Example 4

In Exercises 13–18, solve the system by elimination.

13. $2x + 7y = -11$
$5x + 2y = 19$

14. $x - 3y = 4$
$3x + 2y = 1$

15. $-x + 3y = 11$
$5x + 4y = 21$

16. $-2x + 3y = 30$
$3x + 5y = 69$

17. $2x + 7y = -19$
$-2x + 2y = 10$

18. $3x + 4y = 8$
$4x - y = -2$

Alternate Example 5   Finding No Solution

Solve the system

$$2x - 3y = 9$$
$$-4x + 6y = 10.$$

SOLUTION

Solve Algebraically

We use the elimination method. Multiply the first equation by 2 and add the result to the second equation.

$$4x - 6y = 18 \qquad \text{Multiply first equation by 2.}$$
$$-4x + 6y = 10 \qquad \text{Second equation}$$
$$0 = 28 \qquad \text{Add.}$$

The last equation is true for no values of x and y. That means that the system has no solution.

Support Graphically

Each equation represents a line. Figure 7.3 suggests that the graphs of the two equations are parallel. Solving each equation for y yields

$$y = \frac{2}{3}x - 3$$
$$y = \frac{2}{3}x + \frac{5}{3}.$$

The two lines have the same slope and are therefore parallel. The system has no solution because the graphs of the two equations do not intersect.

Note, however, that when both the slopes and the intercepts of the two lines match, the system has infinitely many solutions because the lines coincide.

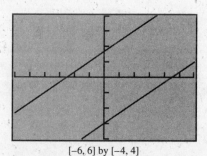

[–6, 6] by [–4, 4]

Figure 7.3 The graphs of the two equations appear to be parallel.

Exercises for Alternate Example 5

In Exercises 19–26, solve the system by elimination.

19. $2x + 5y = 5$
$5x + 12.5y = 8$

20. $x - 3y = 4$
$3x - 9y = 10$

21. $2x + 5y = 10$
$5x + 12.5y = 25$

22. $x - 3y = 4$
$3x - 9y = 12$

23. $-x + 2y = 10$
$4x - 8y = 20$

24. $-3x + 3y = 30$
$x - y = 10$

25. $-x + 2y = 10$
$4x - 8y = -40$

26. $-3x + 3y = 30$
$x - y = -10$

7.2 Matrix Algebra

Alternate Example 2 Using Matrix Addition

The table below represents the state and local expenditures on public works project in 1999–2001 (*Source:U.S. Census Bureau*). Model the data with two matrices—one for state expenditures and one for local expenditures. Then express the combined state and local expenditures in all categories as a single matrix.

State and Local Expenditures on Public Works

Category	State Expenditures			Local Expenditures		
	1999	2000	2001	1999	2000	2001
Highways (HW)	56,242	61,942	66,436	36,776	39,394	40,800
Airport transportation (AT)	1,013	1,106	1,321	11,619	12,054	14,211
Water transportation (WT)	798	863	951	2,273	2,277	5,044
Sewage (SW)	1,128	955	987	25,852	27,098	27,075
Solid waste management (WM)	1,958	2,347	2,601	14,110	14,861	16,055
Water supply (WS)	164	354	347	33,924	35,435	36,410
Mass transit (MT)	6,089	7,407	6,874	22,433	24,476	26,001

SOLUTION

First, rewrite the data as two matrices.

$$A = \begin{array}{c} \\ HW \\ AT \\ WT \\ SW \\ WM \\ WS \\ MT \end{array} \begin{array}{ccc} 99 & 00 & 01 \\ \left[\begin{array}{ccc} 56{,}242 & 61{,}942 & 66{,}436 \\ 1{,}013 & 1{,}106 & 1{,}321 \\ 798 & 863 & 951 \\ 1{,}128 & 955 & 987 \\ 1{,}958 & 2{,}347 & 2{,}601 \\ 164 & 354 & 347 \\ 6{,}089 & 7{,}407 & 6{,}874 \end{array}\right] \end{array} \qquad B = \begin{array}{c} \\ HW \\ AT \\ WT \\ SW \\ WM \\ WS \\ MT \end{array} \begin{array}{ccc} 99 & 00 & 01 \\ \left[\begin{array}{ccc} 36{,}776 & 39{,}394 & 40{,}800 \\ 11{,}619 & 12{,}054 & 14{,}211 \\ 2{,}273 & 2{,}277 & 5{,}044 \\ 25{,}852 & 27{,}098 & 27{,}075 \\ 14{,}110 & 14{,}861 & 16{,}055 \\ 33{,}924 & 35{,}435 & 36{,}410 \\ 22{,}433 & 24{,}476 & 26{,}001 \end{array}\right] \end{array}$$

To add the two matrices, combine the data in the corresponding entries.

$$
A + B = \begin{array}{c} \\ \text{HW} \\ \text{AT} \\ \text{WT} \\ \text{SW} \\ \text{WM} \\ \text{WS} \\ \text{MT} \end{array}
\begin{array}{ccc} 99 & 00 & 01 \end{array}
\left[\begin{array}{ccc}
93{,}018 & 101{,}336 & 107{,}236 \\
12{,}632 & 13{,}160 & 15{,}532 \\
3{,}071 & 3{,}140 & 5{,}995 \\
26{,}980 & 28{,}053 & 28{,}062 \\
16{,}068 & 17{,}208 & 18{,}656 \\
34{,}088 & 35{,}789 & 36{,}757 \\
28{,}522 & 31{,}883 & 32{,}875
\end{array} \right]
$$

Exercises for Alternate Example 1

In Exercises 1–6, find **(a)** $A + B$ and **(b)** $A - B$.

1. $A = \begin{bmatrix} -1 & 2 \\ 0 & -1 \\ 2 & 4 \end{bmatrix}$, $B = \begin{bmatrix} 3 & 1 \\ 5 & 1 \\ 0 & 2 \end{bmatrix}$

2. $A = \begin{bmatrix} -1 \\ 0 \\ 3 \end{bmatrix}$, $B = \begin{bmatrix} -3 \\ 2 \\ 0 \end{bmatrix}$

3. $A = \begin{bmatrix} 11 & -1 \\ 0 & 3 \end{bmatrix}$, $B = \begin{bmatrix} -5 & 0 \\ 3 & 1 \end{bmatrix}$

4. $A = \begin{bmatrix} -7 & 1 & 0 & 4 \\ 0 & 4 & -6 & -3 \end{bmatrix}$, $B = \begin{bmatrix} 2 & 0 & -5 & 3 \\ 1 & -2 & 0 & -1 \end{bmatrix}$

5. $A = \begin{bmatrix} 1 & 2 & 3 \\ 0 & 1 & 4 \\ 0 & 0 & 1 \end{bmatrix}$, $B = \begin{bmatrix} 1 & 0 & 0 \\ -1 & 1 & 0 \\ -2 & -1 & 1 \end{bmatrix}$

6. $A = \begin{bmatrix} 3 & 2 & 1 & 0 \\ -2 & 5 & 0 & 7 \\ 0 & -1 & 2 & -1 \end{bmatrix}$, $B = \begin{bmatrix} -1 & 3 & 0 & 5 \\ 2 & 2 & 2 & 3 \\ 0 & 1 & 7 & -4 \end{bmatrix}$

■

Alternate Example 3 Using Scalar Multiplication

The matrix below shows the sales tax rates for 4 cities.

$$
\begin{array}{c} \\ \text{City A} \\ \text{City B} \\ \text{City C} \\ \text{City D} \end{array}
\begin{array}{cc} \text{State} & \text{City} \end{array}
\left[\begin{array}{cc}
0.06 & 0.025 \\
0.06 & 0.01 \\
0.05 & 0.015 \\
0.05 & 0
\end{array} \right]
$$

Construct a matrix that shows the amount of tax that each state and city collects on a $900 purchase.

SOLUTION

Multiply the original matrix by the scalar 900.

$$
900 \times \begin{bmatrix}
0.06 & 0.025 \\
0.06 & 0.01 \\
0.05 & 0.015 \\
0.05 & 0
\end{bmatrix}
\begin{array}{c} \text{City A} \\ \text{City B} \\ \text{City C} \\ \text{City D} \end{array}
=
\begin{array}{c} \\ \\ \\ \\ \end{array}
\begin{array}{cc} \text{State} & \text{City} \end{array}
\begin{bmatrix}
54 & 22.5 \\
54 & 9 \\
45 & 13.5 \\
45 & 0
\end{bmatrix}
$$

Exercises for Alternate Example 3

In Exercises 7–12, find **(a)** $2A + B$ and **(b)** $2A - 3B$.

7. $A = \begin{bmatrix} -1 & 2 \\ 0 & -1 \\ 2 & 4 \end{bmatrix}$, $B = \begin{bmatrix} 3 & 1 \\ 5 & 1 \\ 0 & 2 \end{bmatrix}$

8. $A = \begin{bmatrix} -1 \\ 0 \\ 3 \end{bmatrix}$, $B = \begin{bmatrix} -3 \\ 2 \\ 0 \end{bmatrix}$

9. $A = \begin{bmatrix} 11 & -1 \\ 0 & 3 \end{bmatrix}$, $B = \begin{bmatrix} -5 & 0 \\ 3 & 1 \end{bmatrix}$

10. $A = \begin{bmatrix} -7 & 1 & 0 & 4 \\ 0 & 4 & -6 & -3 \end{bmatrix}$, $B = \begin{bmatrix} 2 & 0 & -5 & 3 \\ 1 & -2 & 0 & -1 \end{bmatrix}$

11. $A = \begin{bmatrix} 1 & 2 & 3 \\ 0 & 1 & 4 \\ 0 & 0 & 1 \end{bmatrix}$, $B = \begin{bmatrix} 1 & 0 & 0 \\ -1 & 1 & 0 \\ -2 & -1 & 1 \end{bmatrix}$

12. $A = \begin{bmatrix} 3 & 2 & 1 & 0 \\ -2 & 5 & 0 & 7 \\ 0 & -1 & 2 & -1 \end{bmatrix}$, $B = \begin{bmatrix} -1 & 3 & 0 & 5 \\ 2 & 2 & 2 & 3 \\ 0 & 1 & 7 & -4 \end{bmatrix}$

Alternate Example 4   **Finding the Product of Two Matrices**

Find the product AB if possible, where

(a) $A = \begin{bmatrix} 3 & 1 \\ 0 & 4 \end{bmatrix}$ and $B = \begin{bmatrix} 1 & -1 & 3 \\ 2 & 0 & 3 \end{bmatrix}$

(b) $A = \begin{bmatrix} 1 & -1 & 3 \\ 2 & 0 & 3 \end{bmatrix}$ and $B = \begin{bmatrix} 3 & 1 \\ 0 & 4 \end{bmatrix}$

SOLUTION

(a) The number of columns of A is 2 and the number of rows of B is 2, so the product AB is defined. The product $AB = [c_{ij}]$ is a 2×3 matrix, where

$$c_{11} = \begin{bmatrix} 3 & 1 \end{bmatrix} \begin{bmatrix} 1 \\ 2 \end{bmatrix} = 3 \cdot 1 + 1 \cdot 2 = 5$$

$$c_{12} = \begin{bmatrix} 3 & 1 \end{bmatrix} \begin{bmatrix} -1 \\ 0 \end{bmatrix} = 3 \cdot (-1) + 1 \cdot 0 = -3$$

$$c_{13} = \begin{bmatrix} 3 & 1 \end{bmatrix} \begin{bmatrix} 3 \\ 3 \end{bmatrix} = 3 \cdot 3 + 1 \cdot 3 = 12$$

$$c_{21} = \begin{bmatrix} 0 & 4 \end{bmatrix} \begin{bmatrix} 1 \\ 2 \end{bmatrix} = 0 \cdot 1 + 4 \cdot 2 = 8$$

$$c_{22} = \begin{bmatrix} 0 & 4 \end{bmatrix} \begin{bmatrix} -1 \\ 0 \end{bmatrix} = 0 \cdot (-1) + 4 \cdot 0 = 0$$

$$c_{23} = \begin{bmatrix} 0 & 4 \end{bmatrix} \begin{bmatrix} 3 \\ 3 \end{bmatrix} = 0 \cdot 3 + 4 \cdot 3 = 12$$

Thus, $AB = \begin{bmatrix} 5 & -3 & 12 \\ 8 & 0 & 12 \end{bmatrix}$. You can check your result on a grapher. Figure 7.4 supports the computation.

(b) The number of columns of A is 3 and the number of rows of B is 2, so the product AB is not defined.

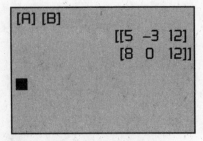

Figure 7.4 The matrix AB is displayed as two 1×3 matrices.

Exercises for Alternate Example 4

In Exercises 13–18, use the definition of matrix multiplication to find **(a)** AB and **(b)** BA, where possible.

13. $A = \begin{bmatrix} 3 & -1 \\ 2 & 1 \end{bmatrix}$, $B = \begin{bmatrix} -5 & 0 \\ 3 & 1 \end{bmatrix}$

14. $A = \begin{bmatrix} -1 & -2 \\ 3 & 4 \end{bmatrix}$, $B = \begin{bmatrix} 2 & 1 \\ 0 & 1 \end{bmatrix}$

15. $A = \begin{bmatrix} -1 & 0 & 2 & -3 \\ 0 & 4 & 1 & 2 \end{bmatrix}$, $B = \begin{bmatrix} 3 & 5 \\ 7 & 1 \\ 0 & -2 \\ -1 & 0 \end{bmatrix}$

16. $A = \begin{bmatrix} 3 & 1 & 0 \\ -1 & 2 & 5 \end{bmatrix}$, $B = \begin{bmatrix} 1 & 2 \\ 0 & 1 \\ -3 & 8 \end{bmatrix}$

17. $A = \begin{bmatrix} 1 & 2 & 3 \\ 0 & 1 & 4 \\ 0 & 0 & 1 \end{bmatrix}$, $B = \begin{bmatrix} 1 & 0 & 0 \\ -1 & 1 & 0 \\ -2 & -1 & 1 \end{bmatrix}$

18. $A = \begin{bmatrix} 1 & -1 & 0 \\ 3 & 2 & -1 \\ 0 & 2 & 1 \end{bmatrix}$, $B = \begin{bmatrix} 5 & 0 & -1 \\ -1 & 1 & 2 \\ -3 & 2 & 2 \end{bmatrix}$

Alternate Example 8   **Finding Inverse Matrices**

Determine whether the matrix has an inverse. If so, find the inverse matrix.

(a) $A = \begin{bmatrix} -1 & -2 \\ 3 & 4 \end{bmatrix}$ **(b)** $B = \begin{bmatrix} 1 & 1 & 0 \\ 3 & 1 & 0 \\ -2 & -1 & 1 \end{bmatrix}$

SOLUTION

(a) The determinant of A is $\det A = ad - bc = (-1)4 - (-2)3 = 2 \neq 0$. We can conclude that A has an inverse. Using the formula for the inverse of a 2×2 matrix, we obtain

$$A^{-1} = \frac{1}{ad - bc}\begin{bmatrix} d & -b \\ -c & a \end{bmatrix} = \frac{1}{2}\begin{bmatrix} 4 & 2 \\ -3 & -1 \end{bmatrix}$$

$$= \begin{bmatrix} 2 & 1 \\ -1.5 & -0.5 \end{bmatrix}$$

We can verify that $AA^{-1} = I_2$.

$$AA^{-1} = \begin{bmatrix} -1 & -2 \\ 3 & 4 \end{bmatrix}\begin{bmatrix} 2 & 1 \\ -1.5 & -0.5 \end{bmatrix}$$

$$= \begin{bmatrix} (-1)2 + (-2)(-1.5) & (-1)1 + (-2)(-0.5) \\ 3(2) + 4(-1.5) & 3(1) + 4(-0.5) \end{bmatrix}$$

$$= \begin{bmatrix} 1 & 0 \\ 0 & 1 \end{bmatrix}$$

You can similarly verify that $A^{-1}A = I_2$.

(b) The determinant of B is -2, so the matrix has an inverse. Use your grapher to compute the inverse matrix (Figure 7.5).

$$B^{-1} = \begin{bmatrix} -0.5 & 0.5 & 0 \\ 1.5 & -0.5 & 0 \\ 0.5 & 0.5 & 1 \end{bmatrix}$$

You can also use your grapher to verify that $BB^{-1} = B^{-1}B = I_3$.

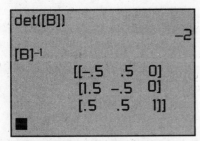

Figure 7.5 The matrix B is nonsingular, so you can use grapher to compute its inverse.

Exercises for Alternate Example 8

In Exercises 19–24, find the inverse of the matrix if it has one, or state that the inverse does not exist.

19. $\begin{bmatrix} 2 & 1 \\ 1 & 3 \end{bmatrix}$

20. $\begin{bmatrix} 4 & 8 \\ 3 & 6 \end{bmatrix}$

21. $\begin{bmatrix} 5 & -2 \\ -3 & 2 \end{bmatrix}$

22. $\begin{bmatrix} -1 & -3 \\ 3 & 9 \end{bmatrix}$

23. $\begin{bmatrix} 1 & 0 & 1 \\ 3 & 1 & 0 \\ -2 & -1 & 1 \end{bmatrix}$

24. $\begin{bmatrix} 0 & 2 & 2 \\ 2 & -2 & 1 \\ 3 & -1 & 1 \end{bmatrix}$

7.3 Multivariate Linear Systems and Row Operations

Alternate Example 1 Solving by Substitution

Solve the system

$$2x - 3y + z = -1$$
$$4y - 2z = 2$$
$$z = 3.$$

SOLUTION

The third equation determines z, namely $z = 3$. Substitute the value of z into the second equation to determine y.

$4y - 2z = 2$	Second equation
$4y - 2(3) = 2$	Substitute $z = 3$.
$4y = 8$	Collect like terms.
$y = 2$	

Finally, substitute the values for y and z into the first equation.

$2x - 3y + z = -1$	First equation
$2x - 3(2) + 3 = -1$	Substitute $y = 2$, $z = 3$.
$2x = 2$	Collect like terms.
$x = 1$	

The solution of the system is $x = 1$, $y = 2$, $z = 3$, or the ordered triple $(1, 2, 3)$.

Exercises for Alternate Example 1

In Exercises 1–6, use substitution to solve the system of equations.

1. $3x + 2y - 2z = 9$
   $4y - 2z = 0$
   $5z = 5$

2. $x + 4y + 3z = -2$
   $2y + 3z = 3$
   $-4z = -12$

3. $6x + y - z = 5$
   $3y + 2z = 16$
   $z = 2$

4. $2x + 5y - z = 16$
   $7y - 2z = 18$
   $4z = -8$

5. $2x - 10y - z = -1$
   $5y + 3z = -1$
   $z = -1$

6. $5x - y + 2z = 9$
   $2y - 5z = -8$
   $z = 0$

Alternate Example 2 **Using Gaussian Elimination**

Solve the system

$$x - y + z = 0$$
$$3x - y - 2z = -1$$
$$-x - y + 2z = -1.$$

SOLUTION

Each time you multiply one of the equations of in a system by a non-zero constant you obtain an equivalent system. When you add two of the equations in a system and replace one of them with the result, you also obtain an equivalent system. So each step in the following process leads to a system of equations equivalent to the original system.

Add the first equation to the third equation and replace the third equation. Leave the first two equations unchanged.

$$x - y + z = 0$$
$$3x - y - 2z = -1$$
$$-2y + 3z = -1 \qquad \begin{array}{l} x - y + z = 0 \\ -x - y + 2z = -1 \end{array}$$

Multiply the first equation by –3 and add to the second equation, replacing it. (Or multiply the first equation by 3 and subtract it from the second equation.)

$$x - y + z = 0$$
$$2y - 5z = -1$$
$$-2y + 3z = -1 \qquad \begin{array}{l} -3x + 3y - 3z = 0 \\ 3x - y - 2z = -1 \end{array}$$

Add the second equation to the third and replace the third equation.

$$x - y + z = 0$$
$$2y - 5z = -1$$
$$-2z = -2 \qquad \begin{array}{l} 2y - 5z = -1 \\ -2y + 3z = -1 \end{array}$$

This system is now in triangular form and can be solved by substitution like the system in Alternate Example 1. The solution of this system is (1, 2, 1). The solution of the original system is also (1, 2, 1).

Exercises for Alternate Example 2

In Exercises 7–12, use Gaussian elimination and substitution to solve the system of equations

7. $x + y + z = 2$
$2x + 3y + z = 7$
$x - y + 2z = -3$

8. $2x - y = 0$
$x + 3y - z = -3$
$2x + 2y + z = 8$

9. $2x + 3y - z = 5$
$x + 3y + 2z = 13$
$3x - 2y - z = -4$

10. $x + y - z = 3$
$2x + 2y - 3z = 5$
$3x - 3y + 4z = 10$

11. $2x - 3y - z = 4$
$5x + 2y + 3z = 27$
$x - y + 4z = 17$

12. $5x - 2y + 2z = 2$
$x - 3y + 5z = 7$
$2x + y - 4z = -7$

Alternate Example 5 **Finding Infinitely Many Solutions**

Solve the system

$$2x + y + 4z = 2$$
$$-2x + 4y + z = 18$$
$$-4x + 3y - 3z = 16.$$

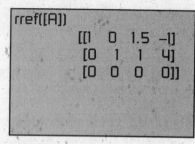

Figure 7.6 The grapher produces the reduced row echelon form for the augmented matrix A for the system of equations.

SOLUTION

Use a grapher to find the reduced row echelon form of the augmented matrix of the system (Figure 7.6).

The matrix shows that the following system of equations is equivalent to the original system.

$$x + 1.5z = -1$$
$$y + z = 4$$
$$0 = 0$$

Solving the first two equations for x and y in terms of z yields:

$$x = -1.5z - 1$$
$$y = 4 - z$$

This system has infinitely many solutions because for every value of z we can use these two equations to find corresponding values for x and y.

Interpret

The solution is a set of all ordered triples of the form $(-1.5z - 1, 4 - z, z)$, where z is any real number.

Exercises for Alternate Example 3

In Exercises 13–18, give the solution of the system in terms of z.

13. $x + y + z = 2$
    $2x + 3y + z = 7$

14. $x + 2y = -2$
    $3x + 2y + 2z = 4$
    $2x - 4y + 4z = 16$

15. $x + y - z = 3$
    $2x - y - 8z = -3$
    $3x + 2y - 5z = 8$

16. $x + 3y + 10z = 2$
    $2x - 3y + 14z = -5$
    $-x + 6y - 4z = 7$

17. $x + y + 2z = 12$
    $2x - y + z = 3$
    $-3x + 2y - z = -1$

18. $x + y + 3z = 3$
    $x - y - z = -1$
    $-2x + 3y + 4z = 4$

Alternate Example 7 Solving a System Using Inverse Matrices

Use the inverse matrix method to solve the system

$$2x - 3y = 9$$
$$3x - 5y = 10.$$

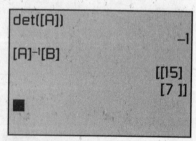

Figure 7.7 If the coefficient matrix is not singular, you can find the solution by multiplying its inverse by the constant matrix.

SOLUTION

First we write the system as a matrix equation.

$$A = \begin{bmatrix} 2 & -3 \\ 3 & -5 \end{bmatrix}, \qquad X = \begin{bmatrix} x \\ y \end{bmatrix}, \qquad \text{and} \qquad B = \begin{bmatrix} 9 \\ 10 \end{bmatrix}$$

Then

$$A \cdot X = \begin{bmatrix} 2 & -3 \\ 3 & -5 \end{bmatrix} \cdot \begin{bmatrix} x \\ y \end{bmatrix} = \begin{bmatrix} 2x - 3y \\ 3x - 5y \end{bmatrix}$$

so that

$$A \cdot X = B,$$

where A is the coefficient matrix of the system. Check that $\det A \neq 0$, so that A^{-1} exists. Then use your grapher to solve the system (Figure 7.7).

$$X = A^{-1}B = \begin{bmatrix} 15 \\ 7 \end{bmatrix}$$

The solution of the system is $x = 15$, $y = 7$, or $(15, 7)$.

Exercises for Alternate Example 7

In Exercises 19–24, solve the system by using an inverse matrix.

19. $3x + 7y = -3$
    $2x + 5y = 8$

20. $4x - 6y = 1$
    $3x - 5y = 5$

21. $x + 5y = 3$
    $2x + 4y = 6$

22. $x + y = 3$
    $-y + z = -2$
    $3x + z = 5$

23. $x - 5y + 3z = 2$
    $2x - 3y - z = 1$
    $-2x + 2y + z = 12$

24. $2x + y - z = -3$
    $x + 3y - 4z = 8$
    $-3x + y + 3z = 9$

7.4 Partial Fractions

Alternate Example 1 Writing the Decomposition Factors

Write the terms of the partial fraction decomposition of the rational function

$$\frac{3x + 1}{x^2(x^2 + x + 1)(x + 2)^3},$$

but do not solve for the corresponding constants.

SOLUTION

Because the degree of the numerator is less than the degree of the denominator, we don't need to divide in order to remove a polynomial part of the function. Start by identifying the factors in the denominator. Two are linear factors to some power—x^2 and $(x + 2)^3$. One is quadratic, $x^2 + x + 1$. The quadratic factor is irreducible.

For each factor of the form $(mx + n)^u$, the partial fraction decomposition is

$$\frac{A_1}{mx + n} + \frac{A_2}{(mx + n)^2} + \dots + \frac{A_u}{(mx + n)^u}.$$

Then, the powers of the linear factors correspond to partial fractions

$$\frac{A_1}{x} + \frac{A_2}{x^2} \quad \text{and} \quad \frac{B_1}{x + 2} + \frac{B_2}{(x + 2)^2} + \frac{B_3}{(x + 2)^3}.$$

The quadratic factor corresponds to

$$\frac{C_1 x + D_1}{x^2 + x + 1}.$$

So the complete partial fraction decomposition of the rational function is

$$\frac{3x + 1}{x^2(x^2 + x + 1)(x + 2)^3} = \frac{A_1}{x} + \frac{A_2}{x^2} + \frac{B_1}{x + 2} + \frac{B_2}{(x + 2)^2} + \frac{B_3}{(x + 2)^3} + \frac{C_1 x + D_1}{x^2 + x + 1}.$$

Exercises for Alternate Example 1

In Exercises 1–6, write the terms of the partial fraction decomposition of the rational function. Do not solve for the corresponding constants.

1. $\dfrac{x}{(x^2+4x+4)(3x-2)^2}$

2. $\dfrac{3x^3+2x^2+1}{x^2(x-1)}$

3. $\dfrac{x^4+1}{x(x-1)(x+2)}$

4. $\dfrac{5}{x^3(x^2+9)(2x+3)}$

5. $\dfrac{5x+2}{(x^2+3x+3)(x+1)^2(2x-1)^2}$

6. $\dfrac{5}{x^3(x^2-9)(2x+3)}$

Alternate Example 2 **Decomposing a Fraction with Distinct Linear Factors**

Find the partial fraction decomposition of

$$\frac{2x-1}{x^2-x-6}.$$

SOLUTION

Solve Algebraically

First factor the denominator of the fraction into linear terms.

$$x^2-x-6=(x-3)(x+2)$$

Following the process in Alternate Example 1, we write

$$\frac{2x-1}{x^2-x-6}=\frac{A_1}{x-3}+\frac{A_2}{x+2}.$$

Now "clear fractions" by multiplying both sides of the equation by the original denominator.

$$2x-1=A_1(x+2)+A_2(x-3)$$
$$2x-1=(A_1+A_2)x+(2A_1-3A_2)$$

Match the coefficients on the left and the right sides of the equation and write the correspondences as a system of two equations in two unknowns.

$$A_1+A_2=2$$
$$2A_1-3A_2=-1$$

We can now rewrite the system as a matrix equation $A \cdot X = B$, .

$$A=\begin{bmatrix}1 & 1\\2 & -3\end{bmatrix}, \qquad X=\begin{bmatrix}x\\y\end{bmatrix}, \qquad \text{and} \qquad B=\begin{bmatrix}2\\-1\end{bmatrix}$$

Then

$$X=A^{-1}B=\begin{bmatrix}1\\1\end{bmatrix},$$

as shown in Figure 7.8.

Then the coefficients are $A_1 = 1$ and $A_2 = 1$, and

$$\frac{2x-1}{x^2-x-6}=\frac{1}{x-3}+\frac{1}{x+2}.$$

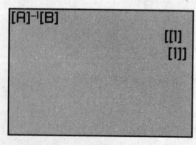

Figure 7.8 The solution of the coefficient matrix equation.

Support Graphically

Figure 7.9 suggests that the following two functions are the same:

$$\frac{2x-1}{x^2-x-6} \qquad \text{and} \qquad \frac{1}{x-3}+\frac{1}{x+2}.$$

Note that both graphs have discontinuities at $x = 3$ and $x = -2$.

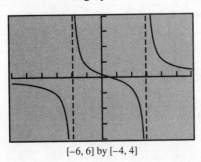

[-6, 6] by [-4, 4]

Figure 7.9 The graphs of $y=\dfrac{2x-1}{x^2-x-6}$ and $y=\dfrac{1}{x-3}+\dfrac{1}{x+2}$ appear to be the same.

Exercises for Alternate Example 2

In Exercises 7–12, use inverse matrices to write partial fraction decompositions.

7. $\dfrac{7x-1}{2x^2-x-1}$ 8. $\dfrac{-1}{x^2+9x+20}$ 9. $\dfrac{x}{2x^2-7x+6}$ 10. $\dfrac{10x-7}{3x^2-5x+2}$ 11. $\dfrac{7x-20}{x^2-7x+10}$

12. $\dfrac{6x}{9x^2-4}$

Alternate Example 3 **Decomposing a Fraction with a Repeated Linear Factor**

Find the partial fraction decomposition of

$$\frac{x^2+10x+8}{x^3+4x^2+4x}.$$

SOLUTION

First factor the denominator of the fraction into linear terms.

$$x^3-4x^2+4x = x(x-2)^2$$

Because the factor $x - 2$ is squared, it contributes two terms to the decomposition. Following the process in Alternate Example 1, we write

$$\frac{x^2+10x+8}{x^3+4x^2+4x}=\frac{A_1}{x}+\frac{A_2}{x+2}+\frac{A_3}{(x+2)^2}.$$

Now clear fractions by multiplying both sides of the equation by the original denominator.

$$x^2+10x+8 = A_1(x+2)^2 + A_2x(x+2)+A_3x$$
$$= A_1(x^2+4x+4 + A_2x^2+2A_2x+A_3x$$
$$= A_1x^2+4A_1x+4A_1 + A_2x^2+2A_2x+A_3x$$

Combining like terms in the equation above we obtain:

$$x^2+10x+8 = (A_1+A_2)x^2+(4A_1+2A_2+A_3)x+4A_1.$$

Comparing the coefficients of the powers of x on the left and the right sides of the equation, we obtain the following system of equations:

$$A_1 + A_2 = 1$$
$$4A_1 + 2A_2 + A_3 = 10$$
$$4A_1 = 8$$

The reduced row echelon form of the augmented matrix

$$\begin{bmatrix} 1 & 1 & 0 & 1 \\ 4 & 2 & 1 & 10 \\ 4 & 0 & 0 & 8 \end{bmatrix}$$

of the above system of equations is

$$\begin{bmatrix} 1 & 0 & 0 & 2 \\ 0 & 1 & 0 & -1 \\ 0 & 0 & 1 & 4 \end{bmatrix}.$$

So $A_1 = 2$, $A_2 = -1$, $A_3 = 4$, and.

$$\frac{x^2 + 10x + 8}{x^3 + 4x^2 + 4x} = \frac{2}{x} + \frac{-1}{x+2} + \frac{4}{(x+2)^2}.$$

Exercises for Alternate Example 3

In Exercises 13–18, find the partial fraction decomposition.

13. $\dfrac{4x^2 + 7x + 2}{x^3 + 2x^2}$

14. $\dfrac{3x^2 - 3x - 1}{x^3 - 2x^2 + x}$

15. $\dfrac{-3x^2 - 4x + 3}{2x^3 + 3x^2}$

16. $\dfrac{-5x^2 - 14x + 5}{x^3 + 5x^2}$

17. $\dfrac{x^2 + 3x - 9}{x^3 + 6x^2 + 9x}$

18. $\dfrac{3x^2 - 4x - 2}{(x-1)^2(x+2)}$

Alternate Example 4 **Decomposing a Fraction with an Irreducible Quadratic Factor**

Find the partial fraction decomposition of

$$\frac{8x^2 + x + 2}{4x^3 - 2x^2 + 2x - 1}.$$

SOLUTION

First factor the denominator of the fraction into linear terms.

$$4x^3 - 2x^2 + 2x - 1 = 2x^2(2x - 1) + (2x - 1) = (2x^2 + 1)(2x - 1)$$

Each factor occurs once, so each one leads to one term in the decomposition. Following the process in Alternate Example 1, we write

$$\frac{8x^2 + x + 2}{4x^3 - 2x^2 + 2x - 1} = \frac{A}{2x - 1} + \frac{Bx + C}{2x^2 + 1}.$$

Now clear fractions by multiplying both sides of the equation by the original denominator.

$$8x^2 + x + 2 = A(2x^2 + 1) + (Bx + C)(2x - 1)$$
$$= 2Ax^2 + A + 2Bx^2 - Bx + 2Cx - C$$

Combining like terms in the equation above we obtain:

$$8x^2 + x + 2 = (2A + 2B)x^2 + (2C - B)x + A - C.$$

Comparing the coefficients of the powers of x on the left and the right sides of the equation, we obtain the following system of equations:

$$2A + 2B = 8$$
$$-B + 2C = 1$$
$$A - C = 2$$

We can use any method for solving a system of linear equations to solve this system. For example, we can multiply the second equation by 2, then add it to the first equation.

$$2A + 4C = 10$$
$$-B + 2C = 1$$
$$A - C = 2$$

Now multiply the first equation by one half and subtract from the third equation.

$$2A + 4C = 10$$
$$-B + 2C = 1$$
$$-3C = -3$$

So $C = 1$, $A = 3$, $B = 1$, and

$$\frac{8x^2 + x + 2}{4x^3 - 2x^2 + 2x - 1} = \frac{3}{2x - 1} + \frac{x + 1}{2x^2 + 1}.$$

Exercises for Alternate Example 4

In Exercises 19–24, find the partial fraction decomposition.

19. $\dfrac{2x^2 - 3x + 1}{x^3 + 1}$ **20.** $\dfrac{5x^2 - 9x + 6}{x^3 - 4x^2 + 6x - 4}$ **21.** $\dfrac{-2x^2 + 4x + 2}{x^3 + x^2 + x + 1}$ **22.** $\dfrac{6x^2 - 2x + 21}{x^3 - x^2 + 4x - 4}$

23. $\dfrac{x^2 - 11x - 6}{x^3 - 8}$ **24.** $\dfrac{3x^2 + x + 1}{x^3 + 1}$

7.5 Systems of Inequalities in Two Variables

Alternate Example 1 **Graphing a Linear Inequality**

Draw the graph of $y \leq -x + 4$.

SOLUTION

Step 1. Because of the "$\leq$", the graph of the line $y = -x + 4$ is part of the graph of the inequality and should be drawn as a solid line.

Step 2. The point $(0, 0)$ is below the line and satisfies the inequality because

$$0 \leq -0 + 4 = 4.$$

Thus the graph of $y \leq -x + 4$ consists of the all the points on or below the line of $y = -x + 4$. The boundary is the graph of $y = -x + 4$. The graph is shown in Figure 7.10.

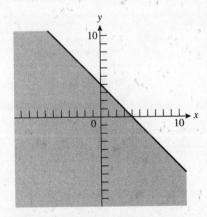

Figure 7.10 The graph of $y \leq -x + 4$.

Exercises for Alternate Example 1

In Exercises 1–6, graph the inequality.

1. $y \leq -3x - 1$ **2.** $y > 2x + 3$ **3.** $y \leq x + 2.5$ **4.** $y \geq -3x + 1$ **5.** $y < 4x - 1$ **6.** $y > -x - 2$

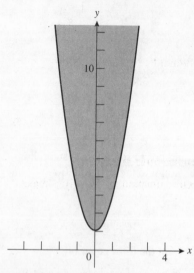

Figure 7.11 The graph of $y \geq 2x^2 + 1$.

Alternate Example 3 **Graphing a Quadratic Inequality**

Draw the graph of $y \geq 2x^2 + 1$.

SOLUTION

Step 1. Replace the "$\geq$" by "$=$" to obtain the equation $y = 2x^2 + 1$ whose graph is a parabola.

Step 2. The point $(0, 2)$ is above the parabola and satisfies the inequality because

$$2 \geq 2(0)^2 + 1 = 1.$$

Thus the graph of $\geq 2x^2 + 1$ consists of the parabola together with the region inside the parabola (Figure 7.11). The parabola is the boundary region.

Exercises for Alternate Example 3

In Exercises 7–12, graph the inequality.

7. $y \leq -x^2 + 1$ **8.** $y \geq x^2 + 3$ **9.** $y < x^2 + 5$ **10.** $y > -2x^2 - 1$ **11.** $y \leq 4x^2$ **12.** $y > -x^2 - 2$ ∎

Alternate Example 4 **Solving a System of Inequalities Graphically**

Solve the system

$$y > x^2 - 1$$
$$3x + 2y \leq 8$$

SOLUTION

The graph of $y = x^2 - 1$ is a parabola. Because of the "$>$", the parabola is not included in the solution. The region is above (or, inside) the parabola, as Figure 7.12a shows.

The graph of the line $3x + 2y = 8$ is included in the solution. The solution region is below the line, as Figure 7.12b shows.

Figure 7.12c shows the intersection of these two graphs. This is the solution of the inequality.

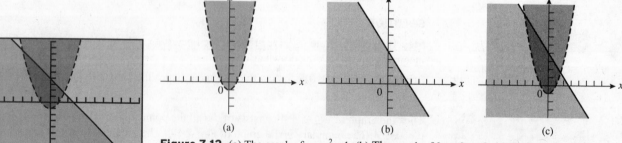

(a)   (b)   (c)

Figure 7.12 (a) The graph of $y = x^2 - 1$. (b) The graph of $3x + 2y \leq 8$. (c) The solution of the system.

[−10, 10] by [−10, 10]

Figure 7.13 The graph of the solution of the system of inequality is the intersection of the two shaded regions

Support with a Grapher

Figure 7.13 shows the solution of the same system of inequalities that our grapher produces. The double shaded portion appears to be identical to the shaded portion in Figure 7.12c.

Exercises for Alternate Example 4

In Exercises 13–18, graph the solution for the system of inequalities.

13. $y \le 3 - x^2$
$y \ge x - 1$

14. $y > x^2 + 2$
$y < 5$

15. $y > x^2 - 3x - 1$
$2x - 3y \le 6$

16. $4 \le x^2 + y^2$
$y > x$

17. $y \ge x^2 - 3$
$x < 2y + 2$

18. $y > x + 1$
$y \ge -x^2 - x - 1$

■

Alternate Example 6 **Solving a Linear Programming Problem**

Find the maximum and the minimum values of the objective function $f = 3x + 4y$, subject to the constraints given by the system of inequalities.

$$y - x \le 2$$
$$2x + y \le 8$$
$$x \ge 0$$
$$y \ge 0$$

SOLUTION

The feasible xy points are in the first quadrant because $x \ge 0$ and $y \ge 0$. Figure 7.14 shows the boundary of the feasible region. The corner points are

$(0, 0)$

$(0, 2)$, the y-intercept of $y - x \le 2$

$(4, 0)$, the x-intercept of $2x + y \le 8$

$(2, 4)$, the point of intersection of the two lines

You can use a table to evaluate f at the corner points of the region.

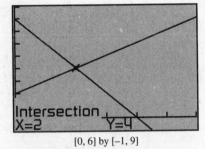

[0, 6] by [–1, 9]

Figure 7.14 The feasible region is bounded by the two lines and the x- and y-axes.

(x, y)	$(0, 0)$	$(0, 2)$	$(4, 0)$	$(2, 4)$
f	0	8	12	22

The maximum value is 22 at $(2, 4)$. The minimum value is 0 at $(0, 0)$.

Exercises for Alternate Example 6

In Exercises 19–24, find the maximum and the minimum values of the objective function subject to the given constraints. Identify points where the objective function takes on these values.

19. $y \ge x - 1$
$y \le 7 - x$
$x \ge 0$
$y \ge 0$
$f(x) = 7x + 3y$

20. $y \le x + 3$
$y \le -2x + 9$
$x \ge 0$
$y \ge 0$
$f(x) = 3x - y - 1$

21. $y \ge x - 4$
$3x + y \le 12$
$x \ge 0$
$f(x) = 2x + 2y + 3$

22. $y \ge \frac{1}{2}x - 3$
$y \le -\frac{1}{2}x + 6$
$y \le -\frac{5}{2}x + 18$
$x \ge 0$
$f(x) = x + y$

23. $y \ge x - 4$
$y \le x + 4$
$x \ge 0$
$y \ge 0$
$y \le 5$
$x \le 5$
$f(x) = 2x - 3y$

24. $2y + x \le 12$
$y \le x + 2$
$x \ge 0$
$y \ge 0$
$x \le 6$
$f(x) = 2y - 3x + 3$

■

Chapter 8

Analytic Geometry in Two and Three Dimensions

8.1 Conic Sections and Parabolas

Alternate Example 2 Finding an Equation of a Parabola

Find an equation in standard form for the parabola whose directrix is the line $x = 3$ and whose focus is the point $(-3, 0)$.

SOLUTION

Because the directrix is $x = 3$ and the focus is $(-3, 0)$, the focal length is $p = -3$ and the parabola opens to the left. The equation of the parabola in standard from is $y^2 = 4px$, or more specifically, $y^2 = -12x$.

Exercises for Alternate Example 2

In Exercises 1–6, find an equation in standard form for the parabola that satisfies the given conditions.

1. directrix: $x = -5$, focus: $(5, 0)$
2. directrix: $y = 4$, focus: $(0, -4)$
3. directrix: $x = 1$, focus: $(-1, 0)$
4. vertex: $(0, 0)$, focus: $(0, -10)$
5. vertex: $(0, 0)$, opens to the left, focal width = 9
6. vertex: $(0, 0)$, opens up, focal width = 7

Alternate Example 3 Finding an Equation of a Parabola

Find the standard form of the equation for the parabola with vertex $(-5, 5)$ and focus $(-3, 5)$.

SOLUTION

The axis of the parabola is the line passing through the vertex $(-5, 5)$ and the focus $(-3, 5)$. This is the line $y = 5$. So the equation has the form

$$(y - k)^2 = 4p(x - h)$$

Because the vertex $(h, k) = (-5, 5)$, $h = -5$ and $k = 5$. The directed distance from the vertex $(-5, 5)$ to the focus $(-3, 5)$ is $p = 5 - 3 = 2$, so $4p = 8$. Thus the equation we seek is

$$(y - 5)^2 = 8(x + 5).$$

Exercises for Alternate Example 3

In Exercises 7–12, find the standard form of the equation for the parabola with given vertex and focus.

7. vertex: $(0, 0)$, focus: $(2, 0)$
8. vertex: $(3, 0)$, focus: $(3, 4)$
9. vertex: $(1, 2)$, focus: $(1, 4)$
10. vertex: $(6, 5)$, focus: $(12, 5)$
11. vertex: $(8, 6)$, focus: $(16, 6)$
12. vertex: $(0, -5)$, focus: $(0, -3)$

Alternate Example 4 Graphing a Parabola

Use a function grapher to graph the parabola $(y - 5)^2 = 8(x + 5)$ of Alternate Example 3.

SOLUTION

$$(y - 5)^2 = 8(x + 5)$$
$$y - 5 = \pm \sqrt{8(x + 5)}$$
$$y = 5 \pm \sqrt{8(x + 5)}$$

Let $Y_1 = 5 + \sqrt{8(x + 5)}$ and $Y_2 = 5 - \sqrt{8(x + 5)}$, and graph the two equations in a window centered at the vertex, as shown in Figure 8.1.

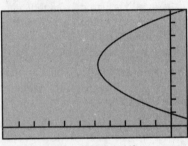

[−11, 1] by [−1, 11]

Figure 8.1 The graphs of $Y_1 = 5 + \sqrt{8(x + 5)}$ and $Y_2 = 5 - \sqrt{8(x + 5)}$ together form the graph of $(y - 5)^2 = 8(x + 5)$.

Exercises for Alternate Example 4

In Exercises 13–18, use a function grapher to graph the parabola.

13. $(y - 1)^2 = 8(x + 1)$ **14.** $(y + 1)^2 = -8(x + 1)$ **15.** $(y - 2)^2 = 4(x + 3)$ **16.** $(y + 3)^2 = 4(x - 3)$

17. $y^2 = -8x$ **18.** $(y + 3)^2 = -8(x + 3)$

8.2 Ellipses

Alternate Example 1 Finding the Vertices and Foci of an Ellipse

Find the vertices and foci of the ellipse $4x^2 + 25y^2 = 100$.

SOLUTION

Dividing both sides of the equation by 100 yields the standard form $x^2/25 + y^2/4 = 1$. Because the larger number is the denominator of x^2, the focal axis is the x-axis. So $a^2 = 25$ and $b^2 = 4$, and $c^2 = a^2 - b^2 = 25 - 4 = 21$. Thus the vertices are $(\pm 5, 0)$, and the foci are $\left(\pm \sqrt{21}, 0 \right)$.

Exercises for Alternate Example 1

In Exercises 1–6, find the vertices and foci of each ellipse.

1. $25x^2 + 4y^2 = 100$ **2.** $9x^2 + 4y^2 = 36$ **3.** $x^2 + 4y^2 = 4$ **4.** $4x^2 + 36y^2 = 144$ **5.** $4x^2 + y^2 = 4$

6. $9x^2 + 25y^2 = 225$

Alternate Example 3   Finding an Equation of an Ellipse

Find the standard form of the equation of the ellipse whose major axis has endpoints (2, 1) and (10, 1), and whose minor axis has length 4.

SOLUTION

Figure 8.2 shows the major axis endpoints, the minor axis, and the center of the ellipse. The standard equation of this ellipse has the form

$$\frac{(x-h)^2}{a^2} + \frac{(y-k)^2}{b^2} = 1,$$

where the center (h, k) is the midpoint (6, 1) of the major axis. The semimajor axis and semiminor axis are

$$a = \frac{10-2}{2} = 4 \ \text{ and } \ b = \frac{4}{2} = 2.$$

So the equation we seek is

$$\frac{(x-6)^2}{16} + \frac{(y-1)^2}{4} = 1.$$

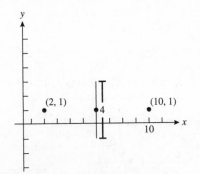

Figure 8.2 Given information for Alternate Example 3.

Exercises for Alternate Example 3

In Exercises 7–12, find the standard form of the equation for the ellipse that satisfies the given conditions.

  7. major axis endpoints: (−4, 0) and (4, 0); length of minor axis: 3
  8. minor axis endpoints: (0, 0) and (0, 4); length of major axis: 8
  9. major axis endpoints: (0, −4) and (0, 6); length of minor axis: 4
 10. minor axis endpoints: (10, −2) and (10, 2); length of major axis: 6
 11. major axis endpoints: (6, 0) and (6, 4); length of minor axis: 1
 12. major axis endpoints: (−8, −2) and (−2, −2); length of minor axis: 3

Alternate Example 4   Locating Key Points of an Ellipse

Find the center, vertices, and foci of the ellipse

$$\frac{(x+3)^2}{16} + \frac{(y-4)^2}{25} = 1.$$

SOLUTION

The standard form of this ellipse has the form

$$\frac{(x-(-3))^2}{16} + \frac{(y-4)^2}{25} = 1.$$

The center (h, k) is (−3, 4). Because the semimajor axis is $a = \sqrt{16} = 4$, the vertices $(h, k \pm a)$ are

$$(h, k + a) = (-3, 4 + 4) = (-3, 8) \text{ and}$$
$$(h, k - a) = (-3, 4 - 4) = (-3, 0)$$

Because $c = \sqrt{a^2 - b^2} = \sqrt{25 - 16} = 3$, the foci $(h, k \pm c)$ are (−3, 4 ± 3), or (−3, 7) and (−3, 1).

Exercises for Alternate Example 4

In Exercises 13–18, find the center, vertices, and foci of each ellipse.

13. $\dfrac{(x+1)^2}{16} + \dfrac{(y-1)^2}{25} = 1$ **14.** $\dfrac{(x+5)^2}{4} + \dfrac{(y-6)^2}{9} = 1$ **15.** $\dfrac{(x-3)^2}{4} + \dfrac{(y-3)^2}{36} = 1$

16. $\dfrac{(x-2)^2}{25} + \dfrac{(y+5)^2}{49} = 1$ **17.** $\dfrac{(x-3)^2}{64} + \dfrac{(y-3)^2}{49} = 1$ **18.** $\dfrac{(x+5)^2}{169} + \dfrac{(y-7)^2}{144} = 1$

8.3 Hyperbolas

Alternate Example 1   **Finding the Vertices and Foci of a Hyperbola**

Find the vertices and foci of the hyperbola $9x^2 - 25y^2 = 225$.

SOLUTION

Dividing both sides of the equation by 225 yields the standard form $x^2/25 - y^2/9 = 1$. So $a^2 = 25$, $b^2 = 9$, and $c^2 = a^2 + b^2 = 25 + 9 = 34$. Thus the vertices are $(\pm 5, 0)$ and the foci are $\left(\pm\sqrt{34}, 0 \right)$.

Exercises for Alternate Example 1

In Exercises 1–6, find the vertices and foci of each hyperbola.

1. $x^2 - 25y^2 = 25$ **2.** $4x^2 - 25y^2 = 100$ **3.** $36x^2 - 4y^2 = 144$ **4.** $4x^2 - 100y^2 = 400$ **5.** $100x^2 - 9y^2 = 900$

6. $25x^2 - 25y^2 = 625$

Alternate Example 2   **Finding an Equation and Graphing a Hyperbola**

Find an equation of the hyperbola with foci $(0, -5)$ and $(0, 5)$ whose conjugate axis has length 4. Sketch the hyperbola and its asymptotes, and support your sketch with a grapher.

SOLUTION

The center is $(0, 0)$. The foci are on the y-axis with $c = 5$. The semiconjugate axis is $b = 4/2 = 2$. Thus $a^2 = c^2 - b^2 = 5^2 - 2^2 = 21$. So the standard form of the equation for the hyperbola is

$$\frac{y^2}{21} - \frac{x^2}{4} = 1.$$

Using $a = \sqrt{21} \approx 4.58$ and $b = 2$, we can sketch the central rectangle, the asymptotes, and the hyperbola itself. Try doing this. To graph the hyperbola using a function grapher, we solve for y in terms of x.

$$\frac{y^2}{21} = 1 + \frac{x^2}{4} \qquad \text{Add } \frac{x^2}{4}.$$

$$y^2 = 21\left(1 + \frac{x^2}{4} \right) \qquad \text{Multiply by 21.}$$

$$y = \pm\sqrt{21\left(1 + \frac{x^2}{4} \right)} \qquad \text{Extract square roots.}$$

The asymptotes are determined by $y = \pm\dfrac{a}{b}x$. They are $y = \pm\dfrac{\sqrt{21}}{2}x$.

Figure 8.3 shows the graphs of

$$Y_1 = \sqrt{21\left(1 + \frac{x^2}{4}\right)} \text{ and } Y_2 = -\sqrt{21\left(1 + \frac{x^2}{4}\right)}$$

together with the asymptotes of the hyperbola

$$Y_3 = \frac{\sqrt{21}}{2}x \text{ and } Y_4 = -\frac{\sqrt{21}}{2}x.$$

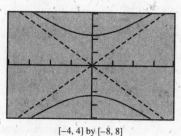

[−4, 4] by [−8, 8]

Figure 8.3 The hyperbola $\dfrac{y^2}{21} - \dfrac{x^2}{4} = 1$, shown with its asymptotes.

Exercises for Alternate Example 2

In Exercises 7–12, find an equation of the hyperbola with the given foci and given conjugate axis length. Sketch the hyperbola and its asymptotes, and support your sketch with a grapher.

7. foci: (0, −6) and (0, 6); conjugate axis length: 4

8. foci: (−4, 0) and (4, 0); conjugate axis length: 4

9. foci: (12, 0) and (12, 0); conjugate axis length: 8

10. foci: (0, −4) and (0, 4); conjugate axis length: 2

11. foci: (0, −5) and (0, 5); conjugate axis length: 6

12. foci: (−9, 0) and (9, 0); conjugate axis length: 6

Alternate Example 4 **Locating Key Points of a Hyperbola**

Find the center, vertices, and foci of the hyperbola

$$\frac{(x-3)^2}{4} - \frac{(y+4)^2}{25} = 1.$$

SOLUTION

The center (h, k) is (3, −4). Because the semitransverse axis $a = \sqrt{4} = 2$, the vertices are

$$(h + a, k) = (3 + 2, -4) = (5, -4) \text{ and}$$
$$(h - a, k) = (3 - 2, -4) = (1, -4)$$

Because $c = \sqrt{a^2 + b^2} = \sqrt{4 + 25} = \sqrt{29}$, the foci $(h \pm c, k)$ are $(3 \pm \sqrt{29}, 4)$, or approximately, (8.39, 4) and (−2.39, 4).

Exercises for Alternate Example 4

In Exercises 13–18, find the center, vertices, and foci of the hyperbola.

13. $\dfrac{(x-1)^2}{9} - \dfrac{(y+2)^2}{25} = 1$

14. $\dfrac{(x+2)^2}{16} - \dfrac{(y+1)^2}{36} = 1$

15. $\dfrac{(y+1)^2}{4} - \dfrac{(x-4)^2}{1} = 1$

16. $\dfrac{(x+1)^2}{36} - \dfrac{(y+1)^2}{9} = 1$

17. $\dfrac{x^2}{81} - \dfrac{(y+2)^2}{49} = 1$

18. $\dfrac{y^2}{100} - \dfrac{(x-4)^2}{49} = 1$

8.4 Translation and Rotation of Axes

Alternate Example 1 **Graphing a Second-Degree Equation**

Solve for y, and use a function grapher to graph

$$4x^2 + 9y^2 - 16x - 18y - 11 = 0.$$

SOLUTION

Rearranging the terms yields the equation:

$$9y^2 - 18y + (4x^2 - 16x - 11) = 0.$$

The quadratic formula gives us

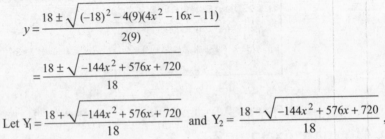

$$y = \frac{18 \pm \sqrt{(-18)^2 - 4(9)(4x^2 - 16x - 11)}}{2(9)}$$

$$= \frac{18 \pm \sqrt{-144x^2 + 576x + 720}}{18}$$

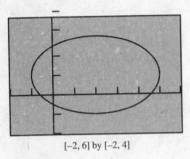

[−2, 6] by [−2, 4]

Figure 8.4 The graph of
$4x^2 + 9y^2 - 16x - 18y - 11 = 0.$

Let $Y_1 = \dfrac{18 + \sqrt{-144x^2 + 576x + 720}}{18}$ and $Y_2 = \dfrac{18 - \sqrt{-144x^2 + 576x + 720}}{18}$,

and graph the two equations in the same viewing window, as shown in Figure 8.4. The combined figure appears to be an ellipse.

Exercises for Alternate Example 1

In Exercises 1–6, solve for y, and use a function grapher to graph each equation.

1. $4x^2 + 9y^2 - 18y - 27 = 0$
2. $4x^2 - 9y^2 + 8x - 32 = 0$
3. $3x^2 - 12y - 12x - 21 = 0$
4. $x^2 + 9y^2 - 2x + 18y + 1 = 0$
5. $25x^2 + 4y^2 + 50x - 8y - 71 = 0$
6. $-25x^2 + 4y^2 + 100x - 16y - 184 = 0$

Alternate Example 2 **Graphing a Second-Degree Equation**

Solve for y, and use a function grapher to graph

$$3xy - 4 = 0.$$

SOLUTION

This equation can be rewritten as $3xy = 4$ or as $y = 4/(3x)$. The graph of this equation is shown in Figure 8.5. It appears to be a hyperbola with a slant focal axis.

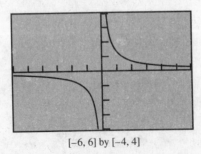

[−6, 6] by [−4, 4]

Figure 8.5 The graph of $3xy - 4 = 0$.

Exercises for Alternate Example 2

In Exercises 7–12, solve for y, and use a function grapher to graph each equation.

7. $xy - 1 = 0$ **8.** $-2xy - 3 = 0$ **9.** $-3xy - 9 = 0$ **10.** $5xy - 15 = 0$ **11.** $2xy - 5 = 0$

12. $16xy + 4 = 0$

Alternate Example 4   **Revisiting Alternate Example 1**

Prove that $4x^2 + 9y^2 - 16x - 18y - 11 = 0$ is the equation of an ellipse. Translate the coordinate axes so that the origin is at the center of the ellipse.

SOLUTION

We complete the square in x and y:

$$4x^2 + 9y^2 - 16x - 18y - 11 = 0$$
$$4(x^2 - 4x + 4) + 9(y^2 - 2x + 1) = 11 + 4(4) + 9(1)$$
$$4(x - 2)^2 + 9(y - 1)^2 = 36$$
$$\frac{(x-2)^2}{9} + \frac{(y-1)^2}{4} = 1$$

This is the standard form of an ellipse. If we let $x' = x - 2$ and $y' = y - 1$, then the equation of the ellipse becomes

$$\frac{(x')^2}{9} + \frac{(y')^2}{4} = 1.$$

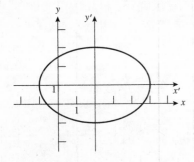

Figure 8.6 The graph of $\frac{(x')^2}{9} + \frac{(y')^2}{4} = 1.$

Figure 8.6 shows the graph of this final equation in the new $x'y'$ coordinate system, with the original xy-axes overlaid. Compare Figures 8.4 and 8.6.

Exercises for Alternate Example 4

In Exercises 13–18, identify the type of conic, write the equation in standard form, translate the conic to the origin, and sketch it in the translated coordinate system.

13. $16x^2 + 9y^2 - 64x - 54y + 1 = 0$ **14.** $-y^2 - 8x + 6y + 23 = 0$ **15.** $-4x^2 + 49y^2 - 8x + 98y - 151 = 0$

16. $x^2 - 2x + 4y + 1 = 0$ **17.** $25x^2 - 4y^2 + 50x + 8y - 79 = 0$ **18.** $25x^2 + 4y^2 - 100x - 16y + 16 = 0$

Alternate Example 6   **Revisiting Example 3**

Prove that $\sqrt{2}x^2 + 2\sqrt{2}xy + \sqrt{2}y^2 - 8x + 8y + 10 = 0$ is the equation of a parabola by rotating the coordinate axes through a suitable angle α.

SOLUTION

The angle of rotation α must satisfy the equation

$$\cot 2\alpha = \frac{A - C}{B} = \frac{\sqrt{2} - \sqrt{2}}{2\sqrt{2}} = 0.$$

So, $2\alpha = \dfrac{\pi}{2}$ and $\alpha = \dfrac{\pi}{4}$. $\cos \alpha = \dfrac{\sqrt{2}}{2}$ and $\sin \alpha = \dfrac{\sqrt{2}}{2}$. (Note, in other circumstances you may need to use Pythagorean relationships and the half-angle cosine formula to find $\sin \alpha$ and $\cos \alpha$.)

The coefficients of the transformed equation are:

$$A' = A\cos^2\alpha + B\cos\alpha\sin\alpha + C\sin^2\alpha$$

$$= \sqrt{2}\left(\frac{\sqrt{2}}{2}\right)^2 + 2\sqrt{2}\cdot\frac{\sqrt{2}}{2}\cdot\frac{\sqrt{2}}{2} + \sqrt{2}\left(\frac{\sqrt{2}}{2}\right)^2$$

$$= \frac{\sqrt{2}}{2} + \frac{2\sqrt{2}}{2} + \frac{\sqrt{2}}{2}$$

$$= 2\sqrt{2}$$

$$B' = B\cos 2\alpha + (C - A)\sin 2\alpha$$

$$= 2\sqrt{2}\cos\left(\frac{\pi}{2}\right) - \left(\sqrt{2} - \sqrt{2}\right)\sin\left(\frac{\pi}{2}\right)$$

$$= 2\sqrt{2}\cdot 0 - 0\cdot 1$$

$$= 0$$

$$C' = C\cos^2\alpha - B\cos\alpha\sin\alpha + C\sin^2\alpha$$

$$= \sqrt{2}\left(\frac{\sqrt{2}}{2}\right)^2 - 2\sqrt{2}\cdot\frac{\sqrt{2}}{2}\cdot\frac{\sqrt{2}}{2} + \sqrt{2}\left(\frac{\sqrt{2}}{2}\right)^2$$

$$= \frac{\sqrt{2}}{2} - \frac{2\sqrt{2}}{2} + \frac{\sqrt{2}}{2}$$

$$= 0$$

$$D' = D\cos\alpha + E\sin\alpha$$

$$= \frac{-8\sqrt{2}}{2} + \frac{8\sqrt{2}}{2} = 0$$

$$E' = E\cos\alpha - D\sin\alpha$$

$$= \frac{8\sqrt{2}}{2} + \frac{8\sqrt{2}}{2} = 8\sqrt{2}$$

$$F' = F$$

$$= 10$$

So the equation $\sqrt{2}x^2 + 2\sqrt{2}xy + \sqrt{2}y^2 - 8x + 8y + 10 = 0$ becomes

$$2\sqrt{2}(x')^2 + 8\sqrt{2}y' + 10 = 0$$

which becomes

$$8\sqrt{2}y' = -2\sqrt{2}(x')^2 - 10$$

$$y' = \frac{-2\sqrt{2}(x')^2}{8\sqrt{2}} - \frac{10}{8\sqrt{2}}$$

$$y' = -\frac{(x')^2}{4} - \frac{5\sqrt{2}}{8}$$

Figure 8.7 shows the graph of the original equation in the original xy coordinate system with the $x'y'$-axes overlaid, along with the graph of the transformed equation.

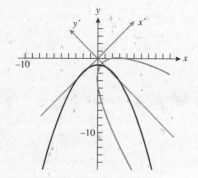

Figure 8.7 The graphs of $\sqrt{2}x^2 + 2\sqrt{2}xy + \sqrt{2}y^2 - 8x + 8y + 10 = 0$ and $y' = -\dfrac{(x')^2}{4} - \dfrac{5\sqrt{2}}{8}$.

Exercises for Alternate Example 6

In Exercises 19–24, identify the type of conic, and rotate the coordinate axes to eliminate the xy term. Write and graph the transformed equation

19. $4xy - 10 = 0$

20. $x^2 - xy + y^2 = 1$

21. $13x^2 - 10xy + 13y^2 = 72$

22. $-5x^2 - 26xy - y^2 = 72$

23. $x^2 - \sqrt{3}xy - 2y^2 = 1$

24. $3x^2 + 2\sqrt{3}xy + y^2 - 8x + 8\sqrt{3}y = 0$

8.5   Polar Equations of Conics

Alternate Example 1 **Writing and Graphing Polar Equations for Conics**

Given that the focus is at the pole, write a polar equation for the specified conic, and graph it.

(a) eccentricity $e = 0.6$, directrix $x = 3$.   **(b)** eccentricity $e = 1$, directrix $x = -3$.

(c) eccentricity $e = 4/3$, directrix $y = 3$.

SOLUTION

(a) Setting $e = 0.6$ and $k = 3$ in $r = \dfrac{ke}{1 + e \cos \theta}$ yields

$$r = \dfrac{3(0.6)}{1 + 0.6 \cos \theta}$$
$$= \dfrac{9}{5 + 3 \cos \theta}$$

Figure 8.8a shows this ellipse and its directrix.

(b) Setting $e = 1$ and $k = 3$ in $r = \dfrac{ke}{1 - e \cos \theta}$ yields

$$r = \dfrac{3(1)}{1 - 1 \cos \theta}$$
$$= \dfrac{3}{1 - \cos \theta}$$

Figure 8.8b shows this parabola and its directrix.

(c) Setting $e = 4/3$, and $k = 3$ in $r = \dfrac{ke}{1 + e \sin \theta}$ yields

$$r = \dfrac{(4/3)(3)}{1 + (4/3) \sin \theta}$$
$$= \dfrac{12}{3 + 4 \sin \theta}$$

Figure 8.8c shows this hyperbola and its directrix.

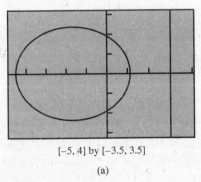

[−5, 4] by [−3.5, 3.5]

(a)

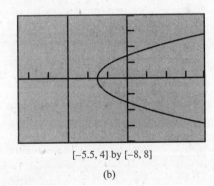

[−5.5, 4] by [−8, 8]

(b)

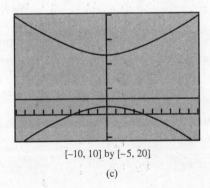

[−10, 10] by [−5, 20]

(c)

Figure 8.8 Graphs for Alternate Example 1.

Exercises for Alternate Example 1

In Exercises 1–6, given that the focus is at the pole, write a polar equation for the specified conic, and graph it

1. $e = 3/2$, and directrix $y = 2$ **2.** $e = 0.4$, and directrix $x = 2$ **3.** $e = 0.5$, and directrix $x = 2$

4. $e = 2$, and directrix $y = 2$ **5.** $e = 0.2$, and directrix $x = 2$ **6.** $e = 1$, and directrix $y = 4$

Alternate Example 2 **Identifying Conics from Their Polar Equations**

Determine the eccentricity, the type of conic, and the directrix.

(a) $r = \dfrac{9}{3 + 6\cos\theta}$ (b) $r = \dfrac{15}{5 - 2\sin\theta}$

SOLUTION

(a) Dividing numerator and denominator by 3 yields $r = 3/(1 + 2\cos\theta)$. So the eccentricity $e = 2$, and thus the conic is a hyperbola. The numerator $ke = 3$, so $k = 1.5$ and thus the equation of the directrix is $x = 1.5$.

(b) Dividing numerator and denominator by 5 yields $r = 3/(1 - 0.4\cos\theta)$. So the eccentricity $e = 0.4$, and thus the conic is an ellipse. The numerator $ke = 3$, so $k = 7.5$ and thus the equation of the directrix is $y = -7.5$.

Exercises for Alternate Example 2

In Exercises 7–12, determine the eccentricity, the type of conic, and the directrix.

7. $r = \dfrac{9}{2 + 5\cos\theta}$ 8. $r = \dfrac{7}{5 + 15\cos\theta}$ 9. $r = \dfrac{3}{3 - 2\sin\theta}$ 10. $r = \dfrac{7}{9 - 7\sin\theta}$ 11. $r = \dfrac{1}{1 - 0.5\sin\theta}$

12. $r = \dfrac{12}{6 + 6\cos\theta}$

Alternate Example 3 **Analyzing a Conic**

Analyze the conic section given by the equation $r = 6/(2 + \sin\theta)$. Include in the analysis the values of e, a, b, and c.

SOLUTION

Dividing the numerator and denominator by 2 yields

$$r = \frac{3}{1 + 0.5\sin\theta}.$$

So the eccentricity $e = 0.5$, and thus the conic is an ellipse. The directrix is $y = 6$. Figure 8.9 shows this ellipse. The chart on page 677 in the text shows that an equation in this form is oriented with y-axis as the major axis. So, the vertices (endpoints of the major axis) are detemined at $\theta = \dfrac{\pi}{2}$ and $\theta = \dfrac{3\pi}{2}$. The vertices have polar coordinates $\left(2, \dfrac{\pi}{2}\right)$ and $\left(6, \dfrac{3\pi}{2}\right)$. So $2a = 2 + 6 = 8$, and thus $a = 4$.

The vertex $\left(2, \dfrac{\pi}{2}\right)$ is 2 units above the pole, and the pole is a focus of the ellipse. So $a - c = 2$, and thus $c = 2$. An alternative way to find c is to use the fact that the eccentricity of an ellipse is $e = c/a$ or $c = ae = 2$.

To find b, we use the Pythagorean relation of an ellipse:

$$b = \sqrt{a^2 - c^2} = \sqrt{16 - 4} = \sqrt{12} = 2\sqrt{3}.$$

To find the center (h, k), use the fact that one focus is at the pole. So, $h = 0$, and $k + c = 0$ or $k = -c$.

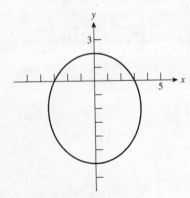

Figure 8.9 The graph of $r = \dfrac{6}{2 + \sin\theta}$.

With all of this information, we can write the Cartesian equation of the ellipse:

$$\frac{x^2}{12} + \frac{(y+2)^2}{16} = 1$$

Exercises for Alternate Example 3

In Exercises 13–18, determine a Cartesian equation for the given polar equation by determining e, a, b, and c along with the directrix d and the polar vertices.

13. $r = \dfrac{2}{1 + 2\sin\theta}$ **14.** $r = \dfrac{2}{1 + \cos\theta}$ (Hint: Use the fact that $k = 2p$.)   **15.** $r = \dfrac{12}{3 - 6\cos\theta}$ **16.** $r = \dfrac{8}{4 - 3\sin\theta}$

17. $r = \dfrac{6}{3 + 2\sin\theta}$ **18.** $r = \dfrac{6}{2 + 3\cos\theta}$

8.6 Three-Dimensional Cartesian Coordinate System

Alternate Example 2 **Calculating a Distance and Finding a Midpoint**

Find the distance between the points $P(-3, 3, 0)$ and $Q(5, 11, -6)$, and find the midpoint of line segment PQ.

SOLUTION

The distance is given by

$$d(P, Q) = \sqrt{(5 - (-3))^2 + (11 - 3)^2 + (-6 - 0)^2}$$
$$= \sqrt{64 + 64 + 36}$$
$$= \sqrt{164} \approx 12.81$$

The midpoint is

$$M = \left(\frac{-3 + 5}{2}, \frac{3 + 11}{2}, \frac{0 + (-6)}{2} \right) = (1, 7, -3)$$

Exercises for Alternate Example 2

In Exercises 1–6, find the distance between the points P and Q, and find the midpoint of line segment PQ.

1. $P(0, 0, 0)$ and $Q(6, -2, -5)$ **2.** $P(3, 0, 0)$ and $Q(7, 2, -5)$ **3.** $P(5, -2, 1)$ and $Q(2, 2, 2)$

4. $P(5, -3, 1)$ and $Q(2, 11, -3)$ **5.** $P(-3, -3, -3)$ and $Q(3, 3, 3)$ **6.** $P(-4, 1, 5)$ and $Q(2, -2, -1)$

Alternate Example 4 **Sketching a Plane in Space**

Sketch the graph of $2x + 3y + 5z = 30$.

SOLUTION

Because this is a first-degree equation, its graph is a plane. Three points determine a plane. To find three points, we first divide both sides of $2x + 3y + 5z = 30$ by 30.

$$\frac{x}{15} + \frac{y}{10} + \frac{z}{6} = 1.$$

In this form, it is easy to see that $(15, 0, 0)$, $(0, 10, 0)$, and $(0, 0, 6)$ satisfy the equation. These are the points where the graph crosses the coordinate axes. Figure 8.10 shows the completed sketch.

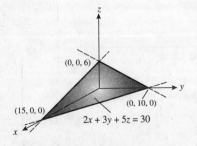

Figure 8.10 The intercepts $(15, 0, 0)$, $(0, 3, 0)$, and $(0, 0, 6)$ determine the plane $2x + 3y + 5z = 30$.

Exercises for Alternate Example 4

In Exercises 7–12, sketch each plane. Label all intercepts.

7. $3x - 2y = 6$ **8.** $2x + 3y + 2z = 6$ **9.** $x + 2y + z = 2$ **10.** $3x - 2y - 4z = 12$ **11.** $6x + 2y - 3z = 6$
12. $5x - 2y + z = 10$

Alternate Example 5 **Computing with Vectors**

Evaluate each expression.

(a) $4\langle -3, 1, 5\rangle$

(b) $\langle -3, 2, -6\rangle + \langle 4, -2, -6\rangle$

(c) $\langle 4, 5, -7\rangle - \langle -4, 3, 0\rangle$

(d) $\left|\langle 4, 3, -3\rangle\right|$

(e) $\langle 2, -5, 2\rangle \cdot \langle -1, 2, 4\rangle$

SOLUTION

(a) $4\langle -3, 1, 5\rangle = \langle 4(-3), 4(1), 4(5)\rangle = \langle -12, 4, 20\rangle$

(b) $\langle -3, 2, -6\rangle + \langle 4, -2, -6\rangle = \langle -3+4, 2+(-2), -6+(-6)\rangle = \langle 1, 0, -12\rangle$

(c) $\langle 4, 5, -7\rangle - \langle -4, 3, 0\rangle = \langle 4-(-4), 5-3, -7-0\rangle = \langle 8, 2, -7\rangle$

(d) $\left|\langle 4, 3, -3\rangle\right| = \sqrt{4^2 + 3^2 + (-3)^2} = \sqrt{34} \approx 5.83$

(e) $\langle 2, -5, 2\rangle \cdot \langle -1, 2, 4\rangle = 2(-1) + (-5)(2) + 2(4) = -2 + (-10) + 8 = -4$

Exercises for Alternate Example 5

In Exercises 13–20, evaluate each expression.

13. $-2\langle -2, 3, 7\rangle$ **14.** $\langle 5, 1, -2\rangle + \langle -4, 2, -1\rangle$ **15.** $\langle -2, 6, -2\rangle - \langle -1, -3, 1\rangle$

16. $\left|\langle -3, -3, -3\rangle\right|$ **17.** $3\langle 1, 0, 0\rangle$ **18.** $\langle 1, 0, 0\rangle \cdot \langle 0, 1, 0\rangle$

19. $\left|\langle 0, 5, 0\rangle\right|$ **20.** $\langle 1, 1, 1\rangle \cdot \langle 1, 1, 1\rangle$

Alternate Example 8 **Finding Equations for a Line**

Using the standard unit vectors **i**, **j**, and **k**, write a vector equation for the line containing the points $A(2, 1, -3)$ and $B(4, 5, -2)$, and compare it to the parametric equations for the line.

SOLUTION

The line is in the direction of
$$\mathbf{v} = \overrightarrow{AB} = \langle 4-2, 5-1, -2-(-3)\rangle = \langle 2, 4, 1\rangle.$$

So using $\mathbf{r} = \overline{OA}$, the vector equation of the line becomes:

$$\mathbf{r} = \mathbf{r}_0 + t\,\mathbf{v}$$

$$\left\langle x, y, z \right\rangle = \left\langle 2, 1, -3 \right\rangle + t \left\langle 2, 4, 1 \right\rangle$$

$$\left\langle x, y, z \right\rangle = \left\langle 2 + 2t, 1 + 4t, -3 + t \right\rangle$$

$$x\mathbf{i} + y\mathbf{j} + z\mathbf{k} = (2 + 2t)\mathbf{i} + (1 + 4t)\mathbf{j} + (-3 + t)\mathbf{k}$$

The parametric equations are the three component equations

$$x = 2 + 2t, \qquad y = 1 + 4t, \qquad \text{and} \qquad z = -3 + t.$$

Exercises for Alternate Example 8

In Exercises 21–26, using the standard unit vectors $\mathbf{i}$, $\mathbf{j}$, and $\mathbf{k}$, write a vector equation for the line containing the points A and B, and then write the parametric equations for the line.

21. $A(0, 0, 0)$ and $B(1, 1, 1)$

22. $A(1, 2, -2)$ and $B(3, 1, -1)$

23. $A(-4, 2, -1)$ and $B(-3, 1, -1)$

24. $A(0, 2, 0)$ and $B(1, 0, 3)$

25. $A(-3, 5, 2)$ and $B(4, -1, 0)$

26. $A(8, -9, -1)$ and $B(2, -1, 2)$

Chapter 9 Discrete Mathematics

9.1 Basic Combinatorics

Alternate Example 2 Using the Multiplication Principle

A New York State license plate consists of three letters of the alphabet followed by four numerical digits (0 through 9). Find the number of different license plates that can be formed

(a) if there is no restriction on the letters or numbers that can be used;

(b) if no letter or digit can be repeated.

SOLUTION

Consider each license plate as having seven blanks to be filled in: three letters followed by four numerical digits.

(a) If there are no restrictions on letters or digits, then we can fill in the first blank 26 ways, the second blank 26 ways, the third blank 26 ways, the fourth blank 10 ways, the fifth blank 10 ways, the sixth blank 10 ways, and the seventh blank 10 ways. By the Multiplication Principle, we can fill in all seven blanks in

$$26 \times 26 \times 26 \times 10 \times 10 \times 10 \times 10 = 175{,}760{,}000 \text{ ways.}$$

There are 175,760,000 possible license plates with no restrictions on letters or digits.

(b) If no letter or digit can be repeated, then we can fill in the first blank 26 ways, the second blank 25 ways, the third blank 24 ways, the fourth blank 10 ways, the fifth blank 9 ways, the sixth blank 8 ways, and the seventh blank 7 ways. By the Multiplication Principle, we can fill in all seven blanks in

$$26 \times 25 \times 24 \times 10 \times 9 \times 8 \times 7 = 78{,}624{,}000 \text{ ways.}$$

There are 78,624,000 possible license plates with no letters or digits repeated.

Exercises for Alternate Example 2

Exercises 1–6 all consist of a license plate with seven blanks. Determine the number of possible license plates

1. if the first four blanks are digits that can be repeated and the last three blanks are letters that can be repeated.

2. if the first two blanks are letters that cannot be repeated and the last five blanks are digits that cannot be repeated.

3. if the first blank is a letter and the remaining six blanks are digits that can be repeated.

4. if all seven blanks are letters that cannot be repeated.

5. if the first six blanks are letters that can be repeated and the last blank is a digit.

6. if the first three blanks are letters that can be repeated and the remaining four blanks are digits that cannot be repeated.

Alternate Example 3 **Distinguishable Permutations**

Count the number of different 7-letter "words" (don't worry about whether they're in the dictionary) that can be formed using the letters in each word.

(a) FLORIDA

(b) ALABAMA

(c) MONTANA

SOLUTION

(a) Each permutation of the 7 letters forms a different word. There are $7! = 5,040$ such permutations.

(b) There are also $7!$ permutations of these letters, but a simple permutation of the four A's does not result in a new word. We correct for the overcount by dividing by $4!$.

There are $\dfrac{7!}{4!} = 210$ *distinguishable* permutations of the letters in ALABAMA.

(c) Again there are $7!$ permutations, but the two N's are indistinguishable, as are the two A's, so we divide by $2!$ twice to correct for the overcount. There are

$\dfrac{7!}{2!2!} = 1,260$ distinguishable permutations of the letters in MONTANA.

Exercises for Alternate Example 3

In Exercises 7–14, count the number of different 6-letter "words" (don't worry about whether they're in the dictionary) that can be formed using the letters in each word.

7. PEANUT   **8.** SUNDAE   **9.** BUTTER   **10.** APPLES   **11.** CHERRY   **12.** COFFEE

13. HAWAII   **14.** KANSAS

Alternate Example 4 **Counting Permutations**

Evaluate each expression without a calculator.

(a) $_6P_2$ **(b)** $_{10}P_6$ **(c)** $_nP_5$

SOLUTION

(a) By the formula, $_6P_2 = 6!/(6-2)! = 6!/4! = (6 \cdot 5 \cdot 4!)/4! = 6 \cdot 5 = 30$.

(b) Although you could use the formula again, you might prefer to apply the Multiplication Principle directly. We have 10 objects and 6 blanks to fill:

$_{10}P_6 = 10 \cdot 9 \cdot 8 \cdot 7 \cdot 6 \cdot 5 = 151,200.$

(c) This time it is definitely easier to use the Multiplication Principle. We have n objects and 5 blanks to fill; so assuming $n \geq 5$,

$_nP_5 = n(n-1)(n-2)(n-3)(n-4).$

Exercises for Alternate Example 4

In Exercises 15–21, evaluate each expression without a calculator.

15. $_8P_6$ **16.** $_9P_6$ **17.** $_5P_3$ **18.** $_7P_4$ **19.** $_{10}P_9$ **20.** $_nP_8$ **21.** $_nP_6$

Alternate Example 7   **Counting Combinations**

In a beauty pageant, 50 contestants must be narrowed down to 15 finalists. In how many possible ways can the fifteen finalists be selected?

SOLUTION

Notice that the *order* of the finalists does not matter at this phase; all that matters is which women are selected. So we count combinations rather than permutations.

$$_{50}C_{15} = \frac{50!}{15!35!} = 2,250,829,575,120.$$

The 15 finalists can be chosen in 2,250,829,575,120 ways.

Exercises for Alternate Example 7

In Exercises 22–29, the number of contestants in a contest is to be narrowed to the given number of finalists. In how many possible ways can the given number of finalists be selected?

22. 50 contestants, 14 finalists

23. 30 contestants, 13 finalists

24. 100 contestants, 10 finalists

25. 25 contestants, 9 finalists

26. 60 contestants, 7 finalists

27. 40 contestants, 4 finalists

28. 75 contestants, 3 finalists

29. 50 contestants, 2 finalists

9.2   The Binomial Theorem

Alternate Example 1   **Using $_nC_r$ to Expand a Binomial**

Expand $(a+b)^7$, using a calculator to compute the binomial coefficients.

SOLUTION

Enter $7\,_nC_r\{0,1,2,3,4,5,6,7\}$ into the calculator to find the binomial coefficients for $n=7$. The calculator returns the list $\{1, 7, 21, 35, 35, 21, 7, 1\}$. Using these coefficients, we construct the expansion:

$$(a+b)^7 = 1a^7 + 7a^6b + 21a^5b^2 + 35a^4b^3 + 35a^3b^4 + 21a^2b^5 + 7ab^6 + 1b^7.$$

Exercises for Alternate Example 1

In Exercises 1–6, expand the binomial using a calculator to find the binomial coefficients.

1. $(a+b)^3$ **2.** $(a+b)^4$ **3.** $(a+b)^5$ **4.** $(a+b)^6$ **5.** $(a+b)^8$ **6.** $(a+b)^9$

Alternate Example 3   **Computing Binomial Coefficients**

Find the coefficient of x^{13} in the expansion of $(x+4)^{17}$.

SOLUTION

The only term in the expansion that we need to deal with is $_{17}C_{13}x^{13}4^4$. This is

$$\frac{17!}{13!4!} \cdot x^{13} \cdot 4^4 = 609,280x^{13}.$$

The coefficient of x^{13} is 609,280.

Exercises for Alternate Example 3

In Exercises 7–14, find the coefficient of the given term in the binomial expansion.

7. x^{13} term, $(x+1)^{15}$ **8.** $x^3 y^{14}$ term, $(x+y)^{17}$ **9.** x^{15} term, $(x+5)^{19}$ **10.** $x^{10} y^6$ term, $(x+y)^{16}$

11. x^{17} term, $(x-3)^{18}$ **12.** xy^{14} term, $(x-y)^{15}$ **13.** $x^7 y^{13}$ term, $(x-y)^{20}$ **14.** x^9 term, $(x-3)^{12}$

Alternate Example 4   **Expanding a Binomial**

Expand $(4x - y^4)^3$.

SOLUTION

We use the Binomial Theorem to expand $(a+b)^3$ where $a = 4x$ and $b = -y^4$.

$$(a+b)^3 = a^3 + 3a^2 b + 3ab^2 + b^3$$
$$(4x - y^4) = (4x)^3 + 3(4x)^2(-y^4) + 3(4x)(-y^4)^2 + (-y^4)^3$$
$$= 64x^3 - 48x^2 y^4 + 12xy^8 - y^{12}$$

Exercises for Alternate Example 4

In Exercises 15–22, expand:

15. $(5x - 3y^2)^3$ **16.** $(x^2 - y^2)^5$ **17.** $(10x - 15y)^4$ **18.** $(2x^3 + 2y^2)^4$ **19.** $(5x^3 - 5y^3)^5$

20. $(3x^3 - y)^3$ **21.** $(x - 3y^3)^3$ **22.** $(x^5 - y^5)^5$

9.3 Probability

Alternate Example 4   **Choosing Chocolates, Sample Space I**

Sal opens a box of two-dozen chocolate crèmes and generously offers three of them to Val. Val likes vanilla crème the best, but all the chocolates look alike on the outside. If 11 of the 24 cremes are vanilla, what is the probability that all three of Val's picks turn out to be vanilla?

SOLUTION

The experiment in question is the selection of three chocolates, without regard to order, from a box of 24. There are $_{24}C_3 = 2{,}024$ outcomes of this experiment, and all of them are equally likely. We can therefore determine the probability by counting.

The event E consists of all possible pairs of 3 vanilla crèmes that can be chosen, without regard to order, from 11 vanilla crèmes available. There are $_{11}C_3 = 165$ ways to form such pairs.

Therefore, $P(E) = 165/2024 = 15/184$.

Exercises for Alternate Example 4

Sal opens a box of two-dozen chocolate crèmes and generously offers some to Val. Val likes vanilla crème the best, but all the chocolates look alike on the outside. In Exercises 1–6, a given number of the two-dozen crèmes are vanilla. Determine the probability that all of Val's picks turn out to be vanilla if:

1. Sal offers Val 4 chocolate crèmes and 4 are vanilla.   **2.** Sal offers Val 10 chocolate crèmes and 12 are vanilla.

3. Sal offers Val 8 chocolate crèmes and 9 are vanilla.   **4.** Sal offers Val 2 chocolate crèmes and 5 are vanilla.

5. Sal offers Val 12 chocolate crèmes and 12 are vanilla.   **6.** Sal offers Val 1 chocolate crème and 4 are vanilla.

Alternate Example 6 Using a Venn Diagram

In a large high school, 43% of the students are girls and 45% of the students play sports. Half of the girls at the school play sports.

(a) What percentage of the students who play sports are boys?

(b) If a student is chosen at random, what is the probability that it is a boy who does not play sports?

SOLUTION

To organize the categories, we draw a large rectangle to represent the sample space (all students at the school) and two overlapping regions to represent "girls" and "sports" (Figure 9.1a). We fill in the percentages (as in Figure 9.1b) using the following logic:

- The overlapping (dark gray) region contains half the girls, or $(0.5)(43\%) = 21.5\%$ of the students.

- The light gray region (the rest of the girls) then contains $(43 - 21.5)\% = 21.5\%$ of the students.

- The medium gray region (the rest of the sports players) then contains $(45 - 21.5)\% = 23.5\%$ of the students.

- The white region (the rest of the students) then contains $(100 - 66.5)\% = 33.5\%$ of the students. These are boys who do not play sports.

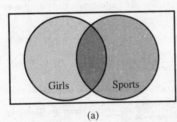 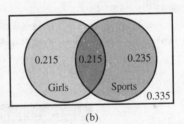

        (a)   (b)

Figure 9.1 (a) The Venn diagram for Alternate Example 6. (b) The Venn diagram with the probabilities filled in.

We can now answer the two questions by looking at the Venn diagram.

(a) We see from the diagram that the ratio of *boys* who play sports to *all students* who play sports is $\dfrac{23.5}{45}$, which is about 52.22%.

(b) We see that 33.5% of the students are boys who do not play sports, so 0.335 is the probability.

Exercises for Alternate Example 6

In Exercises 7–12, determine (a) what percentage of the students who play sports are boys, and (b) if a student is chosen at random, what is the probability that it is a boy who does not play sports, if in a large high school:

7. 60% of the students are girls, 35% of the students play sports, and 30% of the girls at the school play sports.

8. 50% of the students are girls, 45% of the students play sports, and 25% of the girls at the school play sports.

9. 38% of the students are girls, 38% of the students play sports, and 50% of the girls at the school play sports.

10. 62% of the students are girls, 80% of the students play sports, and 75% of the girls at the school play sports.

11. 20% of the students are girls, 50% of the students play sports, and 90% of the girls at the school play sports.

12. 60% of the students are girls, 50% of the students play sports, and 55% of the girls at the school play sports.

Alternate Example 8 **Using the Conditional Probability Formula**

Suppose we have drawn a cookie at random from one of two identical cookie jars (jar A contains 3 chocolate chip and 2 peanut butter cookies, while jar B contains 2 chocolate chip cookies). Given that it is chocolate chip, what is the probability that it came from jar A?

SOLUTION

The Conditional Probability Formula states that if the event B depends on the event A,

then $P\left(B\middle|A\right) = \dfrac{P\left(A \text{ and } B\right)}{P\left(A\right)}$.

Using the formula,

$$P(\text{jar A}\middle|\text{chocolate chip}) = \frac{P(\text{jar A and chocolate chip})}{P(\text{chocolate chip})}$$

$$= \frac{(1/2)(3/5)}{0.8} = \frac{0.3}{0.8} = \frac{3}{8}$$

Exercises for Alternate Example 8

In Exercises 13–18, suppose we have drawn a cookie at random from one of two identical cookie jars (jar A contains 3 chocolate chip and 2 peanut butter cookies, while jar B contains 1 chocolate chip and 4 peanut butter cookies). In Exercises 13–16, what is the probability that:

13. it is peanut butter and came from jar B?

14. it is chocolate chip and came from jar B?

15. it is chocolate chip and came from jar A?

16. it is peanut butter and came from jar A?

17. If one chocolate chip cookie is eaten from jar A, what is the probability that the next cookie drawn will be chocolate chip from jar A?

18. If one peanut butter cookie is eaten from jar A, what is the probability that the next cookie drawn will be peanut butter from jar A?

Alternate Example 9 **Repeating a Simple Experiment**

We roll a fair die four times. Find the probability that we roll:

(a) all 5's. **(b)** no 5's. **(c)** exactly three 5's.

SOLUTION

(a) We have a probability 1/6 of rolling a 5 each time. By the Multiplication Principle, the probability of rolling a 5 all four times is $(1/6)^4 \approx 0.00077$.

(b) There is a probability 5/6 of rolling something other than 5 each time. By the Multiplication Principle, the probability of rolling a non-5 all four times is $(5/6)^4 \approx 0.48225$.

(c) The probability of rolling three 5's followed by one non-5 (again by the Multiplication Principle) is $(1/6)^3(5/6)^1 \approx .003858$. However, that is not the only outcome we must consider. In fact, the three 5's could occur anywhere among the four rolls, in exactly $\dbinom{4}{3} = 4$ ways. That gives 4 outcomes, each with probability

$(1/6)^3(5/6)^1$. The probability of the event "exactly three 5's" is therefore

$$\dbinom{4}{3}(1/6)^3(5/6)^1 \approx .015432.$$

Exercises for Alternate Example 9

In Exercises 19–24, we roll an eight sided die five times. Find the probability that we roll:

19. no 1's   **20.** all 1's   **21.** four 1's   **22.** two 1's   **23.** three 1's   **24.** one 1

9.4 Sequences

Alternate Example 3 **Finding Limits of Sequences**

Determine whether the sequence converges or diverges. If it converges, give the limit.

(a) $\dfrac{1}{2}, \dfrac{1}{4}, \dfrac{1}{6}, \dfrac{1}{8}, ..., \dfrac{1}{2n}, ...$

(b) $\dfrac{11}{1}, \dfrac{12}{2}, \dfrac{13}{3}, \dfrac{14}{4}, ...$

(c) 5, 10, 15, 20, 25, ...

(d) 0.1, 0.2, 0.3, 0.4, 0.5, ...

SOLUTION

(a) $\lim\limits_{n \to \infty} \dfrac{1}{2n} = 0$. So the sequence converges to a limit of 0.

(b) Although the nth term is not explicitly given, we can see that $a_n = \dfrac{n+10}{n}$.

$\lim\limits_{n \to \infty} \dfrac{n+10}{n} = \lim\limits_{n \to \infty} \left(1 + \dfrac{10}{n}\right) = 1 + 0 = 1$. The sequence converges to a limit of 1.

(c) This time we see that $a_n = 5n$. Since $\lim\limits_{n \to \infty} 5n = \infty$, the sequence diverges.

(d) $a_n = 0.1n$ and the $\lim\limits_{n \to \infty} 0.1n = \infty$. This sequence diverges.

Exercises for Alternate Example 3

In Exercises 1–6, determine whether the sequence converges or diverges. If it converges, give the limit.

1. $10n - 10$ **2.** 5^n **3.** $\dfrac{1}{5}, \dfrac{1}{25}, \dfrac{1}{125}, \dfrac{1}{625}, \dfrac{1}{3125}, ...$ **4.** $\dfrac{5}{6}, 1, \dfrac{7}{6}, \dfrac{4}{3}, \dfrac{3}{2}, ...$ **5.** 0.1^n

6. $\left\{ \dfrac{10n - 2}{4 - 5n} \right\}$

Alternate Example 5 **Defining Arithmetic Sequences**

For each of the following arithmetic sequences, find **(a)** the common difference, **(b)** the tenth term, **(c)** a recursive rule for the nth term, and **(d)** an explicit rule for the nth term.

(1) −7, −4, −1, 2, 5, ...

(2) $\ln 3, \ln 9, \ln 27, \ln 81, ...$

SOLUTION

(1) **(a)** The difference between successive terms is 3.

(b) $a_{10} = -7 + (10 - 1)(3) = 20$

(c) The sequence is defined recursively by $a_1 = -7$ and $a_n = a_{n-1} + 3$ for all $n \geq 2$.

(d) The sequence is defined explicitly by $a_n = -7 + (n-1)(3) = 3n - 3 - 7 = 3n - 10$.

(2) (a) This sequence might not look arithmetic at first, but $\ln 9 - \ln 3 = \ln \dfrac{9}{3} = \ln 3$ (by a law of logarithms) and the difference between successive terms continues to be $\ln 3$.

(b) $a_{10} = \ln 3 + (10-1)\ln 3 = \ln 3 + 9\ln 3 = \ln(3 \cdot 3^9) = \ln 3^{10} = \ln 10.9861$

(c) The sequence is defined recursively by $a_1 = \ln 3$ and $a_n = a_{n-1} + \ln 3$ for all $n \geq 2$.

(d) The sequence is defined explicitly by $a_n = \ln 3 + (n-1)\ln 3 = \ln(3 \cdot 3^{n-1}) = \ln(3^n)$.

Exercises for Alternate Example 5

In Exercises 7–13, for each of the following arithmetic sequences, find (a) the common difference, (b) the tenth term, (c) a recursive rule for the nth term, and (d) an explicit rule for the nth term.

7. 31, 36, 41, 46, 51, ...   8. −15, −13, −11, −9, −7, ...   9. −9, −1, 7, 15, 23, ...   10. 1, 11, 21, 31, 41, ...

11. 2, 14, 26, 38, 50, ...   12. 8, 17, 26, 35, 44, ...   13. 4, 19, 34, 49, 64, ...

Alternate Example 6   **Defining Geometric Sequences**

For each of the following geometric sequences, find (a) the common ratio, (b) the tenth term, (c) a recursive rule for the nth term, and (d) an explicit rule for the nth term.

(1) 2, 6, 18, 54, 162, ...

(2) $2^6, 2^9, 2^{12}, 2^{15}, 2^{18}, \ldots$

SOLUTION

(1) (a) The ratio between the successive terms is 3.

(b) $a_{10} = 2 \cdot 3^{10-1} = 2 \cdot 3^9 = 39{,}366$

(c) The sequence is defined recursively by $a_1 = 2$ and $a_n = 3a_{n-1}$ for $n \geq 2$.

(d) The sequence is defined explicitly by $a_n = 2 \cdot 3^{n-1}$.

(2) (a) Applying a law of exponents, $\dfrac{2^9}{2^6} = 2^{9+(-6)} = 2^3$, and the ratio between successive terms continues to be 2^3.

(b) $a_{10} = 2^6 \cdot (2^3)^{10-1} = 2^{6+27} = 2^{33}$

(c) The sequence is defined recursively by $a_1 = 2^6$ and $a_n = 2^3 a_{n-1}$ for $n \geq 2$.

(d) The sequence is defined explicitly by
$$a_n = 2^6 \cdot (2^3)^{n-1} = 2^6 \cdot 2^{3n-3} = 2^{3n-3+6} = 2^{3n+3}.$$

Exercises for Alternate Example 6

In Exercises 14–19, for each of the following geometric sequences, find (a) the common ratio, (b) the tenth term, (c) a recursive rule for the nth term, and (d) an explicit rule for the nth term.

14. 7, 21, 63, 189, 567, ...   15. 2500, 500, 100, 20, 4, ...   16. 10, 20, 40, 80, 160, ...

17. 3, −9, 27, −81, 243, ...   18. 7, 42, 252, 1512, 9072, ...   19. −9, 18, −36, 72, −144, ...

Alternate Example 7 Constructing Sequences

The second and fifth terms of a sequence are 3 and 192, respectively. Find explicit and recursive formulas for the sequence if it is **(a)** arithmetic and **(b)** geometric.

SOLUTION

(a) If the sequence is arithmetic, then $a_2 = a + d = 3$ and $a_5 = a + 4d = 192$. Subtracting, we have

$$(a + 4d) - (a + d) = 192 - 3$$
$$a + 4d - a - d = 189$$
$$3d = 189$$
$$d = 63$$

Then $a + d = a + 63 = 3$ implies $a = -60$.

The sequence is defined explicitly by

$$a_n = -60 + (n - 1) \cdot 63 = -60 + 63n - 63 = 63n - 123.$$

The sequence is recursively by $a_1 = -60$ and $a = a_{n-1} + 63$ for $n \geq 2$.

(b) If the sequence is geometric, then $a_2 = a \cdot r^1 = 3$ and $a_5 = a \cdot r^4 = 192$.

Dividing, we have

$$\frac{a \cdot r^4}{a \cdot r^1} = \frac{192}{3}$$
$$r^3 = 64$$
$$r = 4$$

Then $a \cdot r^1 = 3$ implies $a = \frac{3}{4}$.

The sequence is defined explicitly by $a_n = \frac{3}{4}(4)^{n-1}$, or $a_n = 3 \cdot 4^{n-2}$.

The sequence is defined recursively by $a_1 = \frac{3}{4}$ and $a = 4 \cdot a_{n-1}$.

Exercises for Alternate Example 7

Solve Exercises 20–25.

20. The third and fifth terms of an arithmetic sequence are –1 and 3 respectively. Find the first term and a recursive rule for the nth term.

21. The third and eighth terms of an arithmetic sequence are 9 and 74 respectively. Find the first term and a recursive rule for the nth term.

22. The fifth and tenth terms of an arithmetic sequence are 5 and 41 respectively. Find the first term and a recursive rule for the nth term.

23. The second and fifth terms of a geometric series are 21 and 567 respectively. Find the first term, common ratio, and explicit rule for the nth term

24. The second and fifth terms of a geometric series are 10 and 1250 respectively. Find the first term, common ratio, and explicit rule for the nth term.

25. The fifth and seventh terms of a geometric series are 324 and 2916 respectively. Find the first term, common ratio, and explicit rule for the nth term.

9.5 Series

Alternate Example 1 **Summing the Terms of an Arithmetic Sequence**

A corner section of a stadium has 10 seats along the front row. Each successive row has 5 more seats than the row preceding it. If the top row has 70 seats, how many seats are in the entire section?

SOLUTION

The numbers of seats in the rows form an arithmetic sequence with

$$a_1 = 10, \ a_n = 70, \ \text{and} \ d = 5.$$

Solving $a_n = a_1 + (n-1)d$, we find that

$$70 = 10 + (n-1)5$$
$$70 = 10 + 5n - 5$$
$$70 = 5n + 5$$
$$65 = 5n$$
$$n = 13$$

Applying the Sum of a Finite Arithmetic Sequence Theorem, the total number of seats in the section is $13(10 + 70)/2 = 520$.

We can support this answer numerically by computing the sum on a calculator (Figure 9.2):

$$\text{sum(seq}(10 + (N-1)5, \ N, \ 1, \ 13) = 520.$$

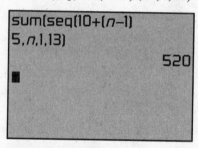

Figure 9.2 The commands necessary to compute the sum of a finite series on a typical calculator, in sequential mode.

Exercises for Alternate Example 1

A corner section of a stadium has seats along the front row. Each successive row has more seats than the row preceding it. In Exercises 1–6, determine how many seats are in the entire section if:

1. there are 10 seats along the front row, 46 seats along the top row, and each successive row has 4 more seats than the row preceding it.

2. there are 19 seats along the front row, 380 seats along the top row, and each successive row has 19 more seats than the row preceding it.

3. there are 3 seats along the front row, 58 seats along the top row, and each successive row has 5 more seats than the row preceding it.

4. there are 9 seats along the front row, 33 seats along the top row, and each successive row has 2 more seats than the row preceding it.

5. there are 15 seats along the front row, 51 seats along the top row, and each successive row has 3 more seats than the row preceding it.

6. there is 1 seat along the front row, 22 seats along the top row, and each successive row has 3 more seats than the row preceding it.

Alternate Example 2 **Summing the Terms of a Geometric Sequence**

Find the sum of the geometric sequence 9, 9/7, 9/49, 9/343, ..., $9(1/7)^7$.

SOLUTION

We can see that $a_1 = 9$ and $r = 1/7$. The nth term is $9(1/7)^7$, which means that $n = 8$. (Remember that the exponent on the nth term is $n-1$, not n.) Applying the Sum of a Finite Geometric Sequence Theorem, we find that

$$\sum_{n=1}^{8} 9\left(\frac{1}{7}\right)^{n-1} = \frac{9(1-(1/7)^8)}{1-(1/7)} \approx 10.499998179.$$

We can support this answer by having the calculator do the actual summing:

$$\text{sum(seq}(9(1/7)\char`\^(N-1), N, 1, 8) = 10.499998179.$$

Exercises for Alternate Example 2

In Exercises 7–12, find the sum of the geometric sequence.

7. $7, \dfrac{7}{5}, \dfrac{7}{25}, \dfrac{7}{125}, ..., 7(1/5)^{10}$

8. 2, 6, 18, 54, ..., 118098

9. 1, 2, 4, 8, 16, ..., 256

10. $1, \dfrac{1}{6}, \dfrac{1}{36}, \dfrac{1}{216}, \dfrac{1}{1296}, \dfrac{1}{7776}$

11. $4, 2, 1, ..., \dfrac{1}{8}$

12. $6, 1, \dfrac{1}{6}, \dfrac{1}{36}, ..., \dfrac{1}{1296}$

Alternate Example 4 **Summing Infinite Geometric Series**

Determine whether the series converges. If it converges, give the sum.

(a) $\displaystyle\sum_{k=1}^{\infty} 6(0.25)^{k-1}$

(b) $\displaystyle\sum_{n=0}^{\infty} \left(\frac{-12}{33}\right)^n$

(c) $\displaystyle\sum_{n=1}^{\infty} \left(\frac{17}{16}\right)^n$

(d) $1 + \dfrac{1}{3} + \dfrac{1}{9} + \dfrac{1}{27} + \cdots$

SOLUTION

(a) Since $|r| = |0.25| < 1$, the series converges. The first term is $6(0.25)^0 = 6$, so the sum is $a/(1-r) = 6/(1-0.25) = 8$.

(b) Since $|r| = |12/33| < 1$, the series converges. The first term is $(-12/33)^0 = 1$, so the sum is $a/(1-r) = 1/(1-(-12/33)) = 11/15$.

(c) Since $|r| = |17/16| > 1$, the series diverges.

(d) Since $|r| = |1/3| < 1$, the series converges. The first term is 1, and so the sum is $a/(1-r) = 1/(1-1/3) = 3/2$.

Exercises for Alternate Example 4

In Exercises 13–18, determine whether the series converges. If it converges, give the sum.

13. $\displaystyle\sum_{n=0}^{\infty} 4\left(\frac{1}{4}\right)^n$

14. 10, 7.5, 5.625, ...

15. 8, 12, 18, 27, ...

16. $\dfrac{1}{2}, \dfrac{3}{4}, \dfrac{9}{8}, ...$

17. $\dfrac{1}{2}, \dfrac{1}{6}, \dfrac{1}{18}, ...$

18. $\displaystyle\sum_{n=0}^{\infty} \frac{1}{3}\left(\frac{1}{2}\right)^n$

9.6 Mathematical Induction

Alternate Example 1 Using Mathematical Induction

Prove that $1 + 2 + 3 + \ldots + n = \dfrac{n^2 + n}{2}$ is true for all positive integers n.

SOLUTION

Call the statement P_n. We could verify P_n by using the formula for the sum of an arithmetic sequence, but here is how we prove it by mathematical induction.

(The Anchor) For $n = 1$, the equation reduces to $P_1 : 1 = 1$, which is true.

(The Inductive Hypothesis) Assume that the equation is true for $n = k$. That is, assume

$$P_k = 1 + 2 + 3 + \ldots + k = \frac{k^2 + k}{2} \text{ is true.}$$

(The Inductive Step) The next term on the left-hand side would be $n + 1$. We add this to both sides of P_k and get

$$1 + 2 + 3 + \ldots + k + (k + 1) = \frac{k^2 + k}{2} + k + 1$$

$$= \frac{k^2 + k + 2(k + 1)}{2}$$

$$= \frac{k^2 + k + 2k + 2}{2}$$

$$= \frac{k^2 + k + 2k + 1 + 1}{2}$$

$$= \frac{(k^2 + 2k + 1) + (k + 1)}{2}$$

$$= \frac{(k + 1)^2 + (k + 1)}{2}$$

This is exactly the statement P_{k+1}, so the equation is true for $n = k + 1$. Therefore, P_n is true for all positive integers, by mathematical induction.

Exercises for Alternate Example 1

Prove that each of the following statements is true for all positive integers n by mathematical induction.

1. $2 + 4 + 6 + \ldots + 2n = n(n + 1)$

2. $1 + 4 + 7 + \ldots + (3n - 2) = \dfrac{n(3n - 1)}{2}$

3. $1 + 7 + 13 + \ldots + (6n - 5) = n(3n - 2)$

4. $1 + 5 + 5^2 + \ldots + 5^{n-1} = \dfrac{1}{4}(5^n - 1)$

5. $1 \cdot 2 + 2 \cdot 3 + 3 \cdot 4 + \ldots + n(n + 1) = \dfrac{n(n + 1)(n + 2)}{3}$

6. $5 + 7 + 9 + \ldots + (5 + 2(n - 1)) = \dfrac{n(8 + 2n)}{2}$

Alternate Example 2 **Using Mathematical Induction**

Prove that $1+8+27+...+n^3 = \dfrac{n^2(n+1)^2}{4}$ is true for all positive integers n.

SOLUTION

Let P_n be the statement $1+8+27+...+n^3 = \dfrac{n^2(n+1)^2}{4}$.

(The Anchor) P_1 is true because $1^3 = \dfrac{1(1+1)^2}{4}$.

(The Inductive Hypothesis) Assume that P_k is true, so that

$$1^3 + 2^3 + ... + k^3 = \frac{k^2(k+1)^2}{4}.$$

(The Inductive Step) The next term on the left-hand side would be $(k+1)^3$. We add this to both sides of P_k and get

$$\begin{aligned}
1^3 + 2^3 + 3^3 + \ ... \ + k^3 + (k+1)^3 &= \frac{k^2(k+1)^2}{4} + (k+1)^3 \\
&= \frac{k^2(k^2+2k+1) + 4(k^3+3k^2+3k+1)}{4} \\
&= \frac{k^4 + 2k^3 + k^2 + 4k^3 + 12k^2 + 12k + 4}{4} \\
&= \frac{k^4 + 2k^3 + k^2 + 4k^3 + 4k^2 + 8k^2 + 4k + 8k + 4}{4} \\
&= \frac{(k^4 + 4k^3 + 4k^2) + (2k^3 + 8k^2 + 8k) + (k^2 + 4k + 4)}{4} \\
&= \frac{k^2(k^2 + 4k + 4) + 2k(k^2 + 4k + 4) + (k^2 + 4k + 4)}{4} \\
&= \frac{(k^2 + 2k + 1)(k^2 + 4k + 4)}{4} \\
&= \frac{(k+1)^2(k+2)^2}{4}
\end{aligned}$$

This is exactly the statement P_{k+1}, so the equation is true for $n = k+1$. Therefore, P_n is true for all positive integers, by mathematical induction.

Exercises for Alternate Example 2

Prove that each of the following statements is true for all positive integers n by mathematical induction.

7. $3+9+15+ \ ... \ +(6n-3) = 3n^2$

8. $3+8+13+ \ ... \ +(5n-2) = \dfrac{n}{2}(5n+1)$

9. $7+11+15+ \ ... \ +(4n+3) = n(2n+5)$

10. $1+6+6^2 + \ ... \ +6^{n-1} = \dfrac{1}{5}(6^n - 1)$

11. $\left(1+\dfrac{1}{1}\right)+\left(1+\dfrac{1}{2}\right)+\left(1+\dfrac{1}{3}\right)+ \ ... \ +\left(1+\dfrac{1}{n}\right) = n+1$

12. $n! > 2^n$, for $n \geq 4$

Alternate Example 3 Proving Divisibility

Prove that $5^n - 1$ is evenly divisible by 2 for all positive integers n.

SOLUTION

Let P_n be the statement that $5^n - 1$ is evenly divisible by 2 for all positive integers n.

(The Anchor) P_1 is true because $5^1 - 1 = 4$ is divisible by 2.

(The Inductive Hypothesis) Assume that P_k is true, so that $5^k - 1$ is divisible by 2.

(The Inductive Step) We need to prove that $5^{k+1} - 1$ is divisible by 2.

Using a little algebra, we see that $5^{k+1} - 1 = 5 \cdot 5^k - 1 = 5(5^k - 1) + 4$.

By the inductive hypothesis, $5^k - 1$ is divisible by 2. Of course, so is 4. Therefore, $5(5^k - 1) + 4$ is a sum of multiples of 2, and hence divisible by 2. This is exactly the statement P_{k+1}, so P_{k+1} is true. Therefore, P_n is true for all positive integers, by mathematical induction.

Exercises for Alternate Example 3

Prove that each of the following statements is true for all positive integers n by mathematical induction.

13. Prove that $18^n - 1$ is evenly divisible by 17 for all positive integers n.

14. Prove that $n(n^2 + 5)$ is evenly divisible by 6 for all positive integers n.

15. Prove that $11^n - 4^n$ is evenly divisible by 7 for all positive integers n.

16. Prove that $n(n^2 + 3n + 8)$ is evenly divisible by 6 for all positive integers n.

17. Prove that $2^{5n} - 1$ is evenly divisible by 31 for all positive integers n.

18. Prove that $(n+1)(n+2) \ldots (2n)$ is evenly divisible by 2^n for all positive integers n.

9.7 Statistics and Data (Graphical)

Alternate Example 3 Making Back-to-Back Stemplots

Mark McGwire and Barry Bonds entered the major leagues in 1986 and had overlapping careers until 2001, the year that McGwire retired. During that period they averaged 101.63 and 144.56 hits per year, respectively. Compare their annual hit totals with a back-to-back stemplot. Can you tell which player was a more consistent hitter?

Major League Hit Totals for Mark McGwire and Barry Bonds through 2001															
Year	86	87	88	89	90	91	92	93	94	95	96	97	98	99	00	01
McGwire	10	161	143	113	123	97	125	28	34	87	132	148	152	145	72	56
Bonds	92	144	152	144	156	149	147	181	122	149	159	155	167	93	147	156

SOLUTION

We form a back-to-back stemplot with McGwire's totals branching off to the left and Bonds' to the right.

Mark McGwire		Barry Bonds
1	0	
	1	
8	2	
4	3	
	4	
6	5	
	6	
2	7	
7	8	
7	9	2 3
	10	
3	11	
3 5	12	2
2	13	
3 5 8	14	4 4 7 7 9 9
2	15	2 5 6 6 9
1	16	7
	17	
	18	1

The low hit years for McGwire can be explained by fewer times at bat (late entry into the league in 1986 and injuries in 1993 and 1994). It appears that Barry Bonds was a more consistent hitter.

The frequency tables below present the same data distributed in intervals. Comparing the data as shown in these tables clearly shows that Barry Bonds was a more consistent hitter because the number of hits is clustered within a smaller number of intervals.

Frequency Table for Barry Bonds Yearly Hits Totals, 1986–2001			
Hits	Frequency	Hits	Frequency
0 – 24	0	125 – 149	6
25 – 49	0	150 – 174	6
50 – 74	0	175 – 199	1
75 – 99	2	Total	16
100 – 124	1		

Frequency Table for Mark McGwire's Yearly Hits Totals, 1986–2001			
Hits	Frequency	Hits	Frequency
0 – 24	1	100 – 124	2
25 – 49	2	125 – 149	5
50 – 74	2	150 – 174	2
75 – 99	2	Total	16

Exercises for Alternate Example 3

For Exercises 1–3, use the table below:

Nutritional Values in Burger King Menu Items			
	Protein (g)	Carbohydrates (g)	Fat (g)
Whopper	27	46	31
Whopper with Cheese	32	48	38
Double Whopper	46	46	48
Double Whopper with Cheese	51	48	55
Cheeseburger	16	28	14
Whopper Jr. with Cheese	16	30	19
Hamburger	14	28	10
Whopper Jr.	14	29	15
Bacon Double Cheeseburger	30	26	28
Bacon Double Cheeseburger Deluxe	30	28	33
Double Cheeseburger	27	29	25

Source: Burger King's Your Guide to Nutrition

1. Make a back-to-back stem plot comparing the carbohydrates and fat in the 12 Burger King burgers. If you were on a low-carbohydrate diet, which is your best burger choice?

2. Make a back-to-back stem plot comparing the protein and the fat in the 12 Burger King burgers. If you were trying to limit your fat, but maintain a high protein diet, which burger might you choose?

3. Make a back-to-back stem plot comparing the protein and the carbohydrates in the 12 Burger King burgers. If you were trying to limit your carbohydrates, but maintain a high protein diet, which burger might you choose?

For Exercises 4–6, use the table below:

Home Run Leaders				
Year	National League	HR	American League	HR
1970	Johnny Bench	45	Frank Howard	44
1971	Willie Stargell	48	Bill Melton	33
1972	Johnny Bench	40	Dick Allen	37
1973	Willie Stargell	44	Reggie Jackson	32
1974	Mike Schmidt	36	Dick Allen	32
1975	Mike Schmidt	38	George Scott Reggie Jackson	36
1976	Mike Schmidt	38	Craig Nettles	32
1977	George Foster	52	Jim Rice	39
1978	George Foster	40	Jim Rice	46
1979	Dave Kingman	48	Gorman Thomas	45
1980	Mike Schmidt	48	Reggie Jackson Ben Oglivie	41
1981	Mike Schmidt	31	Bobby Grich Tony Armas Dwight Evans Eddie Murray	22
1982	Dave Kingman	37	Gorman Thomas Reggie Jackson	39
1983	Mike Schmidt	40	Jim Rice	39
1984	Mike Schmidt Dale Murphy	36	Tony Armas	43
1985	Dale Murphy	37	Darrell Evans	40

Source: www.baseball-reference.com

4. Make a back-to-back stem and leaf plot comparing the number of home runs hit by the American and National League home run hitters each year during this 16-year time period.

5. Does the American League champion or the National League champion tend to hit the most home runs?

6. Do there seem to be any outliers in either data set? What are some possible explanations for these?

∎

Alternate Example 4 **Graphing a Histogram on a Calculator**

Make a histogram of Mickey Mantle's annual home run totals given in the table below, using intervals of width 5.

Annual Home Run Statistics For Mickey Mantle

Year	Home Runs	Year	Home Runs	Year	Home Runs
1951	13	1957	34	1963	15
1952	23	1958	42	1964	35
1953	21	1959	31	1965	19
1954	27	1960	40	1966	23
1955	37	1961	54	1967	22
1956	52	1962	30	1968	18

Source: www.baseball-reference.com

SOLUTION

We first make a frequency table for the data, using intervals of width 5. (This is not needed for the calculator to produce the histogram, but we will compare this with the result.)

Home Runs	Frequency	Home Runs	Frequency	Home Runs	Frequency
$10-14$	1	$25-29$	1	$40-44$	2
$15-19$	3	$30-34$	3	$45-49$	0
$20-24$	4	$35-39$	2	$50-54$	2
				Total	18

To scale the *x*-axis to be consistent with the intervals of the table, let Xmin = 0, Xmax = 60, and Xscl = 5. Notice that the maximum frequency is 4 years (20–24 home runs), so the *y*-axis ought to extend at least to 5. Enter the data from the above table into list L1, and plot a histogram in the window [0, 60] by [−1, 5]. (See Figure 9.3a.) Tracing along the histogram should reveal the same frequencies as in the frequency table we made. (See Figure 9.3b)

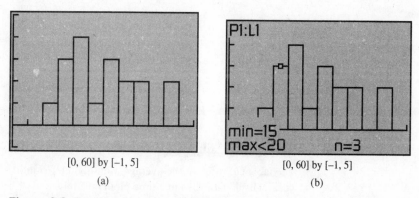

[0, 60] by [−1, 5]   [0, 60] by [−1, 5]
(a)   (b)

Figure 9.3 Calculator histograms of Mickey Mantle's yearly home run totals.

Exercises for Alternate Example 4

In Exercises 7–9, use the data in the table below:

1992 United States Dream Team			
Player	Minutes	FGs	FG Attempts
Charles Barkley	48	14	22
Larry Bird	47	6	12
Clyde Drexler	54	13	21
Patrick Ewing	54	14	24
"Magic" Johnson	72	15	25
Michael Jordan	67	20	45
Christian Laettner	17	1	3
Karl Malone	45	13	21
Chris Mullin	60	16	20
Scottie Pippen	56	11	17
David Robinson	57	13	20
John Stockton	23	3	6

7. Make a histogram of the number of minutes played by members of the Dream Team, using intervals of width 10. Does the distribution appear normal? Support your answer.

8. Make a histogram of the number of field goals by members of the Dream Team, using intervals of width 5. Does the distribution appear normal? Support your answer.

9. Make a histogram of the number of field goal attempts by members of the Dream Team, using intervals of width 5. Do you notice any outliers in your data set? If so, how do they appear graphically?

In Exercises 10–12, use the data in the table below.

International Assessment of Science and Mathematics Achievement of 13-year-old Students			
Country	Percent of Students who Spend 2 Hours or More on Homework Per Day	Percent of Students who Watch Television 5 Hours or More Per Day	Percent Correct for Science on Standardized Assessment
Canada	26	15	69
France	55	4	69
Hungary	61	16	73
Ireland	66	9	63
Israel	49	20	70
Italy	78	7	70
Jordan	54	10	57
Scotland	15	23	68
Slovenia	27	5	70
South Korea	38	10	78
Russia	52	19	71
Spain	62	11	68
Switzerland	21	7	74
Taiwan	44	7	76
United States	31	22	67

Source: National Center of Education Statistics, Learning Mathematics and Learning Science, 1992.

10. Make a histogram of the percent of students who spend 2 hours or more on homework per day, using intervals of width 10. Does the distribution appear normal? Support your answer.

11. Make a histogram of the percent of students who watch television 5 hours or more per day, using intervals of width 5. Does the distribution appear normal? Support your answer.

12. Make a histogram of the percent correct on the standardized assessment, using intervals of width 5. Are there any unusual trends or outliers in the data set?

Alternate Example 5 **Drawing a Time Plot**

The table below gives the average monthly temperature of Albany, New York for the past 50 years. Display the data in a time plot and analyze the 12-month trend.

Average Monthly Temperatures of Albany, New York for the Past 50 Years												
Month	Jan.	Feb.	Mar.	Apr.	May	Jun.	Jul.	Aug.	Sep.	Oct.	Nov.	Dec.
°F	22	24	34	47	58	67	72	70	61	51	40	27

Source: www.weatherbase.com

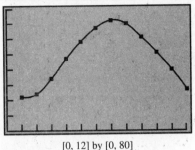

[0, 12] by [0, 80]

Figure 9.4 A time plot comparing average monthly temperatures in Albany, New York over the last 50 years.

SOLUTION

The horizontal axis represents time (in months) from January to December. The vertical axis represents the average temperature of that month for the past 50 years in degrees Fahrenheit. Since the visualization is enhanced by showing both axes in the viewing window, the vertical axis of the time plot is usually translated to cross the *x*-axis at or near the beginning of the time interval in the data. You can create this effect on your grapher by entering the months {1, 2, 3, ..., 12} rather than {January, February, March, ..., December}.

The time plot in Figure 9.4 shows that the average temperature of Albany, New York for the past 50 years steadily increases from January until July, then steadily decreases from July until December.

Exercises for Alternate Example 5

In Exercises 13–16, consider the data set below.

Median Age at First Marriage		
Year	Males	Females
1900	25.9	21.9
1910	25.1	21.6
1920	24.6	21.2
1930	24.3	21.3
1940	24.3	21.5
1950	22.8	20.3
1960	22.8	20.3
1970	23.2	20.8
1980	24.7	22.0
1990	26.1	23.9

Source: United States Bureau of the Census

13. Make a time plot for the median age of the first marriage of males throughout the last century.

14. On the same set of axes, make a time plot for the median age of the first marriage of females throughout the last century. Be sure to distinguish which set of data corresponds with which time plot.

15. During which decade did the largest decrease occur in median age at first marriage for men? for women? Give a possible explanation.

16. Subtract the median female age at first marriage from the median male age for each year given in the data table, and construct a time plot of these differences. Describe any patterns you see in this time series plot.

In Exercises 17–20, consider the following data set.

Average SAT Scores		
Year	Verbal	Mathematics
1980	424	466
1981	424	466
1982	426	467
1983	425	468
1984	426	471
1985	431	475
1986	431	475
1987	430	476
1988	428	476
1989	427	476
1990	424	476
1991	422	474
1992	423	476
1993	424	478

Source: The College Board

17. On the same set of axes, make a time plot for the annual math and verbal SAT scores. Overall, which scores have increased more, mathematics or verbal?

18. Calculate the average math score minus the average verbal score for each year. Construct a time plot over time of these differences.

19. Write a sentence summarizing the pattern over time of the difference between the scores.

20. Calculate the slope of the fitted line of your time plot from Question 18. Write a sentence describing what it represents in terms of the SAT scores

9.8 Statistics and Data (Algebraic)

Alternate Example 5 **Using a Frequency Table**

A teacher gives a 10-point quiz and records the scores in a frequency table as shown below. Find the mode, median, and mean of the data.

Quiz Scores											
Score	10	9	8	7	6	5	4	3	2	1	0
Frequency	4	4	4	4	2	2	5	3	0	2	2

SOLUTION

The total of the frequencies is 32, so there are 32 scores.

The *mode* is 4, since that is the score with the highest frequency.

The *median* of 32 numbers will be the mean of the 16th and 17th numbers. The table is already arranged in descending order, so we count the frequencies from left to right until we come to 16. We see that the 16th number is a 7 and the 17th number is a 6. The median, therefore, is 6.5.

To find the *mean*, we multiply each number by its frequency, add the products, and divide the total by 32:

$$\bar{x} = \frac{[10(4) + 9(4) + 8(4) + 7(4) + 6(2) + 5(2) + 4(5) + 3(3) + 2(0) + 1(2) + 0(2)]}{32}$$

$$= \frac{189}{32} \approx 5.91$$

Exercises for Alternate Example 5

Students in a class were given a weekly, 10-point exam. In Exercises 1–6, use the data given in the frequency tables to find the mean scores.

1.

Scores	Number of students
10	4
9	5
8	5
7	5
6	
5	
4	
3	
2	
1	1

2.

Scores	Number of students
10	2
9	10
8	2
7	5
6	1
5	2
4	
3	
2	
1	

3.

Scores	Number of students
10	5
9	3
8	2
7	10
6	
5	
4	
3	
2	
1	

4.

Scores	Number of students
10	4
9	10
8	3
7	1
6	1
5	
4	1
3	
2	
1	

5.

Scores	Number of students
10	5
9	2
8	3
7	10
6	
5	
4	
3	
2	
1	

6.

Scores	Number of students
10	
9	
8	
7	10
6	5
5	5
4	5
3	
2	
1	

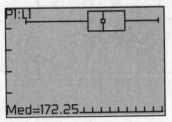

[60, 240] by [0, 5]

FIGURE 9.5 A boxplot for the five-number summary for the male cancer death rates, with the median highlighted.

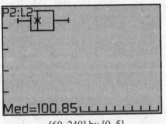

[60, 240] by [0, 5]

FIGURE 9.6 A boxplot for the five-number summary for the female cancer death rates, with the median highlighted.

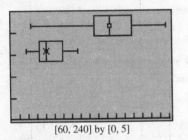

[60, 240] by [0, 5]

FIGURE 9.7 A single graph showing boxplots for male and female cancer death rates gives a good visualization of the difference in the two data sets.

Alternate Example 8 **Comparing Boxplots**

Draw boxplots for the male and female data given below and describe briefly the information displayed in the visualization.

Male and Female Cancer Death Rates Per 100,000 Population Among 12 Countries from 1986–88		
Country	Male	Female
Hungary	235.4	129.4
Italy	193.7	99.7
United States	163.2	109.7
Australia	162.8	102.0
Cuba	129.0	95.9
Japan	149.8	78.1
Canada	170.6	111.5
France	204.7	89.9
Scotland	200.9	136.1
England & Wales	181.9	126.9
Switzerland	173.9	99.5
Ecuador	83.9	86.1

Source: World Health Organization Data as adapted by the American Cancer Society, 1992

SOLUTION

Here are the lists in ascending order.

Males:

{83.9, 129.0, 149.8, 162.8, 163.2, 170.6, 173.9, 181.9, 193.7, 200.9, 204.7, 235.4}

Females:

{78.1, 86.1, 89.9, 95.9, 99.5, 99.7, 102.0, 109.7, 111.5, 126.9, 129.4, 136.1}

The five-number summaries are:

Males: {83.9, 156.3, 172.25, 197.3, 235.4}

Females: {78.1, 92.9, 100.85, 119.20, 136.1}

The boxplots are shown separately in Figures 9.5 and 9.6. They can be graphed simultaneously (Figure 9.7).

From this graph we can see that the cancer death rate for the men is greater than the cancer death rate for the women among the 12 countries.

Exercises for Alternate Example 8

In Exercises 7–12, draw boxplots for the given data and describe briefly the information displayed in the visualization.

7. Males: {135, 142, 151, 159, 168}
Females: {101, 109, 120, 131, 143}

8. Males: {153, 162, 169, 179, 183}
Females: {130, 141, 153, 161, 172}

9. Males: {143, 152, 161, 169, 173}
Females: {141, 157, 163, 168, 176}

10. Males: {80, 92, 101, 107, 115}
Females: {79, 83, 90, 94, 102}

11. Males: {131, 151, 163, 170, 179}
Females: {135, 141, 144, 150, 161}

12. Males: {150, 163, 168, 171, 180}
Females: {163, 172, 178, 181, 193}

Alternate Example 10 **Finding Standard Deviation with a Calculator**

A researcher measured 30 newly hatched loon chicks and recorded their weights in grams as shown in the table below.

Weights in Grams of 30 Loon Chicks									
80	80.1	79.1	77.1	78.2	79.3	78.6	80.1	81.2	83
82.3	84.5	82.3	82.6	82.7	82.9	83.7	85.6	82.7	81.6
80.7	78.9	77.9	83.6	84.5	85.4	82.7	82.9	82.6	82.2

Based on the sample, estimate the mean and standard deviation for the weights of all newly hatched loon chicks. Are these measures useful in this case, or should we use the five-number summary?

SOLUTION

We enter the list of data into a calculator and choose the command that will produce statistics of a single variable. The output from one such calculator is shown in Figure 9.8

The mean is $\bar{x} = 81.63333$ grams. For standard deviation, we choose $Sx = 2.269032$ grams because the calculations are based on a sample of loon chicks, not the entire population of loon chicks.

A histogram in the window [75, 88] by [0, 12] (Figure 9.9) shows that the distribution is skewed slightly to the right. Therefore, the mean and standard deviation may not be appropriate measures of the entire population.

```
1-Var Stats
 x̄=81.63333333
 Σx=2449
 Σx²=200069.34
 Sx=2.269031896
 σx=2.23089419
↓n=30
■
```

Figure 9.8 Single-variable statistics in a typical calculator display.

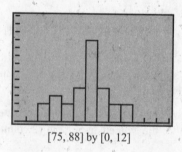

[75, 88] by [0, 12]

Figure 9.9 The weight of the loon chicks.

Exercises for Alternate Example 10

In Exercises 13–18, using the number sets given below, estimate the **(a)** mean and **(b)** standard deviation.

13. {100, 96, 100, 75, 56, 82, 89, 86, 87}

14. {82.3, 86.5, 79.9, 90, 82.6, 87, 85, 84, 86}

15. {9, 8, 3, 7, 5, 9, 9, 9, 7}

16. {15, 17, 18, 19, 20, 15, 17, 16, 19}

17. (4, 3.5, 3.6, 3, 2.9, 3.15, 3.25, 3.46, 3.79}

18. {108, 120, 135, 90, 95, 112, 125, 116, 114}

Alternate Example 11 **Using the 68-95-99.7 Rule**

Based on the research data presented in Alternate Example 10, would a loon chick weighing 86.2 grams be in the top 2.5% of all newly hatched loon chicks?

SOLUTION

We assume that the weights of newly hatched loon chicks are normally distributed in the whole population. Since we do not know the mean and standard deviation for the whole population (the parameters μ and σ), we use $\bar{x} = 81.63$ and $Sx = 2.27$ as estimates.

Look at Figure 9.10. The shaded region contains 95% of the area, so the two identical white regions at either end must each contain 2.5% of the area. That is, to be in the top 2.5%, a loon chick will have to weigh at least 2 standard deviations more than the mean:

$$\bar{x} + 2Sx = 81.63 + 2(2.27) = 86.17 \text{ grams.}$$

Since $86.2 > 86.17$, an 86.2-gram loon chick is indeed in the top 2.5%.

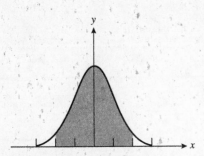

Figure 9.10 A normal curve showing three standard deviations from the mean.

Exercises for Alternate Example 11

Solve Exercises 19–24.

19. If a class's average score on a math final exam is 80.5% with a standard deviation of 5.32, approximately what percentage of the scores were higher than 91.14%?

20. If a class's average score on a math final exam is 85.1% with a standard deviation of 3.31, approximately what percentage of the scores were higher than 88.41%?

21. If a class's average score on a math final exam is 7% with a standard deviation of 1.3, approximately what percentage of the scores were higher than 9.6%?

22. If a class's average score on a math final exam is 76% with a standard deviation of 5.5, approximately what percentage of the scores were higher than 81.5%?

23. If a class's average score on a math final exam is 94% with a standard deviation of 1.21, approximately what percentage of the scores were higher than 96.42%?

24. If a class's average score on a math final exam is 16% with a standard deviation of 0.73, approximately what percentage of the scores were higher than 17.46%?

Chapter 10

An Introduction to Calculus: Limits, Derivatives and Integrals

10.1 Limits and Motion: The Tangent Problem

Alternate Example 1 **Computing Average Velocity**

An airplane travels 1500 miles in 3 hours and 45 minutes. What is the average velocity of the plane over the entire 3.75-hour time interval?

SOLUTION

The average velocity is the change in position (1500 miles) divided by the change in time (3.75 hours). If we denote position by s and time by t, we have

$$v_{ave} = \frac{\Delta s}{\Delta t} = \frac{1500 \text{ miles}}{3.75 \text{ hours}} = 400 \text{ miles per hour}$$

Exercises for Alternate Example 1

In Exercises 1–6, find the average velocity traveled given the distances traveled over the entire given time interval, using appropriate units.

1. 800 meters in 40 seconds
2. 900 kilometers in 2 hours and 15 minutes
3. 50 yards in 2.5 seconds
4. 1200 feet in 5.2 minutes
5. 90 miles in 2 hours and 10 minutes
6. 840 kilometers in 5.7 hours

Alternate Example 3 **Finding the Slope of a Tangent Line**

Use limits to find the slope of the tangent line to the graph of $s = t^2 - 4$ at the point $(-1, -3)$.

SOLUTION

We take the limit as t approaches -1 of the average rate of change of the function (Figure 10.1).

$$\lim_{t \to -1} \frac{\Delta s}{\Delta t} = \lim_{t \to -1} \frac{s(t) - s(-1)}{t - (-1)}$$

$$= \lim_{t \to -1} \frac{(t^2 - 4) - (-3)}{t - (-1)}$$

$$= \lim_{t \to -1} \frac{(t^2 - 1)}{t + 1}$$

$$= \lim_{t \to -1} \frac{(t-1)(t+1)}{t+1} \qquad \text{Factor the numerator.}$$

$$= \lim_{t \to -1} (t-1)\frac{t+1}{t+1}$$

$$= \lim_{t \to -1} (t-1) \qquad \text{Since } t \ne -1, \frac{t+1}{t+1} = 1.$$

$$= -2$$

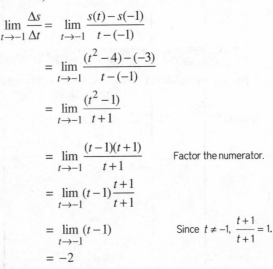

Figure 10.1 A line tangent to the graph of $s = t^2 - 4$ at the point $(-1, -3)$.

Exercises for Alternate Example 3

In Exercises 7–12, use limits to find the slope of the tangent line to the graph of the function at the given point.

7. $s = t^2 - 9$ at $(2, -5)$ **8.** $s = 2t^2$ at $(0, 0)$ **9.** $s = 4t^2 - 9$ at $(2, 7)$ **10.** $s = t^4 - 1$ at $(-1, 0)$

11. $s = t^3 + 8$ at $(-1, 7)$ **12.** $s = t^3$ at $(1, 1)$

Alternate Example 4 **Finding a Derivative at a Point**

Find $f'(-1)$ if $f(x) = 3x^2 + 2$.

SOLUTION

$$f'(-1) = \lim_{h \to 0} \frac{f(-1+h) - f(-1)}{h}$$

$$= \lim_{h \to 0} \frac{3(-1+h)^2 + 2 - [3(-1)^2 + 2]}{h}$$ Replace x with $(-1 + h)$ and -1 in function rule.

$$= \lim_{h \to 0} \frac{3(1 - 2h + h^2) + 2 - 5}{h}$$

$$= \lim_{h \to 0} \frac{3 - 6h + 3h^2 - 3}{h}$$

$$= \lim_{h \to 0} \frac{-6h + 3h^2}{h}$$

$$= \lim_{h \to 0} \frac{h(-6 + 3h)}{h}$$

$$= \lim_{h \to 0} (-6 + 3h)$$ Since $h \neq 0$, $\frac{h}{h} = 1$.

$$= -6$$

Exercises for Alternate Example 4

In Exercises 13–18, find the numerical value of the derivative for each of the following functions.

13. $f'(2)$ if $f(x) = 6x + 5$ **14.** $f'(3)$ if $f(x) = \frac{2}{3}x^2 + 4$ **15.** $f'(7)$ if $f(x) = \sqrt{x+2}$ **16.** $f'(3\pi)$ if $f(x) = \pi$

17. $f'(7)$ if $f(x) = \frac{5}{x+3}$ **18.** $f'(-3)$ if $f(x) = 2x^3$

Alternate Example 5 **Finding the Derivative of a Function**

(a) Find $f'(x)$ if $f(x) = \sqrt{x}$.

(b) Find $\frac{dy}{dx}$ if $y = \frac{2}{x+5}$.

SOLUTION

(a) $f'(x) = \lim_{h \to 0} \dfrac{f(x+h) - f(x)}{h}$

$= \lim_{h \to 0} \dfrac{\sqrt{x+h} - \sqrt{x}}{h}$

$= \lim_{h \to 0} \dfrac{\sqrt{x+h} - \sqrt{x}}{h} \cdot \dfrac{\sqrt{x+h} + \sqrt{x}}{\sqrt{x+h} + \sqrt{x}}$ Rationalize the numerator.

$= \lim_{h \to 0} \dfrac{(x+h) - x}{h\left(\sqrt{x+h} + \sqrt{x}\right)}$

$= \lim_{h \to 0} \dfrac{h}{h\left(\sqrt{x+h} + \sqrt{x}\right)}$

$= \lim_{h \to 0} \dfrac{1}{\left(\sqrt{x+h} + \sqrt{x}\right)}$ Since $h \neq 0$, $\dfrac{h}{h} = 1$.

$= -6$

$= \dfrac{1}{2\sqrt{x}}$

(b) $\dfrac{dy}{dx} = \lim_{h \to 0} \dfrac{f(x+h) - f(x)}{h}$

$= \lim_{h \to 0} \dfrac{\dfrac{2}{x+h+5} - \dfrac{2}{x+5}}{h}$

$= \lim_{h \to 0} \dfrac{\dfrac{2(x+5) - 2(x+h+5)}{(x+h+5)(x+5)}}{h}$ Found common denominator in numerator

$= \lim_{h \to 0} \dfrac{2x + 10 - 2x - 2h - 10}{(x+h+5)(x+5)} \cdot \dfrac{1}{h}$ Division by h changed to multiplication by $\dfrac{1}{h}$.

$= \lim_{h \to 0} \dfrac{-2h}{(x+h+5)(x+5)} \cdot \dfrac{1}{h}$

$= \lim_{h \to 0} \dfrac{-2}{(x+h+5)(x+5)}$

$= -\dfrac{2}{(x+5)^2}$

Exercises for Alternate Example 5

In Exercises 19–24, find the derivatives of each of the following functions.

19. Find $f'(x)$ if $f(x) = -2x + 7$.

20. Find $f'(x)$ if $f(x) = \dfrac{1}{3}x^2 - 9$.

21. Find $f'(x)$ if $f(x) = \sqrt{x-7}$.

22. Find $\dfrac{dy}{dx}$ if $y = \dfrac{3}{x-1}$.

23. Find $\dfrac{dy}{dx}$ if $y = 6\pi$.

24. Find $\dfrac{dy}{dx}$ if $y = -5x^3 + 7x$.

10.2 Limits and Motion: The Area Problem

Alternate Example 2 **Computing Distance Traveled**

An automobile travels at an *average* rate of 54 miles per hour for 3 hours and 30 minutes. How far does the automobile travel?

SOLUTION

The distance traveled is Δs, the time interval has length Δt, and $\Delta s/\Delta t$ is the average velocity.

Therefore, $\Delta s = \dfrac{\Delta s}{\Delta t} \cdot \Delta t = (54 \text{ mph})(3.5 \text{ hr}) = 189$ miles.

Exercises for Alternate Example 2

In Exercises 1–6, find the distance traveled during the given time interval. Include appropriate units.

1. Average rate of 420 miles per hour for 2 hours and 15 minutes
2. Average rate of 9 yards per second for 4.5 seconds
3. Average rate of 80 kilometers per hour for 3 hours
4. Average rate of 20 meters per minute for 15 minutes and 30 seconds
5. Average rate of 8 inches per second for 2.25 seconds
6. Average rate of 30 centimeters per minute for $6\frac{1}{3}$ minutes ■

Alternate Example 3 **Approximating an Area with Rectangles**

Use the six rectangles in Figure 10.2 to approximate the area of the region below the graph of $f(x) = x^3$ over the interval [2, 5].

SOLUTION

The width of the interval is $5 - 2 = 3$. Since we have 6 subintervals, the width of the base of each approximating rectangle is $3/6 = 1/2$. The height is determined by the function value at the right-hand endpoint of each subinterval. The areas of the six rectangles and the total area are computed in the table below:

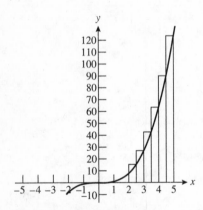

Figure 10.2 The area under the graph of $f(x) = x^3$ is approximated by six rectangles, each with base 1/2. The height of each rectangle is the function value at the right-hand endpoint of the subinterval.

Subinterval	Base of Rectangle	Height of Rectangle	Area of Rectangle
[2, 5/2]	1/2	$f(5/2) = (5/2)^3 = 125/8$	$(1/2)(125/8) = 7.8125$
[5/2, 3]	1/2	$f(3) = (3)^3 = 27$	$(1/2)(27) = 13.5$
[3, 7/2]	1/2	$f(7/2) = (7/2)^3 = 343/8$	$(1/2)(343/8) = 21.4375$
[7/2, 4]	1/2	$f(4) = (4)^3 = 64$	$(1/2)(64) = 32$
[4, 9/2]	1/2	$f(9/2) = (9/2)^3 = 729/8$	$(1/2)(729/8) = 45.5625$
[9/2, 5]	1/2	$f(5) = (5)^3 = 125$	$(1/2)(125) = 62.5$
		Total Area:	182.8125

The six rectangles give an approximation of 182.8125 square units for the area under the curve from 2 to 5.

Exercises for Alternate Example 3

In Exercises 7–9, use the rectangles shown to approximate the area of the region below the graph in the indicated interval.

7. Use 6 rectangles to approximate the area of the region below the graph of $f(x) = \sqrt[3]{x}$ over the interval $[2, 8]$.

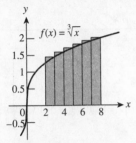

8. Use 6 rectangles to approximate the area of the region below the graph of $f(x) = \cos(x/3)$ over the interval $[0, \pi]$.

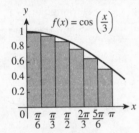

9. Use 4 rectangles to approximate the area of the region below the graph of $f(x) = \sin x$ over the interval $[0, 2]$.

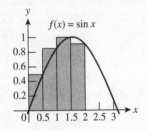

In Exercises 10–12, use the number of rectangles given to approximate the area of the region below the graph of the function over the given interval.

10. Use 4 rectangles to approximate the area of the region below the graph of $f(x) = x^4$ over the interval $[0, 2]$.

11. Use 4 rectangles to approximate the area of the region below the graph of $f(x) = \sqrt{x}$ over the interval $[0, 4]$.

12. Use 4 rectangles to approximate the area of the region below the graph of $f(x) = 2x^2 + 3$ over the interval $[1, 2]$.

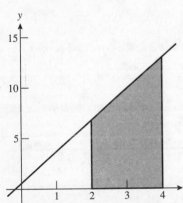

Figure 10.3 The area of the trapezoid equals $\displaystyle\int_2^4 (3x+1)\,dx$.

Alternate Example 4   **Computing an Integral**

Find $\displaystyle\int_2^4 3x+1\,dx$.

SOLUTION

This will be the area under the line $y = 3x + 1$ over the interval $[2, 4]$. The graph in Figure 10.3 shows that this is the area of a trapezoid.

Using the formula $A = h\left(\dfrac{b_1 + b_2}{2}\right)$, we find that

$$\int_2^4 3x+1\,dx = 2\left(\frac{[3(2)+1]+[3(4)+1]}{2}\right) = 2\left(\frac{20}{2}\right) = 20$$

Exercises for Alternate Example 4

In Exercises 13–18, find the indicated definite integrals.

13. Find $\int_{-2}^{3} 5\,dx$.

14. Find $\int_{-2}^{6} \left(\frac{1}{2}x + 2\right) dx$.

15. Find $\int_{0}^{10} 2x\,dx$.

16. Find $\int_{0}^{2} (-2x + 4)\,dx$.

17. Find $\int_{-3}^{0} (-3x + 9)\,dx$.

18. Find $\int_{-5}^{-1} (2x + 11)\,dx$.

Alternate Example 5   **Computing an Integral**

Suppose a ball rolls down a ramp so that its velocity after t seconds is always $5t$ feet per second. How far does it fall during the first 4 seconds?

SOLUTION

The distance traveled will be the same as the area under the velocity graph, $v(t) = 5t$, over the interval $[0, 4]$. The graph is shown in Figure 10.4. Since the region is triangular, we can find its area: $A = (1/2)(4)(20) = 40$. The distance traveled in the first 4 seconds, therefore, is $\Delta s = (1/2)(4\,\text{sec})(20\,\text{feet/sec}) = 40$ feet.

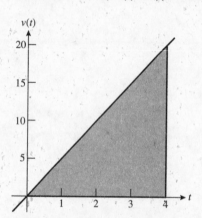

Figure 10.4 The area under the velocity graph $v(t) = 5t$ over the interval $[0, 4]$ is the distance traveled by the ball in Alternate Example 5 during the first 4 seconds.

Exercises for Alternate Example 5

In Exercises 19–24, suppose a ball rolls down a ramp so that its velocity is given by $v(t)$. How far does it fall during the given time period?

19. $v(t) = 3$ meters per second for 10 seconds

20. $v(t) = 3t$ feet per second for 8 seconds

21. $v(t) = 3t + 3$ feet per second for 4 seconds

22. $v(t) = -2t + 10$ meters per second for 3 seconds

23. $v(t) = \sqrt{36 - t^2}$ meters per second for 6 seconds

24. $v(t) = \sqrt{16 - t^2}$ meters per second for 4 seconds

10.3   More on Limits

Alternate Example 3   **Finding Limits by Substitution**

Find each of the following limits:

(a) $\lim_{x \to 0} \dfrac{\sin 3x + 2x}{5}$

(b) $\lim_{x \to 1/2} \dfrac{\ln 2x + \cos \pi x}{x - \sin \pi x}$

SOLUTION

You might not recognize these functions as being continuous, but you can use the limit properties to write the limits in terms of limits of basic functions.

(a) $\displaystyle\lim_{x\to 0}\frac{\sin 3x + 2x}{5} = \frac{\displaystyle\lim_{x\to 0}(\sin 3x + 2x)}{\displaystyle\lim_{x\to 0}(5)}$ Quotient Rule

$\displaystyle = \frac{\displaystyle\lim_{x\to 0}(\sin 3x) + \lim_{x\to 0}(2x)}{\displaystyle\lim_{x\to 0}(5)}$ Sum Rule

$\displaystyle = \frac{\sin(3(0)) + 2(0)}{5}$ Limits of continuous functions

$\displaystyle = \frac{0+0}{5}$

$= 0$

(b) $\displaystyle\lim_{x\to 1/2}\frac{\ln 2x + \cos \pi x}{x - \sin \pi x} = \frac{\displaystyle\lim_{x\to 1/2}(\ln 2x + \cos \pi x)}{\displaystyle\lim_{x\to 1/2}(x - \sin \pi x)}$ Quotient Rule

$\displaystyle = \frac{\displaystyle\lim_{x\to 1/2}(\ln 2x) + \lim_{x\to 1/2}(\cos \pi x)}{\displaystyle\lim_{x\to 1/2}(x) - \lim_{x\to 1/2}(\sin \pi x)}$ Sum and Difference Rules

$\displaystyle = \frac{\ln\left(2\left(\frac{1}{2}\right)\right) + \cos\left(\pi\left(\frac{1}{2}\right)\right)}{\frac{1}{2} - \sin\left(\pi\left(\frac{1}{2}\right)\right)}$ Limits of continuous functions

$\displaystyle = \frac{\ln 1 + \cos\left(\frac{\pi}{2}\right)}{\frac{1}{2} - \sin\left(\frac{\pi}{2}\right)}$

$\displaystyle = \frac{0+0}{\frac{1}{2}-1}$

$= 0$

Exercises for Alternate Example 3

In Exercises 1–6, find the indicated limit.

1. $\displaystyle\lim_{n\to 0}\frac{e^n - \ln(n+1)}{\sqrt{n+9}}$

2. $\displaystyle\lim_{t\to 10}\frac{(\sqrt{4t})(t^2 - 5t)}{6t}$

3. $\displaystyle\lim_{m\to -2}\frac{(m^2 + 4m)}{\sqrt{m+3}}$

4. $\displaystyle\lim_{x\to 2}\frac{\ln(x^2) - x^3}{3x}$

5. $\displaystyle\lim_{x\to 3}\frac{\sin(\pi x) + \sqrt{6x}}{2x^2}$

6. $\displaystyle\lim_{x\to -\pi}\frac{\cos x + \pi x^2}{(x/3)}$

Alternate Example 4 **Finding Left- and Right-Hand Limits**

Find $\lim_{x \to 1^-} f(x)$ and $\lim_{x \to 1^+} f(x)$ where $f(x) = \begin{cases} x^3 - x + 1 & \text{if } x \le 1 \\ 5x + 1 & \text{if } x > 1 \end{cases}$

SOLUTION

Figure 10.5 suggests that the left- and right-hand limits of f exist but are not equal. Using algebra we find:

$$\lim_{x \to 1^-} f(x) = \lim_{x \to 1^-} (x^3 - x + 1) \quad \text{Definition of } f$$

$$= 1^3 - 1 + 1$$

$$= 1$$

$$\lim_{x \to 1^+} f(x) = \lim_{x \to 1^+} (5x + 1) \quad \text{Definition of } f$$

$$= 5 \cdot 1 + 1$$

$$= 6$$

You can use trace or tables to support the above results.

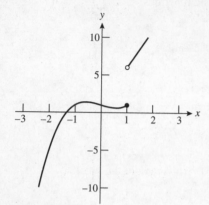

Figure 10.5 A graph of the piecewise-defined function

$$f(x) = \begin{cases} x^3 - x + 1 & \text{if } x \le 1 \\ 5x + 1 & \text{if } x > 1 \end{cases}$$

Exercises for Alternate Example 4

In Exercises 7–12, find the left- and right-hand limits for each of the following piecewise-defined functions.

7. Find $\lim_{x \to 4^-} f(x)$ and $\lim_{x \to 4^+} f(x)$ where

$$f(x) = \begin{cases} x^3 - x^2 & \text{if } x \le 4 \\ 2x^2 + 1 & \text{if } x > 4 \end{cases}$$

8. Find $\lim_{x \to 0^-} f(x)$ and $\lim_{x \to 0^+} f(x)$ where

$$f(x) = \begin{cases} \cos(4x) & \text{if } x \le 0 \\ 3\sin(2x) & \text{if } x > 0 \end{cases}$$

9. Find $\lim_{x \to e^-} f(x)$ and $\lim_{x \to e^+} f(x)$ where

$$f(x) = \begin{cases} \ln x & \text{if } x \le e \\ 5x - 4e & \text{if } x > e \end{cases}$$

10. Find $\lim_{x \to 1^-} f(x)$ and $\lim_{x \to 1^+} f(x)$ where

$$f(x) = \begin{cases} e^x - \ln x & \text{if } x \le 1 \\ e & \text{if } x > 1 \end{cases}$$

11. Find $\lim_{x \to \pi^-} f(x)$ and $\lim_{x \to \pi^+} f(x)$ where

$$f(x) = \begin{cases} \sin x + \cos x & \text{if } x \le \pi \\ \sin 2x + \cos 2x & \text{if } x > \pi \end{cases}$$

12. Find $\lim_{x \to \sqrt{5}^-} f(x)$ and $\lim_{x \to \sqrt{5}^+} f(x)$ where

$$f(x) = \begin{cases} x^4 - 10 & \text{if } x \le \sqrt{5} \\ 5x^2 & \text{if } x > \sqrt{5} \end{cases}$$

Alternate Example 8 **Using Tables to Investigate Limits as**
$$\lim_{x \to \infty} f(x) \text{ and } \lim_{x \to -\infty} f(x)$$

Let $f(x) = x^2 e^x$. Find $\lim_{x \to \infty} f(x)$ and $\lim_{x \to -\infty} f(x)$.

SOLUTION

The tables in Figure 10.6 suggest that $\lim_{x \to \infty} f(x) = \infty$ and $\lim_{x \to -\infty} f(x) = 0$.

The graph of f in Figure 10.7 supports these results.

X	Y1	
0	0	
10	2.2E6	
20	1.9E11	
30	9.6E15	
40	3.8E20	
50	1.3E25	
60	4.1E29	
Y1=X²e^(X)		

(a)

X	Y1	
0	0	
-10	.00454	
-20	8.2E-7	
-30	8E-11	
-40	7E-15	
-50	5E-19	
-60	3E-23	
Y1≡X²e^(X)		

(b)

Figure 10.6 The table in (a) suggests that the values of $f(x) = x^2 e^x$ approach ∞ as $x \to \infty$ and the table in (b) suggests that the values of $f(x) = x^2 e^x$ approach 0 as $x \to -\infty$.

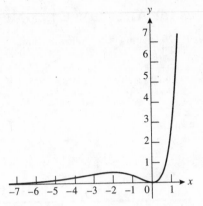

Figure 10.7 The graph of the function $f(x) = x^2 e^x$.

Exercises for Alternate Example 8

In Exercises 13–18, use tables to investigate the limits of the functions as $x \to \infty$ and $x \to -\infty$.

13. Let $f(x) = \dfrac{\ln|x|}{x}$. Find $\lim\limits_{x \to \infty} f(x)$ and $\lim\limits_{x \to -\infty} f(x)$.

14. Let $f(x) = \dfrac{e^{4x}}{\ln|x|}$. Find $\lim\limits_{x \to \infty} f(x)$ and $\lim\limits_{x \to -\infty} f(x)$.

15. Let $f(x) = \dfrac{x}{e^{3x}}$. Find $\lim\limits_{x \to \infty} f(x)$ and $\lim\limits_{x \to -\infty} f(x)$.

16. Let $f(x) = 5x^2 e^{-x} + 4$. Find $\lim\limits_{x \to \infty} f(x)$ and $\lim\limits_{x \to -\infty} f(x)$.

17. Let $f(x) = \dfrac{e^{2x}}{4x^2} + 1$. Find $\lim\limits_{x \to \infty} f(x)$ and $\lim\limits_{x \to -\infty} f(x)$.

18. Let $f(x) = \dfrac{5}{3 + e^{-5x}}$. Find $\lim\limits_{x \to \infty} f(x)$ and $\lim\limits_{x \to -\infty} f(x)$.

Alternate Example 9 **Investigating Unbounded Limits**

Find $\lim\limits_{x \to -5} \dfrac{3}{(x+5)^2}$.

SOLUTION

The graph of $f(x) = \dfrac{3}{(x+5)^2}$ in Figure 10.8 suggests that

$$\lim_{x \to -5^+} \frac{3}{(x+5)^2} = \infty \quad \text{and} \quad \lim_{x \to -5^-} \frac{3}{(x+5)^2} = \infty$$

This means that the limit of f as x approaches -5 does not exist. The table of values in Figure 10.9 agrees with this conclusion. The graph of f has a vertical asymptote at $x = -5$.

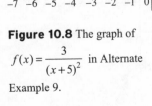

Figure 10.8 The graph of $f(x) = \dfrac{3}{(x+5)^2}$ in Alternate Example 9.

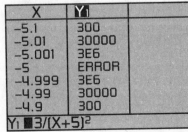

X	Y1	
-5.1	300	
-5.01	30000	
-5.001	3E6	
-5	ERROR	
-4.999	3E6	
-4.99	30000	
-4.9	300	
Y1 ∎3/(X+5)²		

Figure 10.9 A table of values for $f(x) = \dfrac{3}{(x+5)^2}$.

Exercises for Alternate Example 9

In Exercises 19–24, investigate the following limits.

19. $\displaystyle\lim_{x\to-1/4}\frac{-7}{(4x+1)^2}$ **20.** $\displaystyle\lim_{x\to0}\ \ln|x|$ **21.** $\displaystyle\lim_{x\to1}\frac{5}{(x-1)^3}$ **22.** $\displaystyle\lim_{x\to0}\frac{3}{\sin x}$ **23.** $\displaystyle\lim_{x\to0}\frac{-2}{(\cos x)-1}$

24. $\displaystyle\lim_{x\to0}\frac{e^x}{\sin 2x}$

10.4 Numerical Derivatives and Integrals

Alternate Example 1 Computing a Numerical Derivative

Let $f(x)=2x^3$. Compute NDER $f(1)$ by calculating the symmetric difference quotient with $h=0.001$. Compare it to the actual value of $f'(1)$. (See Figure 10.10)

SOLUTION

$$\text{NDER } f(1) = \frac{f(1+0.001)-f(1-0.001)}{2(0.001)} \qquad \text{Definition of numerical derivative}$$

$$= \frac{f(1.001)-f(0.999)}{0.002}$$

$$= 6.000002$$

The actual value is

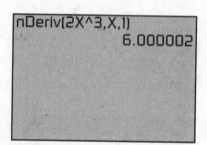

$$f'(1) = \lim_{h\to0}\frac{f(1+h)-f(1)}{h} \qquad \text{Definition of Derivative}$$

$$= \lim_{h\to0}\frac{2(1+h)^3-2(1)^3}{h}$$

$$= \lim_{h\to0}\frac{2(1+3h+3h^2+h^3)-2}{h}$$

$$= \lim_{h\to0}\frac{h(6+6h+2h^2)}{h}$$

$$= \lim_{h\to0}\ 6+6h+2h^2$$

$$= 6$$

Figure 10.10 Applying the numerical derivative command on a graphing calculator.

Exercises for Alternate Example 1

In Exercises 1–6, find the numerical derivatives by calculating the symmetric difference quotient with $h=0.001$. Compare it to the actual value of the derivative at the point.

1. Find $f'(3)$ when $f(x)=4x^3$. **2.** Find $f'(4)$ when $f(x)=\sqrt{x}$. **3.** Find $f'(2)$ when $f(x)=\dfrac{6}{x+2}$.

4. Find $f'(0)$ when $f(x)=x^2+3x+5$. **5.** Find $f'(1/2)$ when $f(x)=2x^3$.

6. Find $f'(\pi)$ when $f(x)=\sin x$. (Hint: Recall the formula: $\sin(\alpha+\beta)=\sin\alpha\cos\beta+\sin\beta\cos\alpha$)

Alternate Example 2 **Finding a Numerical Integral**

Use NINT to find the area of the region R enclosed between the x-axis and the graph of $y = \ln x$ from $x = 2$ to $x = 6$.

SOLUTION

The region is shown in Figure 10.11

The area can be written as the definite integral $\int_2^6 \ln x \, dx$, which we find on a graphing calculator: NINT ($\ln x$, x, 2, 6) = 5.364262454. The exact answer (as you will learn in a calculus course) is $6 \ln 6 - 2 \ln 2 - 4$, which agrees in every displayed digit with the NINT value. Figure 10.12 shows the syntax for numerical integration and verification of this value on one type of calculator.

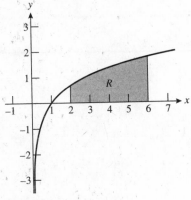

Figure 10.11 The graph of $f(x) = \ln x$ with the area under the curve between $x = 2$ and $x = 6$ shaded.

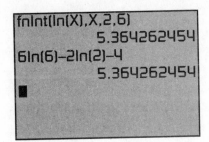

Figure 10.12 A numerical integral approximation for $\int_2^6 \ln x \, dx$.

Exercises for Alternate Example 2

In Exercises 7–12, use NINT to find the area of the region R enclosed between the x-axis and the graph of $y = f(x)$ from $x = a$ (called the lower limit of integration) to $x = b$ (called the upper limit of integration)

7. Find the area of the region R enclosed between the x-axis and the graph of $y = 2x^2$ from $x = 3$ to $x = 8$.

8. Find the area of the region R enclosed between the x-axis and the graph of $y = \cos x$ from $x = 0$ to $x = \pi/2$.

9. Find the area of the region R enclosed between the x-axis and the graph of $y = \sin x$ from $x = 0$ to $x = \pi/2$.

10. Find the area of the region R enclosed between the x-axis and the graph of $y = \ln x$ from $x = 1$ to $x = 2$.

11. Find the area of the region R enclosed between the x-axis and the graph of $y = |x|$ from $x = 0$ to $x = 3$.

12. Find the area of the region R enclosed between the x-axis and the graph of $y = \sqrt{2x}$ from $x = 6$ to $x = 9$.

Alternate Example 3 **Finding Distance Traveled**

An automobile is driven at a variable rate along a test track for 4 hours so that its velocity at any time t ($0 \le t \le 4$) is given by $v(t) = 20 + 5 \cos 3t$ miles per hour. How far does the automobile travel during the 4-hour test?

SOLUTION

According to the analysis found in Section 10.2, the distance traveled is given by

$\int_0^4 (20 + 5 \cos 3t) \, dt$. We use a calculator to find the numerical integral:

NINT($20 + 5 \cos 3t$, t, 0, 4) ≈ 79.11

Interpreting the answer, we conclude that the automobile travels 79.11 miles.

Exercises for Alternate Example 3

In Exercises 13–18, given a velocity function $v(t)$, find the total distance traveled over the given time interval by using NINT to evaluate the appropriate definite integral.

13. $v(t) = 2t + 5$ miles per hour; from $t = 1$ hour to $t = 5$ hours

14. $v(t) = 3t^2 + 7$ miles per hour; from $t = 2$ hour to $t = 4.5$ hours

15. $v(t) = 410t - 110$ miles per hour; from $t = 0$ hour to $t = 3$ hours

16. $v(t) = \sqrt{32t^3 - 6}$ miles per hour; from $t = 1$ hour to $t = 5$ hours

17. $v(t) = 15\cos t + t^2$ miles per hour; from $t = 3$ hour to $t = 7$ hours

18. $v(t) = \dfrac{60}{t + 7}$ miles per hour; from $t = 5$ hour to $t = 9$ hours ■

Answers to Exercises

Chapter P
Prerequisites

P.1 Real Numbers

1. $[-2, 3$; endpoints: -2 and 3; bounded and closed

2. $(-\infty, 0)$; endpoint: 0; unbounded and open

3. $[-4, 4]$; endpoints: -4 and 4; bounded and closed

4. $(2, 5)$; endpoints: 2 and 5; bounded and open

5. $(-3, 5)$; endpoints: -3 and 5; bounded and open

6. $x \geq 3$; endpoint: 3; unbounded and closed

7. $bx^2 - 3b$

8. $6x - 2xc$

9. $2x^2n + 5n$

10. $12r + 6rn$

11. $3n(1 + 2m)$

12. $7c(2 + b)$

13. $g(15h + 7z)$

14. $n^3(p + 2t)$

15. $28x^5y^7$

16. $-15a^7b^{10}$

17. $\dfrac{1}{cd^5}$

18. $\dfrac{x}{y^2}$

19. $\dfrac{1}{64k^6}$

20. $9h^4$

21. 2×10^2

22. 3.5×10^2

23. 4.8×10^1

24. 1.08×10^{-2}

25. 9×10^2

26. 1.5×10^{-7}

P.2 Cartesian Coordinate System

1. $x^2 + (y - 4)^2 = 16$

2. $(x - 1)^2 + (y - 1)^2 = 100$

3. $x^2 + y^2 = 1$

4. $(x + 4)^2 + y^2 = 6.25$

5. $\left(x - \dfrac{1}{2}\right)^2 + \left(y - \dfrac{1}{2}\right)^2 = 16$

6. $(x + 3)^2 + (y + 3)^2 = 64$

7. $a = \sqrt{(1 - 0)^2 + (0 - 0)^2} = 1$

$b = \sqrt{(0 - 0)^2 + (1 - 0)^2} = 1$

$c = \sqrt{(1 - 0)^2 + (0 - 1)^2} = \sqrt{2}$

$a^2 + b^2 = 1 + 1 = 2 = \left(\sqrt{2}\right)^2 = c^2$

8. $a = \sqrt{(13 - 1)^2 + (1 - 1)^2} = 12$

$b = \sqrt{(1 - 1)^2 + (6 - 1)^2} = 5$

$c = \sqrt{(13 - 1)^2 + (1 - 6)^2} = 13$

$a^2 + b^2 = 144 + 25 = 169 = 13^2 = c^2$

9. $a = \sqrt{(0 - (-4))^2 + (0 - 0)^2} = 4$

$b = \sqrt{(-4 - (-4))^2 + (0 - (-3))^2} = 3$

$c = \sqrt{(0 - (-4))^2 + (0 - (-3))^2} = 5$

$a^2 + b^2 = 16 + 9 = 25 = 5^2 = c^2$

10. $a = \sqrt{(-3 - 2)^2 + (4 - 4)^2} = 5$

$b = \sqrt{(2 - 2)^2 + (10 - 4)^2} = 6$

$c = \sqrt{(2 - (-3))^2 + (10 - 4)^2} = \sqrt{61}$

$a^2 + b^2 = 25 + 36 = 61 = \left(\sqrt{61}\right)^2 = c^2$

11. $a = \sqrt{(0-(-1))^2 + (0-2)^2} = \sqrt{5}$

$b = \sqrt{(0-6)^2 + (0-3)^2} = \sqrt{45}$

$c = \sqrt{(-1-6)^2 + (2-3)^2} = \sqrt{50}$

$a^2 + b^2 = 5 + 45 = 50 = \left(\sqrt{50}\right)^2 = c^2$

12. $a = \sqrt{(6-4)^2 + (2-0)^2} = \sqrt{8}$

$b = \sqrt{(6-4)^2 + (-2-0)^2} = \sqrt{8}$

$c = \sqrt{(6-6)^2 + (2-(-2))^2} = 4$

$a^2 + b^2 = \left(\sqrt{8}\right)^2 + \left(\sqrt{8}\right)^2$

$\quad = 16 = 4^2 = c^2$

13. The square is shown below.

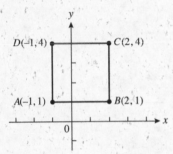

The midpoint of segment AC is $\left(\dfrac{-1+2}{2}, \dfrac{1+4}{2}\right)$, or

$\left(\dfrac{1}{2}, \dfrac{5}{2}\right)$. The midpoint of segment BD is

$\left(\dfrac{-1+2}{2}, \dfrac{1+4}{2}\right)$, or $\left(\dfrac{1}{2}, \dfrac{5}{2}\right)$.

14. The rectangle is shown below.

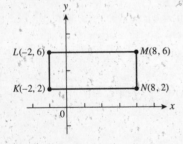

The midpoint of segment KM is $\left(\dfrac{-2+8}{2}, \dfrac{2+6}{2}\right)$, or

$(3, 4)$. The midpoint of segment LN is $\left(\dfrac{-2+8}{2}, \dfrac{6+2}{2}\right)$,

or $(3, 4)$.

15. The parallelogram is shown below.

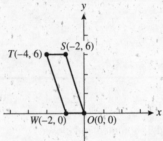

The midpoint of segment OT is $\left(\dfrac{0+(-4)}{2}, \dfrac{0+6}{2}\right)$, or

$(-2, 3)$. The midpoint of segment WS is

$\left(\dfrac{-2+(-2)}{2}, \dfrac{0+6}{2}\right)$, or $(-2, 3)$.

16. The parallelogram is shown below.

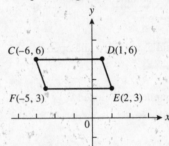

The midpoint of segment EC is $\left(\dfrac{2+(-6)}{2}, \dfrac{3+6}{2}\right)$,

or $(-2, 4.5)$. The midpoint of segment FD is

$\left(\dfrac{-5+1}{2}, \dfrac{3+6}{2}\right)$, or $(-2, 4.5)$.

17. The rectangle is shown below.

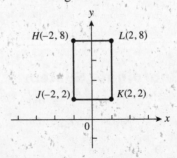

The midpoint of segment JL is $\left(\dfrac{-2+2}{2}, \dfrac{2+8}{2} \right)$,

or $(0, 5)$. The midpoint of segment HK is

$\left(\dfrac{-2+2}{2}, \dfrac{8+2}{2} \right)$, or $(0, 5)$.

18. The square is shown below.

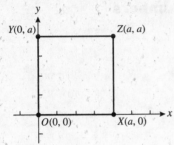

The midpoint of segment OZ is $\left(\dfrac{0+a}{2}, \dfrac{0+a}{2} \right)$,

or $\left(\dfrac{a}{2}, \dfrac{a}{2} \right)$. The midpoint of segment XY is

$\left(\dfrac{0+a}{2}, \dfrac{0+a}{2} \right)$, or $\left(\dfrac{a}{2}, \dfrac{a}{2} \right)$.

P.3 Linear Equations and Inequalities

1. $x = 9$ **2.** $x = -5$

3. $x = -6$ **4.** $x = 0$

5. $x = -9$ **6.** $x = 4$

7. $x = 2$ **8.** $x = 1.5$

9. $x = 3$ **10.** $x = 2.5$

11. $x = -0.5$ **12.** $x = 2.4$

13. $x < 1$ **14.** $x \leq -2$

15. $x > 2$ **16.** $x \leq 3$

17. $x \geq 5$ **18.** $x > -5$

19. $-2 < x < 6$

20. $2 \leq x \leq 5$

21. $-4 < x \leq 2$

22. $0 \leq x < 7$

23. $3 < x \leq 6$

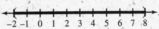

24. $-2 < x < 8$

P.4 Lines in the Plane

1. $y - 6 = 2(x - 4)$ **2.** $y + 2 = 2x$

3. $y - 4 = 1.5(x - 4)$ **4.** $y = 5(x - 6)$

5. $y + 2 = -(x + 3)$ **6.** $y + 2 = -10(x - 5)$

7.

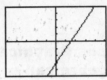

$[-5.1, 5.1]$ by $[-4.7, 4.7]$

8.

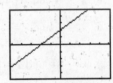

$[-5.1, 5.1]$ by $[-4.7, 4.7]$

9.

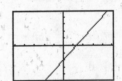

$[-5.1, 5.1]$ by $[-4.7, 4.7]$

10.

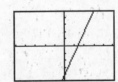

$[-5.1, 5.1]$ by $[-4.7, 4.7]$

11.

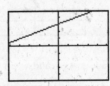

$[-5.1, 5.1]$ by $[-4.7, 4.7]$

12.

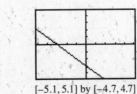

$[-5.1, 5.1]$ by $[-4.7, 4.7]$

13. $y = \frac{3}{2}x + 1$ **14.** $y = x - 3$

15. $y = \frac{1}{3}x + 1$ **16.** $y = -\frac{3}{4}x - \frac{11}{4}$

17. $y = 2x + 1$ **18.** $y = \frac{1}{2}x - \frac{11}{2}$

19. $y = \frac{2}{3}x - 2$ **20.** $y = -2x + 5$

21. $y = \frac{2}{5}x$ **22.** $y = 4x - 12$

23. $y = -x + 4$ **24.** $y = -2x + 17$

P.5 Solving Equations Graphically, Numerically, and Algebraically

1. $x = -2$ and $x = 2$ **2.** $x = 0$ and $x = 4$

3. $x = -2$ and $x = 3$ **4.** $x = -1$ and $x = 1.5$

5. $x = \frac{1}{2}$ and $x = 3$ **6.** $x = -3$ and $x = -\frac{4}{5}$

7. $x = 0$ and $x = \frac{5}{2}$ **8.** $x = -\frac{2}{3}$ and $x = 1$

9. $x = \frac{7}{6} \pm \frac{\sqrt{37}}{6}$ **10.** $x = \frac{3}{4} \pm \frac{\sqrt{21}}{4}$

11. $x = 3 \pm \sqrt{14}$ **12.** $x = -\frac{2}{5}$ and $x = \frac{1}{2}$

13. -1.52138 **14.** 0

15. 1.3788 **16.** 0.165906

17. 1.89329 **18.** -0.32748

19. 1.77 **20.** -1.34

21. $0.59, 3.41$ **22.** $-8.47, 0.47$

23. $-3.87, -0.79, 0.66$ **24.** 4.12

25. $x = -1$ and $x = 5$ **26.** $x = -3$ and $x = 3$

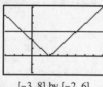
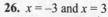

$[-3, 8]$ by $[-2, 6]$ 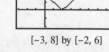 $[-5, 5]$ by $[-2, 8]$

27. $x = -3$ **28.** $x = -1$ and $x = 3$

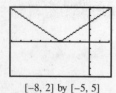   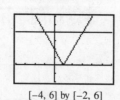

$[-8, 2]$ by $[-5, 5]$ $[-4, 6]$ by $[-2, 6]$

29. $x = 0.25$ **30.** $x = 0$

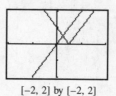   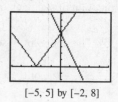

$[-2, 2]$ by $[-2, 2]$ $[-5, 5]$ by $[-2, 8]$

P.6 Complex Numbers

1. $-12 - 2i$ **2.** $-9 + 3i$

3. $9 + 9i$ **4.** $17 + 6i$

5. $-7 + i$ **6.** $-29 + 6i$

7. $13 + 25i$ **8.** $-10 + 92i$

9. $-3 + 4i$ **10.** $117 + 44i$

11. $-936 - 352i$ **12.** $8 - 6i$

13. $46 - 9i$ **14.** $-16 + 30i$

15. $-\frac{3}{2}i$ **16.** $\frac{1}{2} + i$

17. $-\frac{1}{5} - \frac{8}{5}i$ **18.** $\frac{4}{5} - \frac{3}{5}i$

19. $\frac{24}{25} - \frac{7}{25}i$ **20.** 2

21. $x = -\frac{5}{2} \pm \frac{\sqrt{3}}{2}i$ **22.** $x = \frac{1}{6} \pm \frac{\sqrt{47}}{6}i$

23. $x = \frac{3}{8} \pm \frac{\sqrt{71}}{8}i$ **24.** $x = \pm \frac{\sqrt{3}}{2}i$

25. $x = \frac{1}{2} \pm \frac{\sqrt{51}}{2}i$ **26.** $x = \frac{5}{2} \pm \frac{1}{2}i$

27. $x = -\frac{3}{2} \pm \frac{\sqrt{3}}{2}i$ **28.** $x = \pm 2i$

P.7 Solving Inequalities Algebraically and Graphically

1. $(-\infty, -1) \cup (1, \infty)$ **2.** $(-\infty, -4] \cup [0, \infty)$

3. $[-1, 5]$ **4.** $(-\infty, 0) \cup (6, \infty)$

5. $[2, 3]$ **6.** $\left(-\infty, -1\right] \cup \left[\frac{1}{3}, \infty\right)$

7. $(-\infty, -3] \cup [3, \infty)$ **8.** $[1, 4]$

9. $(-\infty, -4] \cup [1, \infty)$ **10.** $(-\infty, 3) \cup (3, \infty)$

11. $(-\infty, -4] \cup [-3, \infty)$ **12.** $\left[-3, -\frac{1}{2}\right]$

13. $[1.21, \infty)$

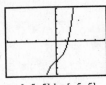

[−5, 5] by [−5, 5]

14. $(-1.73, 0) \cup (1.73, \infty)$

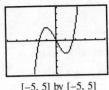

[−5, 5] by [−5, 5]

15. $(-4.79, -0.21) \cup (0, \infty)$

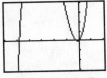

[−6, 2] by [−3, 4]

16. $(-\infty, -0.62] \cup [0, 1.6]$

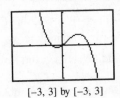

[−3, 3] by [−3, 3]

17. $(-\infty, -0.45) \cup (0, 4.45)$

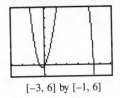

[−3, 6] by [−1, 6]

18. $[-0.83, 1] \cup [4.83, \infty)$

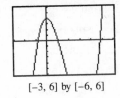

[−3, 6] by [−6, 6]

Chapter 1
Functions and Graphs

1.1 Modeling and Equation Solving

1. square pizza 10 inches on a side
2. square pizza 10 inches on a side
3. rectangular pizza 12 inches long and 7 inches wide
4. circular pizza with an 9-inch diameter
5. square pizza 15 inches on a side
6. rectangular pizza 10 inches long and 8 inches wide
7. $x = 0$, $x = 1$, and $x = 3$
8. $x = -4$, $x = 0$, and $x = 2$
9. $x = 0$ and $x = 4$
10. $x = 0$, $x = 2.5$, and $x = 4$
11. $x = 0$, and $x = 2.5$
12. $x = 0$, $x = \dfrac{1}{3}$, and $x = \dfrac{5}{2}$
13. $x = 2 + \sqrt{2} \approx 3.41421$
    $x = 2 - \sqrt{2} \approx 0.585786$
14. $x = \dfrac{5 + \sqrt{17}}{2} \approx 4.56155$
    $x = \dfrac{5 - \sqrt{17}}{2} \approx 0.438447$

15. $x = \dfrac{1 + \sqrt{29}}{2} \approx 3.19258$
    $x = \dfrac{1 - \sqrt{29}}{2} \approx -2.19258$
16. $x = \dfrac{-3 + \sqrt{13}}{2} \approx 0.302776$
    $x = \dfrac{-3 - \sqrt{13}}{2} \approx -3.30278$
17. $x = -2 + \sqrt{10} \approx 1.16228$
    $x = -2 - \sqrt{10} \approx -5.16228$
18. $x = 2 + \sqrt{11} \approx 5.31662$
    $x = 2 - \sqrt{11} \approx -1.31662$
19. $100   **20.** 175 miles   **21.** 325
22. 170 gallons   **23.** 3.5 minutes
24. 65 minutes, 1 hour 5 minutes

1.2 Functions and Their Properties

1. $x \geq 1$
2. $x \geq 0$
3. $s > 0$
4. $x > 0$
5. $x \geq \dfrac{1}{2}$
6. $r > 0$
7. $x > 0$
8. $x \geq 0$
9. decreasing on $(-\infty, -2]$ and increasing on $[-2, \infty)$
10. increasing on $(-\infty, -3]$ and decreasing on $[-3, \infty)$
11. increasing on $(-\infty, -4)$, increasing again on $(-4, 0]$, decreasing on $[0, 4)$, and decreasing again on $(4, \infty)$
12. increasing on $(-\infty, -5)$, increasing again on $(-5, 0]$, decreasing on $[0, 5)$, and decreasing again on $(5, \infty)$
13. decreasing on $(-\infty, 2]$ and increasing on $[2, \infty)$
14. increasing on $(-\infty, 0]$, decreasing on $[0, 2]$, increasing again on $[2, \infty)$
15. even   **16.** neither   **17.** even
18. even   **19.** neither   **20.** odd
21. vertical asymptotes: $x = -2$ and $x = 2$; horizontal asymptote: $y = 0$
22. vertical asymptote: $x = 5$; horizontal asymptote: $y = 0$
23. vertical asymptote: $x = 5$; horizontal asymptote: $y = 0$
24. vertical asymptote: $x = 2$; horizontal asymptote: $y = 0$
25. vertical asymptote: none; horizontal asymptote: $y = 0$
26. vertical asymptote: $x = -1$ and $x = 4$; horizontal asymptote: $y = 2$

1.3 Twelve Basic Functions

1. increasing on $(-\infty, 0]$; decreasing on $[0, \infty)$; even function; $(0, 0)$ is a local maximum; reflection of the graph of $y = x^2$ in the x-axis

2. increasing on $[-1, \infty)$; decreasing on $(-\infty, -1]$; neither even nor odd; $(-1, 0)$ is a local minimum; translation of the graph of $y = x^2$ one unit to the left

3. increasing on $(-\infty, 3]$; decreasing on $[3, \infty)$; neither even nor odd; $(3, 0)$ is a local maximum; translation of the graph of $y = x^2$ three units to the right and reflection of the graph of $y = (x - 3)^2$ in the x-axis

4. increasing on $[0, \infty)$; decreasing on $(-\infty, 0]$; even function; $(0, 1)$ is a local minimum; translation of the graph of $y = x^2$ one unit up

5. increasing on $[0, \infty)$; decreasing on $(-\infty, 0]$; even function; $(0, -1)$ is a local minimum; translation of the graph of $y = x^2$ one unit down

6. increasing on $[2, \infty)$; decreasing on $(-\infty, 2]$; neither even nor odd; $(2, 0)$ is a local minimum; translation of the graph of $y = x^2$ two units to the right

7. $f(x) = \begin{cases} |x| & \text{if } x \leq 2 \\ x^2 & \text{if } x \geq 2 \end{cases}$

8. $f(x) = \begin{cases} |x| & \text{if } x \leq 0 \\ \sqrt{x} & \text{if } x \geq 0 \end{cases}$ or $f(x) = \begin{cases} -x & \text{if } x \leq 0 \\ \sqrt{x} & \text{if } x > 0 \end{cases}$

9. $f(x) = \begin{cases} x & \text{if } x \leq 0 \\ \sqrt{x} & \text{if } x \geq 0 \end{cases}$

10. $f(x) = \begin{cases} x & \text{if } x \leq 0 \\ \text{int } x & \text{if } x \geq 0 \end{cases}$

11. $f(x) = \begin{cases} x & \text{if } x \leq 0 \\ x^2 & \text{if } x \geq 0 \end{cases}$

12. $f(x) = \begin{cases} \text{int } x & \text{if } x \leq 0 \\ x & \text{if } x \geq 0 \end{cases}$ or $f(x) = \begin{cases} \text{int } x & \text{if } x \leq 0 \\ |x| & \text{if } x \geq 0 \end{cases}$

13. Domain: all real numbers

    Range: $(-\infty, \infty)$

    Continuous

    Increasing on $(-\infty, \infty)$

    Symmetric with respect to origin

    Not bounded below or above

    No local extrema

    No horizontal asymptotes

    No vertical asymptotes

    End behavior: $\lim\limits_{x \to \infty} x^3 = \infty$ and $\lim\limits_{x \to -\infty} x^3 = -\infty$

14. Domain: all positive numbers

    Range: $(-\infty, \infty)$

    Continuous

    Increasing on $(0, \infty)$

    Not symmetric

    Not bounded below or above

    No local extrema

    No horizontal asymptotes

    Vertical asymptote: $x = 0$

    End behavior: $\lim\limits_{x \to \infty} \ln x = \infty$ and $\lim\limits_{x \to 0} \ln x = -\infty$

15. Domain: all real numbers

    Range: $(0, \infty)$

    Continuous

    Increasing on $(-\infty, \infty)$

    Not symmetric

    Bounded below

    No local extrema

    Horizontal asymptote: $y = 0$

    Vertical asymptote: none

    End behavior: $\lim\limits_{x \to \infty} e^x = \infty$ and $\lim\limits_{x \to -\infty} e^x = 0$

16. Domain: all real numbers

    Range: all integers

    Continuous for all real numbers but integers

    Increasing on $(-\infty, \infty)$

    Not symmetric

    Not bounded below or above

    No local extrema

    Horizontal asymptote: none

    Vertical asymptote: none

    End behavior: $\lim\limits_{x \to \infty} \text{int } x = \infty$ and $\lim\limits_{x \to -\infty} \text{int } x = -\infty$

17. Domain: all real numbers but 0

    Range: all real numbers but 0

    Continuous for all real numbers but $x = 0$

    Decreasing on $(-\infty, 0)$; decreasing on $(0, \infty)$

    Symmetric with respect to the origin

    Not bounded below or above

    No local extrema

    Horizontal asymptote: $y = 0$

    Vertical asymptote: $x = 0$

    End behavior: $\lim\limits_{x \to \infty} \dfrac{1}{x} = 0$ and $\lim\limits_{x \to -\infty} \dfrac{1}{x} = 0$

18. Domain: all real numbers

Range: all nonnegative real numbers

Continuous for all real numbers

Increasing on $[0, \infty)$; decreasing on $(-\infty, 0]$

Symmetric with respect to the y-axis

Bounded below

Local minimum at $x = 0$

Horizontal asymptote: none

Vertical asymptote: none

End behavior: $\lim\limits_{x \to \infty} |x| = \infty$ and $\lim\limits_{x \to -\infty} |x| = \infty$

1.4 Building Functions from Functions

1. $(f \circ g)(x) = \dfrac{2}{x^2}$; $(g \circ f)(x) = \dfrac{1}{2x^2}$

2. $(f \circ g)(x) = 2x^2$; $(g \circ f)(x) = 4x^2$

3. $(f \circ g)(x) = \sqrt{x} + 1$; $(g \circ f)(x) = \sqrt{x + 1}$

4. $(f \circ g)(x) = \left(\sqrt{x} - 1\right)^3$; $(g \circ f)(x) = \sqrt{x^3 - 1}$

5. $(f \circ g)(x) = \left(\dfrac{1}{x + 1}\right)^2$; $(g \circ f)(x) = \dfrac{1}{x^2 + 1}$

6. $(f \circ g)(x) = 2^{x^2}$; $(g \circ f)(x) = 2^{2x}$

7. $f(x) = 2x^2$ and $g(x) = x^2 + 5$

8. $f(x) = \dfrac{1}{x + 1}$ and $g(x) = x^2$

9. $f(x) = x^3$ and $g(x) = \dfrac{1}{x^2 + 1}$

10. $f(x) = -2x^3 + x$ and $g(x) = x^2 - 1$

11. $f(x) = \dfrac{1}{x} + 3x^3$ and $g(x) = x^2 - 9$

12. $f(x) = \sqrt[3]{x} + x$ and $g(x) = x^2$

13. the parallel lines whose graphs are the graphs of
$x - y = \pm 3$

14. the perpendicular lines whose graphs are the graphs of
$y = -x$ and $y = x$

15. the intersecting lines whose graphs are the graphs of
$y = -x$ and $y = 0.25x$

16. the intersecting lines whose graphs are the graphs of
$y = \dfrac{1}{2}x$ and $y = \dfrac{1}{3}x$

17. the parallel lines whose graphs are the graphs of
$y = \dfrac{3}{2}x \pm 2$

18. the intersecting lines whose graphs are the graphs of
$y = 3x$ and $y = -x$

1.5 Parametric Relations and Inverses

1. (a)

t	(x, y)
-3	$(-6, -4)$
-2	$(-4, -3)$
-1	$(-2, -2)$
0	$(0, -1)$
1	$(2, 0)$
2	$(4, 1)$
3	$(6, 2)$

(b) $y = \dfrac{1}{2}x - 1$; yes

(c)

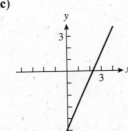

2. (a)

t	(x, y)
-3	$(-1, -7)$
-2	$(0, -5)$
-1	$(1, -3)$
0	$(2, -1)$
1	$(3, 1)$
2	$(4, 3)$
3	$(5, 5)$

(b) $y = 2x - 5$; yes

(c)

3. (a)

t	(x, y)
–3	(–4, 9)
–2	(–3, 4)
–1	(–2, 1)
0	(–1, 0)
1	(0, 1)
2	(1, 4)
3	(2, 9)

(b) $y = x^2 + 2x + 1$; yes

(c)

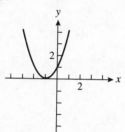

4. (a)

t	(x, y)
–3	(–6, 8)
–2	(–4, 3)
–1	(–2, 0)
0	(0, –1)
1	(2, 0)
2	(4, 3)
3	(6, 8)

(b) $y = \dfrac{1}{4}x^2 - 1$; yes

(c)

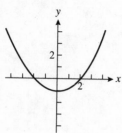

5. (a)

t	(x, y)
–3	(–6, 11)
–2	(–4, 6)
–1	(–2, 3)
0	(0, 2)
1	(2, 3)
2	(4, 6)
3	(6, 11)

(b) $y = \dfrac{1}{4}x^2 + 2$; yes

(c)

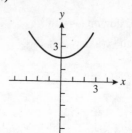

6. (a)

t	(x, y)
–3	(–2, 6)
–2	(–1, 1)
–1	(0, –2)
0	(1, –3)
1	(2, –2)
2	(3, 1)
3	(4, 6)

(b) $y = x^2 - 2x - 2$; yes

(c)

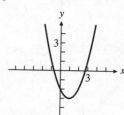

7. $f^{-1}(x) = \dfrac{-x}{x - 1}$

8. $f^{-1}(x) = \dfrac{x}{x - 1}$

9. $f^{-1}(x) = \dfrac{4x}{3x - 2}$

10. $f^{-1}(x) = \dfrac{x + 1}{x - 1}$

11. $f^{-1}(x) = \dfrac{3x + 3}{x - 2}$

12. $f^{-1}(x) = x^3 + 2$

13. $f(x) = 2x + 3$, domain of f: all real numbers;

$f^{-1}(x) = \dfrac{1}{2}x - \dfrac{3}{2}$, domain of f^{-1}: all real numbers

14. $f(x) = \dfrac{1}{2x + 5}$, domain of f: all real numbers but –2.5;

$f^{-1}(x) = \dfrac{-5x + 1}{2x}$, domain of f^{-1}: all real numbers but 0

15. $f(x) = \dfrac{1}{-3x + 2}$, domain of f: all real numbers but $\dfrac{2}{3}$;

$f^{-1}(x) = \dfrac{2x - 1}{3x}$, domain of f^{-1}: all real numbers but 0

16. $f(x) = \sqrt{5x+1}$, domain of f: all real numbers such that $x \geq -\dfrac{1}{5}$; $f^{-1}(x) = \dfrac{1}{5}x^2 - \dfrac{1}{5}$, domain of f^{-1}: all real numbers such that $x \geq 0$

17. $f(x) = \sqrt{4x-8}$, domain of f: all real numbers such that $x \geq 2$; $f^{-1}(x) = \dfrac{1}{4}x^2 + 2$, domain of f^{-1}: all real numbers such that $x \geq 0$

18. $f(x) = 2\sqrt{x-1}$, domain of f: all real numbers such that $x \geq 1$; $f^{-1}(x) = \dfrac{1}{4}x^2 + 1$, domain of f^{-1}: all real numbers such that $x \geq 1$

1.6 Graphical Transformations

1. across the x-axis: $f(x) = \dfrac{-3x-2}{3x^2-5}$

across the y-axis: $f(x) = \dfrac{-3x+2}{3x^2-5}$

2. across the x-axis: $f(x) = \dfrac{-1}{2x+5}$

across the y-axis: $f(x) = \dfrac{1}{-2x+5}$

3. across the x-axis: $f(x) = \dfrac{1}{3x-2}$

across the y-axis: $f(x) = \dfrac{1}{3x+2}$

4. across the x-axis: $f(x) = -\sqrt{x-2} - 1$

across the y-axis: $f(x) = \sqrt{-x-2} + 1$

5. across the x-axis: $f(x) = \dfrac{-x}{x^2+7}$

across the y-axis: $f(x) = \dfrac{-x}{x^2+7}$

6. across the x-axis: $f(x) = \dfrac{-3x^2}{4x^2-1}$

across the y-axis: $f(x) = \dfrac{3x^2}{4x^2-1}$

7. $y = 3x^2 + 6x$

8. $y = 8x^2 - 6x$

9. $y = x^3 + 2x^2$

10. $y = -\dfrac{1}{8}x^3 + \dfrac{5}{2}x$

11. $y = -6x^3 + 3x^2$

12. $y = \dfrac{1}{4}x^2 - \dfrac{1}{2}x$

13.(a) $y = \dfrac{x^2}{2} - 2x + 4$

(b) $y = \dfrac{x^2}{2} - 2x + 3$

14.(a) $y = 3x^2 - 6x + 2$

(b) $y = 3x^2 - 6x$

15.(a) $y = \dfrac{x^2}{4} + 3$

(b) $y = \dfrac{x^2}{4} + 3$

16.(a) $y = x^2 + 6x + 8$

(b) $y = x^2 + 6x + 8$

17.(a) $y = 3x^2 - 24x + 52$

(b) $y = 3x^2 - 24x + 60$

18.(a) $y = 16x^2 + 16x - 1$

(b) $y = 16x^2 + 64x + 59$

19.

20.

21.

22.

23.

24.

1.7 Modeling with Functions

1. $V = 10s^2$; $V = \dfrac{5}{8}P^2$

2. $V = 10\pi r^2$; $V = \dfrac{5\pi d^2}{2}$; $V = \dfrac{2.5C^2}{\pi}$

3. $A = \pi \cdot \dfrac{s^2}{4}$

4. $a = 2\pi s^2$

5. $A = 4w; A = 2(P - 8)$

6. $V = 3\pi r^2$; $V = \dfrac{3\pi d^2}{4}$; $V = \dfrac{3C^2}{4\pi}$

7. 6 inches

8. 6.75 inches

9. $5\dfrac{5}{8}$ inches

10. 7.5 inches

11. $3\dfrac{3}{4}$ inches

12. $9\dfrac{3}{8}$ inches

13. 15.61 inches

14. 12.31 inches

15. 12.90 inches

16. 9.09 inches

17. 24.37 inches

18. 24.65 inches

Chapter 2
Polynomial, Power, and Rational Functions

2.1 Linear and Quadratic Functions and Modeling

1. $f(x) = 3$

2. $f(x) = \dfrac{2}{3}x + 2$

3. $f(x) = \dfrac{4}{3}x + \dfrac{4}{3}$

4. $f(x) = -\dfrac{1}{4}x - \dfrac{7}{2}$

5. $f(x) = -\dfrac{7}{5}x + \dfrac{9}{5}$

6. $f(x) = \dfrac{11}{2}x - \dfrac{15}{2}$

7. $f(x) = (x + 1)^2 - 1$; vertex: $(-1, -1)$; axis of symmetry: $x = -1$

8. $f(x) = (x - 3)^2 - 7$; vertex: $(3, -7)$; axis of symmetry: $x = 3$

9. $f(x) = (x + 3)^2 - 4$; vertex: $(-3, -4)$; axis of symmetry: $x = -3$

10. $f(x) = (x + 5)^2$; vertex: $(-5, 0)$; axis of symmetry: $x = -5$

11. $f(x) = x^2 - 3$; vertex: $(0, -3)$; axis of symmetry: $x = 0$

12. $f(x) = -(x + 6)^2 - 7$; vertex: $(-6, -7)$; axis of symmetry: $x = -6$

13. $f(x) = (x - 1)^2 - 7$; The graph of f is an upward-opening parabola with vertex $(1, -7)$, axis of symmetry $x = 1$ and intersects the x-axis at $x = 1 - \sqrt{7} \approx -1.64575$ and $x = 1 + \sqrt{7} \approx 3.64575$.

14. $f(x) = (x + 3)^2 - 11$; The graph of f is an upward-opening parabola with vertex $(-3, -11)$, axis of symmetry $x = -3$ and intersects the x-axis at $x = -3 - \sqrt{11} \approx -6.31662$ and $x = -3 + \sqrt{11} \approx 0.316625$.

15. $f(x) = -(x + 1)^2 + 4$; The graph of f is a downward-opening parabola with vertex $(-1, 4)$, axis of symmetry $x = -1$ and intersects the x-axis at $x = -3$ and $x = 1$.

16. $f(x) = (x - 4)^2 - 4$; The graph of f is an upward-opening parabola with vertex $(4, -4)$, axis of symmetry $x = 4$ and intersects the x-axis at $x = 2$ and $x = 6$.

17. $f(x) = 2(x - 2)^2 - 3$; The graph of f is an upward-opening parabola with vertex $(2, -3)$, axis of symmetry $x = 2$ and intersects the x-axis at $x = 2 - \dfrac{\sqrt{6}}{2} \approx 0.775255$ and $x = 2 + \dfrac{\sqrt{6}}{2} \approx 3.22474$.

18. $f(x) = 3(x + 6)^2 - 12$; The graph of f is an upward-opening parabola with vertex $(-6, -12)$, axis of symmetry $x = -6$ and intersects the x-axis at $x = -8$ and $x = -4$.

2.2 Power Functions with Modeling

1. power $= \dfrac{1}{2}$; constant of variation $= 2$

Domain: all nonnegative real numbers

Range: $[0, \infty)$

Continuous

Increasing on $[0, \infty)$

Not symmetric with respect to the y-axis or the origin;

Neither odd nor even

Bounded below

Local minimum at $(0, 0)$

No horizontal asymptotes

No vertical asymptotes

End behavior: $\lim\limits_{x \to \infty} 2\sqrt{x} = \infty$

2. power $= -\dfrac{1}{3}$; constant of variation $= -\dfrac{1}{2}$

Domain: all real numbers except 0

Range: $(-\infty, 0) \cup (0, \infty)$

Discontinuous at $x = 0$

Increasing on $(-\infty, 0)$; increasing on $(0, \infty)$

Symmetric with respect to the origin

Not bounded above or below

No local extrema

Horizontal asymptote: $y = 0$

Vertical asymptote: $x = 0$

End behavior: $\lim\limits_{x \to -\infty} -\dfrac{1}{2\sqrt[3]{x}} = 0$ and $\lim\limits_{x \to \infty} -\dfrac{1}{2\sqrt[3]{x}} = 0$

3. power = –4; constant of variation = 1

Domain: all real numbers except 0

Range: $(0, \infty)$

Continuous except at $x = 0$

Increasing on $(-\infty, 0)$; decreasing on $(0, \infty)$

Symmetric with respect to the y-axis

Bounded below

No local extrema

Horizontal asymptote: $y = 0$

Vertical asymptote: $x = 0$

End behavior: $\lim\limits_{x \to -\infty} x^{-4} = 0$ and $\lim\limits_{x \to \infty} x^{-4} = 0$

4. power = $\dfrac{1}{5}$; constant of variation = 1

Domain: all real numbers

Range: $(-\infty, \infty)$

Continuous

increasing on $(-\infty, \infty)$

Symmetric with respect to origin

Not bounded below or above

No local extrema

Horizontal asymptote: none

Vertical asymptote: none

End behavior: $\lim\limits_{x \to -\infty} \sqrt[5]{x} = -\infty$ and $\lim\limits_{x \to \infty} \sqrt[5]{x} = \infty$

5. power = $\dfrac{1}{6}$; constant of variation = 1

Domain: all nonnegative real numbers

Range: $(0, \infty)$

Continuous

increasing on $(0, \infty)$

Not symmetric with respect to either axis or the origin

Bounded below

Local minimum at $(0, 0)$

Horizontal asymptote: none

Vertical asymptote: none

End behavior: $\lim\limits_{x \to \infty} \sqrt[6]{x} = \infty$

6. power = –5; constant of variation = 1

Domain: all real numbers except 0

Range: $(-\infty, 0) \cup (0, \infty)$

Continuous except at $x = 0$

Decreasing on $(-\infty, 0)$; decreasing on $(0, \infty)$

Symmetric with respect to the origin

Not bounded below or above

No local extrema

Horizontal asymptote: $y = 0$

Vertical asymptote: $x = 0$

End behavior: $\lim\limits_{x \to -\infty} x^{-5} = 0$ and $\lim\limits_{x \to \infty} x^{-5} = 0$

7. Vertically stretch the graph of $g(x) = x^2$ by a factor of 2.

8. Vertically stretch the graph of $g(x) = x^4$ by a factor of 2 then reflect it across the x-axis.

9. Reflect the graph of $g(x) = x^3$ across the x-axis.

10. Vertically stretch the graph of $g(x) = x^4$ by a factor of 2.

11. Vertically shrink the graph of $g(x) = x^6$ by a factor of 0.5.

12. Vertically stretch the graph of $g(x) = x^5$ by a factor of 2 then reflect it across the x-axis.

13. $k = 2$ and $a = -2$; In the first quadrant, the graph decreases for all $x > 0$.; The graph is symmetric about the y-axis, (f is even). The graph does not exist in Quadrant II or III.

14. $k = 0.5$ and $a = -5$; In the first quadrant, the graph decreases for all $x > 0$; The graph is symmetric about origin, (f is odd). The graph does not exist in Quadrant II or IV.

15. $k = -1$ and $a = 2.5$; the graph contains $(0, 0)$ and passes through $(1, -1)$. In the fourth quadrant, it is decreasing. The function is undefined for $x < 0$.

16. $k = -2$ and $a = 1.4$; the graph contains $(0, 0)$ and passes through $(1, -1)$. In the second and fourth quadrants, it is decreasing. The function does not exist in Quadrants I and III. The function is symmetric about the origin; f is odd.

17. $k = -1$ and $a = 0.1$, the graph contains $(0, 0)$ and passes through $(1, -1)$. In the fourth quadrant, it is decreasing. The graph does not exist in Quadrants I, II, or III. The function is undefined for $x < 0$.

18. $k = -1$ and $a = 0.2$, the graph contains $(0, 0)$ and passes through $(1, -1)$. In the fourth quadrant, it is decreasing. The graph is symmetric about the origin, (f is odd). The graph does not exist in Quadrant I or III.

2.3 Polynomial Functions of Higher Degree with Modeling

1. Shift the graph of $f(x) = 3x^3$ one unit to the left.

2. Shift the graph of $f(x) = -2x^4$ one unit to the left.

3. Shift the graph of $f(x) = 2x^3$ one unit to the right and one unit down.

4. Shift the graph of $f(x) = -3x^4$ one unit to the right and one unit up.

5. Shift the graph of $f(x) = 2x^3$ one unit down.

Not bounded below or above

No local extrema

Horizontal asymptote: $y = 0$

Vertical asymptote: $x = 0$

6. Shift the graph of $f(x) = -x^4$ two units to the left and two units up.

7. -4, 0, and 4

8. -1, 0, and 10

9. -8, -1, and 0

10. 0, 1, and 7

11. -7, 0, and 5

12. -8, -7, and 0

13. degree 3; zeros: 0 (multiplicity 1) crosses the x-axis, -2 (multiplicity 2) does not cross the x-axis

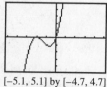

[−5.1, 5.1] by [−4.7, 4.7]

14. degree 4; zeros: -2 (multiplicity 1) crosses the x-axis, 1 (multiplicity 3) crosses the x-axis

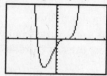

[−5.1, 5.1] by [−10, 10]

15. degree 4; zeros: 1 (multiplicity 2) does not cross the x-axis, 3 (multiplicity 2) does not cross the x-axis

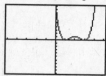

[−5.1, 5.1] by [−10, 10]

16. degree 6; zeros: -3 (multiplicity 3) crosses the x-axis, -1 (multiplicity 3) crosses the x-axis

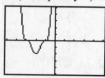

[−5.1, 5.1] by [−2.5, 2.5]

17. degree 5; zeros: -1 (multiplicity 3) crosses the x-axis, 2 (multiplicity 2) does not cross the x-axis

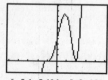

[−5.1, 5.1] by [−2, 10]

18. degree 4; zeros: 0 (multiplicity 1) crosses the x-axis, -2 (multiplicity 1) crosses the x-axis, 3 (multiplicity 2) does not cross the x-axis

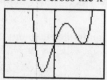

[−5.1, 5.1] by [−20, 20]

19. about -0.2, -0.3, and -2

20. about -1.7, -1.6, and 3

21. about 0, and 0.996

22. about -1, 0, 1.9, and 2.1

23. about 0 and 4.1

24. about 3.8, 3.9, 4, and 4.1

2.4 Real Zeros of Polynomial Functions

1. 19

2. 7

3. 38

4. -10

5. 32

6. -4.25

7. $-\dfrac{1}{2}$ and $\dfrac{1}{3}$

8. $-\dfrac{5}{2}$

9. $-\dfrac{1}{2}$ and 1

10. $-\dfrac{1}{3}$

11. $\dfrac{2}{3}$

12. none

13.

5⌋	1	−4	3
		5	5
	1	1	8

0⌋	1	−4	3
		0	0
	1	−4	3

14.

−2⌋	1	2	−3
		2	8
	1	4	5

−4⌋	1	2	−3
		−4	8
	1	−2	5

15.

7⌋	1	−6	11	6
		7	7	126
	1	1	18	120

0⌋	1	−6	11	−6
		0	0	0
	1	−6	11	−6

16.

2⌋	1	−1	−2	2
		2	2	0
	1	1	0	2

−2⌋	1	−1	−2	2
		−2	6	8
	1	−3	4	10

17.

$$
\begin{array}{r|rrrrr}
2) & 1 & 0 & -1 & 0 & -2 \\
& & 2 & 4 & 6 & 12 \\
\hline
& 1 & 2 & 3 & 6 & 10
\end{array}
$$

$$
\begin{array}{r|rrrrr}
-2) & 1 & 0 & -1 & 0 & -2 \\
& & -2 & 4 & -6 & 12 \\
\hline
& 1 & -2 & 3 & -6 & 10
\end{array}
$$

18.

$$
\begin{array}{r|rrrrr}
2) & 1 & 0 & -2 & 0 & -2 \\
& & 2 & 4 & 4 & 8 \\
\hline
& 1 & 2 & 2 & 4 & 6
\end{array}
$$

$$
\begin{array}{r|rrrrr}
-2) & 1 & 0 & -2 & 0 & -2 \\
& & -2 & 4 & -4 & 8 \\
\hline
& 1 & -2 & 2 & -4 & 6
\end{array}
$$

19. $\dfrac{1}{2}, \pm\sqrt{2}$ **20.** $\dfrac{1}{2}, 1, \pm\sqrt{3}$ **21.** 1 and 2

22. $\dfrac{1}{2}, -\dfrac{1}{2} \pm \dfrac{\sqrt{29}}{2}$ **23.** $\dfrac{2}{3}, \dfrac{1}{2} \pm \dfrac{\sqrt{5}}{2}$ **24.** $\pm 1, \pm\sqrt{7}$

2.5 Complex Zeros and the Fundamental Theorem of Algebra

1. $f(x) = x^2 - 4x + 13$

2. $f(x) = x^3 - 4x^2 + 9x - 36$

3. $f(x) = x^4 - 5x^3 + 7x^2 + 3x - 10$

4. $f(x) = x^4 + x^3 + 20x^2 + 78x$

5. $f(x) = x^4 + x^3 + 10x^2 - 52x + 40$

6. $f(x) = x^6 + 14x^4 + 49x^2 + 36$

7. $f(x) = x(x-1)(x+i)(x-i)$

8. $f(x) = (x-2)(x+2)(x+2i)(x-2i)$

9. $f(x) = (x-3)(x+2)(x+3i)(x-3i)$

10. $f(x) = x(x-2)(x-3)(x-(2+i))(x-(2-i))$

11. $f(x) = (x-3)(x-1)(x-(2+3i))(x-(2-3i))$

12. $f(x) = (x+2)(x-3)(x-1)(x-(2+2i))(x-(2-2i))$

13. $f(x) = (x^2+2)\left(x-\sqrt{3}\right)\left(x+\sqrt{3}\right)$

14. $f(x) = (x-2)(x^2+5)\left(x-\sqrt{3}\right)\left(x+\sqrt{3}\right)$

15. $f(x) = x(x^2+1)(x-1)(x+1)$

16. $f(x) = x(x-1)(x+1)$

17. $f(x) = (2x+1)\left(x-\sqrt{7}\right)\left(x+\sqrt{7}\right)$

18. $f(x) = (2x+1)(x^2-6x+13)$

2.6 Graphs of Rational Functions

1. Shift the graph of $f(x) = \dfrac{1}{x}$ one unit to the left, then vertically stretch the graph by a factor of 2; horizontal asymptote: $y = 0$; vertical asymptote: $x = -1$; $\displaystyle\lim_{x \to \infty} h(x) = \lim_{x \to -\infty} h(x) = 0$, $\displaystyle\lim_{x \to -1^+} h(x) = \infty$, and $\displaystyle\lim_{x \to -1^-} h(x) = -\infty$

2. Shift the graph of $f(x) = \dfrac{1}{x}$ one unit to the left, stretch by a factor of 3, follow by a reflection across the x-axis, and then translate two units up; horizontal asymptote: $y = 2$; vertical asymptote: $x = -1$; $\displaystyle\lim_{x \to \infty} h(x) = \lim_{x \to -\infty} h(x) = 2$, $\displaystyle\lim_{x \to -1^+} h(x) = -\infty$, and $\displaystyle\lim_{x \to -1^-} h(x) = \infty$.

3. Shift the graph of $f(x) = \dfrac{1}{x}$ two units to the right, then vertically stretch the graph by a factor of 2; horizontal asymptote: $y = 0$; vertical asymptote: $x = 2$; $\displaystyle\lim_{x \to \infty} h(x) = \lim_{x \to -\infty} h(x) = 0$, $\displaystyle\lim_{x \to 2^+} h(x) = \infty$, and $\displaystyle\lim_{x \to 2^-} h(x) = -\infty$

4. Shift the graph of $f(x) = \dfrac{1}{x}$ two units to the left, stretch by a factor of 4, follow by a reflection across the x-axis, and then translate three units up; horizontal asymptote: $y = 3$; vertical asymptote: $x = -2$; $\displaystyle\lim_{x \to \infty} h(x) = \lim_{x \to -\infty} h(x) = 3$, $\displaystyle\lim_{x \to -2^+} h(x) = -\infty$, and $\displaystyle\lim_{x \to -2^-} h(x) = \infty$

5. Shift the graph of $f(x) = \dfrac{1}{x}$ three units to the right; horizontal asymptote: $y = 0$; vertical asymptote: $x = 3$; $\displaystyle\lim_{x \to \infty} h(x) = \lim_{x \to -\infty} h(x) = 0$, $\displaystyle\lim_{x \to 3^+} h(x) = \infty$, and $\displaystyle\lim_{x \to 3^-} h(x) = -\infty$

6. Shift the graph of $f(x) = \dfrac{1}{x}$ two units to the right, stretch by a factor of 8, and then translate three units up; horizontal asymptote: $y = 3$; vertical asymptote: $x = 2$; $\displaystyle\lim_{x \to \infty} h(x) = \lim_{x \to -\infty} h(x) = 3$, $\displaystyle\lim_{x \to 2^+} h(x) = \infty$, and $\displaystyle\lim_{x \to 2^-} h(x) = -\infty$

7. slant asymptote: $y = x$; vertical asymptote: $x = \pm 1$;
   x-intercept: 0; y-intercept: 0

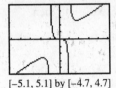

[−5.1, 5.1] by [−4.7, 4.7]

8. slant asymptote: $y = 2x$; vertical asymptote: $x = \pm 2$;
   x-intercept: 0; y-intercept: 0

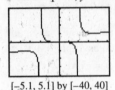

[−5.1, 5.1] by [−40, 40]

9. slant asymptote: $y = x$; vertical asymptote: $x = \pm 4$;
   x-intercept: 0; y-intercept: 0

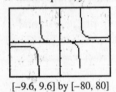

[−9.6, 9.6] by [−80, 80]

10. slant asymptote: $y = x + 4$; vertical asymptote: $x = 2$;
    x-intercepts: −3, 1; y-intercept: $\dfrac{3}{2}$

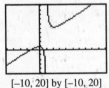

[−10, 20] by [−10, 20]

11. slant asymptote: $y = x - 8$; vertical asymptote: $x = -2$;
    x-intercepts: 1, 5; y-intercept: $\dfrac{5}{2}$

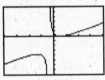

[−20, 20] by [−40, 30]

12. slant asymptote: $y = x - 8$; vertical asymptote: $x = -4$;
    x-intercept: 2; y-intercept: 1

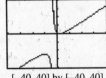

[−40, 40] by [−40, 40]

13. Domain: $(-\infty, -2) \cup (-2, 3) \cup (3, \infty)$
    Range: all reals
    x-intercept: 0
    y-intercept: 0
    Continuity: all $x \neq -2, 3$
    Decreasing on $(-\infty, -2) \cup (-2, 3) \cup (3, \infty)$
    Not symmetric
    Unbounded
    No local extrema
    Horizontal asymptotes: $y = 0$
    Vertical asymptotes: $x = -2$ and $x = 3$
    End behavior: $\lim\limits_{x \to \infty} f(x) = \lim\limits_{x \to -\infty} f(x) = 0$

14. Domain: $(-\infty, -5) \cup (-5, 1) \cup (1, \infty)$
    Range: all reals
    x-intercept: 0
    y-intercept: 0
    Continuity: all $x \neq -5, 1$
    Decreasing on $(-\infty, -5) \cup (-5, 1) \cup (1, \infty)$
    Not symmetric
    Unbounded
    No local extrema
    Horizontal asymptotes: $y = 0$
    Vertical asymptotes: $x = -5$ and $x = 1$
    End behavior: $\lim\limits_{x \to \infty} f(x) = \lim\limits_{x \to -\infty} f(x) = 0$

15. Domain: $(-\infty, 2) \cup (2, 4) \cup (4, \infty)$
    Range: $(-\infty, -3.936) \cup (0, \infty)$
    x-intercept: −1
    y-intercept: 1/8
    Continuity: all $x \neq 2, 4$
    Increasing on $(-\infty, 2) \cup (2, 2.873)$
    Decreasing on $(2.873, 4) \cup (4, \infty)$
    Not symmetric
    Unbounded
    Local maximum at $(2.873, -3.936)$
    Horizontal asymptotes: $y = 0$
    Vertical asymptotes: $x = 2$ and $x = 4$
    End behavior: $\lim\limits_{x \to \infty} f(x) = \lim\limits_{x \to -\infty} f(x) = 0$

16. Domain: $(-\infty, -6) \cup (-6, 1) \cup (1, \infty)$
    Range: all reals
    x-intercept: 0
    y-intercept: 0
    Continuity: all $x \neq -6, 1$

Decreasing on $(-\infty, -6) \cup (-6, 1) \cup (1, \infty)$

Not symmetric

Unbounded

No local extrema

Horizontal asymptotes: $y = 0$

Vertical asymptotes: $x = -6$ and $x = 1$

End behavior: $\lim\limits_{x \to \infty} f(x) = \lim\limits_{x \to -\infty} f(x) = 0$

17. Domain: $(-\infty, -3) \cup (-3, 2) \cup (2, \infty)$

Range: all reals

x-intercept: 1

y-intercept: 1/6

Continuity: all $x \neq -3, 2$

Decreasing on $(-\infty, -3) \cup (-3, 2) \cup (2, \infty)$

Not symmetric

Unbounded

No local extrema

Horizontal asymptotes: $y = 0$

Vertical asymptotes: $x = -3$ and $x = 2$

End behavior: $\lim\limits_{x \to \infty} f(x) = \lim\limits_{x \to -\infty} f(x) = 0$

18. Domain: $(-\infty, -5) \cup (-5, -2) \cup (-2, \infty)$

Range: all reals

x-intercept: -3

y-intercept: 3/10

Continuity: all $x \neq -5, -2$

Decreasing on $(-\infty, -5) \cup (-5, -2) \cup (-2, \infty)$

Not symmetric

Unbounded

No local extrema

Horizontal asymptotes: $y = 0$

Vertical asymptotes: $x = -5$ and $x = -2$

End behavior: $\lim\limits_{x \to \infty} f(x) = \lim\limits_{x \to -\infty} f(x) = 0$

19. Domain: $(-\infty, -4) \cup (-4, 4) \cup (4, \infty)$

Range: $(-\infty, 1/16) \cup (1, \infty)$

x-intercept: ± 1

y-intercept: 1/16

Continuity: all $x \neq -4, 4$

Decreasing on $(0, 4) \cup (4, \infty)$

Increasing on $(-\infty, -4) \cup (-4, 0)$

Symmetric with respect to the y-axis (an even function)

Unbounded

Local maximum of 1/16 at $x = 0$

Horizontal asymptote: $y = 1$

Vertical asymptotes: $x = -4$ and $x = 4$

End behavior: $\lim\limits_{x \to \infty} f(x) = \lim\limits_{x \to -\infty} f(x) = 1$

20. Domain: $(-\infty, -5) \cup (-5, 5) \cup (5, \infty)$

Range: $(-\infty, 1/25) \cup (1, \infty)$

x-intercept: ± 1

y-intercept: 1/25

Continuity: all $x \neq -5, 5$

Decreasing on $(0, 5) \cup (5, \infty)$

Increasing on $(-\infty, -5) \cup (-5, 0)$

Symmetric with respect to the y-axis (an even function)

Unbounded

Local maximum of 1/25 at $x = 0$

Horizontal asymptote: $y = 1$

Vertical asymptotes: $x = -5$ and $x = 5$

End behavior: $\lim\limits_{x \to \infty} f(x) = \lim\limits_{x \to -\infty} f(x) = 1$

21. Domain: $(-\infty, -3) \cup (-3, 3) \cup (3, \infty)$

Range: $(-\infty, 2/9) \cup (1, \infty)$

x-intercepts: $\pm\sqrt{2}$

y-intercept: 2/9

Continuity: all $x \neq -3, 3$

Decreasing on $(0, 3) \cup (3, \infty)$

Increasing on $(-\infty, -3) \cup (-3, 0)$

Symmetric with respect to the y-axis (an even function)

Unbounded

Local maximum of 2/9 at $x = 0$

Horizontal asymptote: $y = 1$

Vertical asymptotes: $x = -3$ and $x = 3$

End behavior: $\lim\limits_{x \to \infty} f(x) = \lim\limits_{x \to -\infty} f(x) = 1$

22. Domain: $(-\infty, -5) \cup (-5, 5) \cup (5, \infty)$

Range: $(-\infty, 0.16) \cup (2, \infty)$

x-intercept: $\pm\sqrt{2}$

y-intercept: 0.16

Continuity: all $x \neq -5, 5$

Decreasing on $(0, 5) \cup (5, \infty)$

Increasing on $(-\infty, -5) \cup (-5, 0)$

Symmetric with respect to the y-axis (an even function)

Unbounded

Local maximum of 0.16 at $x = 0$

Horizontal asymptote: $y = 2$

Vertical asymptotes: $x = -5$ and $x = 5$

End behavior: $\lim\limits_{x \to \infty} f(x) = \lim\limits_{x \to -\infty} f(x) = 2$

Vertical asymptotes: $x = -4$ and $x = 4$

End behavior: $\lim\limits_{x \to \infty} f(x) = \lim\limits_{x \to -\infty} f(x) = 1$

23. Domain: $(-\infty, -4) \cup (-4, 4) \cup (4, \infty)$

Range: $(-\infty, 5/16) \cup (3, \infty)$

x-intercept: $\pm\frac{1}{3}\sqrt{15}$

y-intercept: $5/16$

Continuity: all $x \neq -4, 4$

Decreasing on $(0, 4) \cup (4, \infty)$

Increasing on $(-\infty, -4) \cup (-4, 0)$

Symmetric with respect to the y-axis (an even function)

Unbounded

Local maximum of $5/16$ at $x = 0$

Horizontal asymptote: $y = 3$

Vertical asymptotes: $x = -4$ and $x = 4$

End behavior: $\lim\limits_{x \to \infty} f(x) = \lim\limits_{x \to -\infty} f(x) = 3$

24. Domain: $(-\infty, -3) \cup (-3, 3) \cup (3, \infty)$

Range: $(-\infty, -5/9) \cup (2, \infty)$

x-intercept: none

y-intercept: $-5/9$

Continuity: all $x \neq -3, 3$

Decreasing on $(0, 3) \cup (3, \infty)$

Increasing on $(-\infty, -3) \cup (-3, 0)$

Symmetric with respect to the y-axis (an even function)

Unbounded

Local maximum of $-5/9$ at $x = 0$

Horizontal asymptote: $y = 2$

Vertical asymptotes: $x = -3$ and $x = 3$

End behavior: $\lim\limits_{x \to \infty} f(x) = \lim\limits_{x \to -\infty} f(x) = 2$

2.7 Solving Equations in One Variable

1. $x = -\frac{1}{3}$

2. $x = 4$

3. $x = -1$ and $x = \frac{1}{4}$

4. $x = 1$ and $x = -\frac{1}{3}$

5. $x = -1$ and $x = 5$

6. $x = 0$ and $x = 4$

7. $x = -\frac{3}{2} - \frac{\sqrt{21}}{2} \approx -3.79129$ and $x = -\frac{3}{2} + \frac{\sqrt{21}}{2} \approx 0.791288$

8. $x = \frac{1}{2} - \frac{\sqrt{5}}{2} \approx -0.618034$ and $x = \frac{1}{2} + \frac{\sqrt{5}}{2} \approx 1.61803$

9. $x = -\frac{1}{2} - \frac{\sqrt{5}}{2} \approx -1.61803$ and $x = -\frac{1}{2} + \frac{\sqrt{5}}{2} \approx 0.618034$

10. $x = \frac{5}{6} + \frac{\sqrt{13}}{6} \approx 1.43426$ and $x = \frac{5}{6} - \frac{\sqrt{13}}{6} \approx 0.232408$

11. $x = -1$ and $x = 0.2$

12. $x = 1$ and $x = -0.25$

13. $x = 12$

14. $x = 0.5$

15. $x = 5$

16. $x = -1$

17. $x = 1\frac{3}{7}$

18. $x = 13$

19. 160 mL

20. 62.5 mL

21. 1,100 mL

22. 10 mL

23. 60 mL

24. 262.5 mL

2.8 Solving Inequalities In One Variable

1. $[-3, 2]$

2. $(-\infty, -2] \cup [4, \infty)$

3. $(-\infty, -3) \cup (0, 1)$

4. $[-2, \infty)$

5. $(-\infty, 2)$

6. $[-2, 2] \cup [3, \infty)$

7. (a) $r(x) = 0$ if $x = 0$

(b) $r(x)$ is undefined if $x = 2$

(c) $r(x) > 0$ if $x > 2$ or $x < 0$

(d) $r(x) < 0$ if $0 < x < 2$

8. (a) $r(x) = 0$ if $x = 1$

(b) $r(x)$ is undefined if $x = 2$

(c) $r(x) > 0$ if $x < 1$ or $x > 2$

(d) $r(x) < 0$ if $1 < x < 2$

9. (a) $r(x) = 0$ if $x = 3$

(b) $r(x)$ is undefined if $x = -1$

(c) $r(x) > 0$ if $x < -1$ or $x > 3$

(d) $r(x) < 0$ if $-1 < x < 3$

10. (a) $r(x) = 0$ if $x = -1$

(b) $r(x)$ is undefined if $x = -2$ or $x = 2$

(c) $r(x) > 0$ if $-2 < x < -1$ or $x > 2$

(d) $r(x) < 0$ if $x < -2$ or $-1 < x < 2$

11. (a) $r(x) = 0$ if $x = -2$

(b) $r(x)$ is undefined if $x = -1$ or $x = 4$

(c) $r(x) > 0$ if $-2 < x < -1$ or $x > 4$

(d) $r(x) < 0$ if $x < -2$ or $-1 < x < 4$

12. (a) $r(x) = 0$ if $x = 0$

(b) $r(x)$ is undefined if $x = -1$ or $x = 2$

(c) $r(x) > 0$ if $x < -1$ or $x > 2$

(d) $r(x) < 0$ if $-1 < x < 2$

13. $x < 2$ or $x \geq 3.5$

14. $3 < x \leq 5$

15. $x < -0.5$ or $0 < x < 1$

16. $-1 < x \leq -\frac{1}{3}$ or $x > 1$

17. $x < -1.5$ or $0 < x < 3$

18. $-\frac{3}{7} < x < 0$ or $x > 3$

Chapter 3
Exponential, Logistic, and Logarithmic Functions

3.1 Exponential and Logistic Functions

1. 16

2. 1

3. $\dfrac{1}{64}$

4. 2

5. $\dfrac{1}{2}$

6. $\dfrac{1}{8}$

7. $g(x) = 2 \cdot 5^x$

8. $g(x) = 7 \cdot 2^x$

9. $g(x) = 5 \cdot 3^x$

10. $h(x) = 3 \cdot \left(\dfrac{1}{4}\right)^x$

11. $h(x) = 4 \cdot \left(\dfrac{1}{5}\right)^x$

12. $h(x) = 2 \cdot \left(\dfrac{1}{10}\right)^x$

13. Shift the graph of $f(x) = 5^x$ three units to the left.

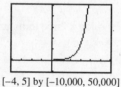

[−4, 5] by [−10,000, 50,000]

14. Shift the graph of $f(x) = 5^x$ one unit to the right.

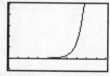

[−1, 9] by [−10,000, 50,000]

15. Reflect the graph of $f(x) = 5^x$ across the x-axis.

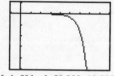

[−1, 9] by [−50,000, 10,000]

16. Vertically stretch the graph of $f(x) = 5^x$ by a factor of 4.

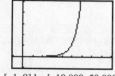

[−1, 9] by [−10,000, 50,000]

17. Vertically stretch the graph of $f(x) = 5^x$ by a factor of 7.

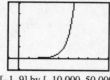

[−1, 9] by [−10,000, 50,000]

18. Shift the graph of $f(x) = 5^x$ five units to the left.

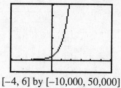

[−4, 6] by [−10,000, 50,000]

19. y-intercept: $\dfrac{5}{4}$; asymptotes: $y = 0$ and $y = 5$

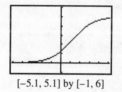

[−5.1, 5.1] by [−1, 6]

20. y-intercept: $\dfrac{20}{3}$; asymptotes: $y = 0$ and $y = 20$

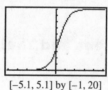

[−5.1, 5.1] by [−1, 20]

21. y-intercept: $\dfrac{1}{5}$; asymptotes: $y = 0$ and $y = 1$

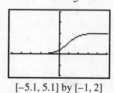

[−5.1, 5.1] by [−1, 2]

22. y-intercept: 3; asymptotes: $y = 0$ and $y = 12$

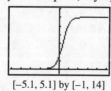

[−5.1, 5.1] by [−1, 14]

23. y-intercept: 2; asymptotes: $y = 0$ and $y = 8$

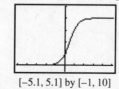

[−5.1, 5.1] by [−1, 10]

24. y-intercept: 4; asymptotes: $y = 0$ and $y = 20$

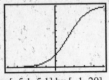

[−5.1, 5.1] by [−1, 20]

3.2 Exponential and Logistic Modeling

1. about 9.6 hr

2. about 8.6 hr

3. about 10.0 hr

4. about 7.3 hr

5. about 8.2 hr

6. about 9.5 hr

7. about 42.5 days

8. 10 days

9. about 10.4 days

10. 120 days

11. about 18.7 days

12. about 56.6 days

13. 100

14. $t \approx 2.7$; toward the end of day 3

15. $t \approx 3.7$; toward the end of day 4

16. 48

17. $t \approx 2.7$; toward the end of day 3

18. $t \approx 4.2$; a little after the beginning of Day 5

3.3 Logarithmic Functions and Their Graphs

1. 4

2. $\frac{1}{2}$

3. −2

4. 0

5. 1

6. $\frac{1}{2}$

7. 4

8. $\frac{1}{3}$

9. 4

10. $\frac{1}{4}$

11. −1

12. 7

13. 5

14. $\frac{1}{7}$

15. −4

16. 11

17. $\frac{1}{5}$

18. 11

19. e

20. $\frac{1}{7}$

21. 6

22. 5

23. 12

24. $\frac{1}{9}$

25. 8

26. $\frac{1}{4}$

27. Translate the graph of $y = \ln x$ one unit to the right.

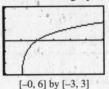

[−0, 6] by [−3, 3]

28. Reflect the graph of $y = \ln x$ across the y-axis then translate the result one unit to the right.

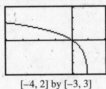

[−4, 2] by [−3, 3]

29. Vertically stretch the graph of $y = \log x$ by a factor of 3.

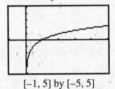

[−1, 5] by [−5, 5]

30. Vertically translate the graph of $y = \log x$ three units up.

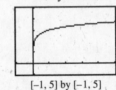

[−1, 5] by [−1, 5]

31. Translate the graph of $y = \ln x$ one unit to the left.

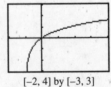

[−2, 4] by [−3, 3]

32. Reflect the graph of $y = \ln x$ across the y-axis then translate the result four units to the right.

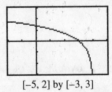

[−5, 2] by [−3, 3]

33. Vertically stretch the graph of $y = \log x$ by a factor of 3.5.

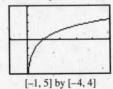

[−1, 5] by [−4, 4]

34. Vertically translate the graph of $y = \log x$ 1.5 units up.

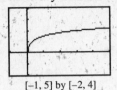

[−1, 5] by [−2, 4]

3.4 Properties of Logarithmic Functions

1. $2 \log 5 + \log x + 3 \log y$

2. $2 + 3 \log x + 4 \log y$

3. $2 \log 12 + 5 \log x + \log y$

4. $3 \log 2 + \log x + \log y$

5. $6 \log 2 + 2 \log x + 3 \log y$

6. $4 \log 3 + \log x + \log y$

7. $2 \log x + 3 \log y$

8. $2 + \log x + \log y$

9. $\ln (x^2 + 1) - \ln x$

10. $3 \ln (x + 3) - 2 \ln x$

11. $\frac{1}{2} \ln (x + 1) - 2 \ln x$

12. $\frac{1}{2} \ln (x^3 + 1) - \ln x$

13. $2 \ln (x - 2) - 2 \ln x$

14. $2 \ln (x - 2) - 2 \ln (x + 1)$

15. $\frac{1}{2} \ln (x^2 - 100) - 6 \ln x$

16. $\frac{1}{2} \ln (x^2 + 1) - \frac{1}{2} \ln (x^2 - 1)$

17. $\ln \dfrac{x}{y}$

18. $\ln \dfrac{1}{x^2 y^4}$

19. $\ln \dfrac{1}{x^2 y^3}$

20. $\ln \dfrac{x^5}{y}$

21. $\ln \dfrac{1}{x^2 y^5}$

22. $\ln \dfrac{1}{y^2}$

23. $\ln \dfrac{x^2}{y^4}$

24. $\ln \dfrac{x^4}{y^2}$

25. Vertically stretch the graph of $f(x) = \ln x$ by a factor of $1/\ln 2 \approx 1.44$.

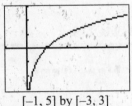

[−1, 5] by [−3, 3]

26. Reflect the graph of $f(x) = \ln x$ across the x-axis and vertically stretch the result by a factor of $1/\ln 2 \approx 1.44$.

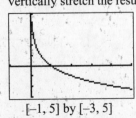

[−1, 5] by [−3, 5]

27. Vertically shrink the graph of $f(x) = \ln x$ by a factor of $1/\ln 4 \approx 0.72$.

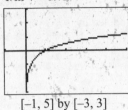

[−1, 5] by [−3, 3]

28. Reflect the graph of $f(x) = \ln x$ across the x-axis and vertically shrink the result by a factor of $1/\ln 5 \approx 0.62$.

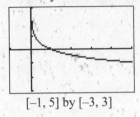

[−1, 5] by [−3, 3]

29. Vertically shrink the graph of $f(x) = \ln x$ by a factor of $1/\ln 7 \approx 0.51$.

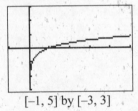

[−1, 5] by [−3, 3]

30. Reflect the graph of $f(x) = \ln x$ across the x-axis and vertically shrink the result by a factor of $1/\ln 6 \approx 0.56$.

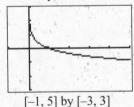

[−1, 5] by [−3, 3]

3.5 Equation Solving and Modeling

1. $x = 3$

2. $x = 4$

3. $x = 8$

4. $x = 12$

5. $x = 6$

6. $x = 9$

7. $x = 8$

8. $x = 9$

9. $x = 100$

10. $x = 1000$

11. $x = 5$

12. $x = 250$

13. $x = \pm 10^4$

14. $x = \pm 10 \sqrt{10}$

15. $x = \pm 5$

16. $x = \pm \sqrt{50} = \pm 5\sqrt{2}$

17. about 11.8 min

18. about 8 min

19. about 12 min

20. about 16 min

21. about 20 min

22. about 19.79 min

23. exponential; the data can be modeled by $y = 3 \cdot 2^x$

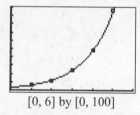

[0, 6] by [0, 100]

24. logarithmic; the data can be modeled by $y = 2 + 3\ln x$

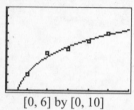

[0, 6] by [0, 10]

25. power; the data can be modeled by $y = 3x^{1.5}$

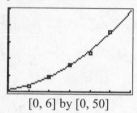

[0, 6] by [0, 50]

26. linear; the data can be modeled by $y = 3x + 2$

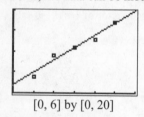

[0, 6] by [0, 20]

27. logarithmic; the data can be modeled by $y = 3 - \ln x$

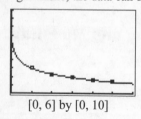

[0, 6] by [0, 10]

28. power; the data can be modeled by $y = 3\sqrt{x}$

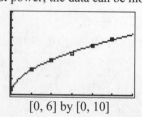

[0, 6] by [0, 10]

3.6 Mathematics of Finance

1. $t \approx 5.81$; 6 years

2. $t \approx 7.4$; 8 years

3. $t \approx 4.5$; 5 years

4. $t \approx 7.7$; 8 years

5. $t \approx 17.2$; 18 years

6. $t \approx 6.4$; 7 years

7.

X	Y1
1	866.63
2	938.01
3	1017
4	1101.7
5	1193.5
6	1292.9
7	1400.5

Y1 ≣ 800e^(.08X)

8.

X	Y1
1	1592.8
2	1691.2
3	1795.8
4	1906.9
5	2024.8
6	2150
7	2282.9

Y1 ≣ 1500e^(.06X)

9.

X	Y1
1	530.92
2	563.75
3	598.61
4	635.62
5	674.93
6	716.66
7	760.98

Y1 ≣ 500e^(.06X)

10.

X	Y1
1	1969.5
2	2155
3	2357.9
4	2580
5	2823
6	3088.8
7	3379.7

Y1 ≣ 1800e^(.09X)

11.

X	Y1
1	965.26
2	1035.2
3	1110.3
4	1190.8
5	1277.2
6	1369.8
7	1469.1

Y1 ≣ 900e^(.07X)

12.

X	Y1
1	1592.8
2	1691.2
3	1795.8
4	1906.9
5	2024.8
6	2150
7	2282.9

Y1 ≣ 1500e^(.06X)

13. $68,640.91

14. $85,896.91

15. $89,201.56

16. $63,857.27

17. $341,017.18

18. $49,018.59

19. $387.35

20. $539.48

21. $386.71

22. $273.55

23. $328.90

24. $250.43

Chapter 4
Trigonometric Functions

4.1 Angles and Their Measures

1. 34°30′

2. 60°18′

3. 50°7′30″

4. 10.5125°

5. 24.205°

6. 29.6°

7. $\frac{2\pi}{5}$ radians

8. $\frac{\pi}{2}$ radians

9. $\frac{2\pi}{3}$ radians

10. 30°

11. 36°

12. 10°

13. 5 inches

14. 14 inches

15. 15.2 inches

16. 18.3 inches

17. 21.3 inches

18. 16.3 inches

19. 19.5 inches

20. 16.4 inches

21. 64.3 miles per hour

22. 57.8 miles per hour

23. 66.4 miles per hour

24. 37.5 miles per hour

25. 51.4 miles per hour **26.** 26.8 miles per hour

27. 53.5 miles per hour **28.** 58.3 miles per hour

4.2 Trigonometric Functions of Acute Angles

1. $\sin \theta = \frac{4}{5}$ $\cos \theta = \frac{3}{5}$ $\tan \theta = \frac{4}{3}$

$\csc \theta = \frac{5}{4}$ $\sec \theta = \frac{5}{3}$ $\cot \theta = \frac{3}{4}$

2. $\sin \theta \approx 0.745356$ $\cos \theta \approx 0.666667$
$\tan \theta \approx 1.11803$ $\csc \theta \approx 1.34164$
$\sec \theta \approx 1.5$ $\cot \theta \approx 0.894427$

3. $\sin \theta \approx 0.166667$ $\cos \theta \approx 0.986013$
$\tan \theta \approx 0.169031$ $\csc \theta \approx 6$
$\sec \theta \approx 1.01419$ $\cot \theta \approx 5.91608$

4. $\sin \theta \approx 0.699854$ $\cos \theta \approx 0.714286$
$\tan \theta \approx 0.979796$ $\csc \theta \approx 1.428869$
$\sec \theta \approx 7/5 = 1.4$ $\cot \theta \approx 1.02062$

5. $\sin \theta \approx 0.444444$ $\cos \theta \approx 0.895806$
$\tan \theta \approx 0.496139$ $\csc \theta \approx 2.25$
$\sec \theta \approx 1.11631$ $\cot \theta \approx 2.01556$

6. $\sin \theta \approx 0.625$ $\cos \theta \approx 0.780625$
$\tan \theta \approx 0.800641$ $\csc \theta \approx 1.6$
$\sec \theta \approx 1.28103$ $\cot \theta \approx 1.249$

7. angles: 90° and 25°; lengths of sides: 9.06 and 4.23

8. angles: 90° and 48°; lengths of sides: 13.38 and 14.86

9. angles: 90° and 52°; lengths of sides: 4.93 and 6.30

10. angles: 90° and 20°; lengths of sides: 23.49 and 8.55

11. angles: 90° and 30°; length of hypotenuse: 100; length of leg: 86.60

12. angles: 90° and 65°; lengths of sides: 19.02 and 40.78

13. about 165 feet **14.** about 261 feet

15. about 111 feet **16.** about 56 feet

17. about 143 feet **18.** about 142 feet

4.3 Trigonometry Extended: The Circular Functions

1. $\sin \theta = \frac{7}{\sqrt{58}} \approx 0.919$; $\cos \theta = \frac{3}{\sqrt{58}} \approx 0.394$;

$\tan \theta = \frac{7}{3} \approx 2.333$; $\csc \theta = \frac{\sqrt{58}}{7} \approx 1.088$;

$\sec \theta = \frac{\sqrt{58}}{3} \approx 2.539$; $\cot \theta = \frac{3}{7} \approx 0.429$

2. $\sin \theta = -\frac{1}{\sqrt{2}} \approx -0.707$; $\cos \theta = -\frac{1}{\sqrt{2}} \approx -0.707$;

$\tan \theta = 1$; $\csc \theta = -\sqrt{2} \approx -1.414$;

$\sec \theta = -\sqrt{2} \approx -1.414$; $\cot \theta = 1$

3. $\sin \theta = \frac{3}{\sqrt{34}} \approx 0.514$; $\cos \theta = -\frac{5}{\sqrt{34}} \approx -0.857$;

$\tan \theta = -\frac{3}{5} = -0.6$; $\csc \theta = \frac{\sqrt{34}}{3} \approx 1.944$;

$\sec \theta = -\frac{\sqrt{34}}{5} \approx -1.166$; $\cot \theta = -\frac{5}{3} \approx -1.667$

4. $\sin \theta = -\frac{5}{\sqrt{29}} \approx -0.928$; $\cos \theta = -\frac{2}{\sqrt{29}} \approx -0.371$;

$\tan \theta = \frac{5}{2} = 2.5$; $\csc \theta = -\frac{\sqrt{29}}{5} \approx -1.077$;

$\sec \theta = -\frac{\sqrt{29}}{2} \approx -2.693$; $\cot \theta = \frac{2}{5} = 0.4$

5. $\sin \theta = \frac{2}{\sqrt{13}} \approx 0.555$; $\cos \theta = \frac{3}{\sqrt{13}} \approx 0.832$;

$\tan \theta = \frac{2}{3} \approx 0.666$; $\csc \theta = \frac{\sqrt{13}}{2} \approx 1.803$;

$\sec \theta = \frac{\sqrt{13}}{3} \approx 1.202$; $\cot \theta = \frac{3}{2} = 1.5$

6. $\sin \theta = -\frac{5}{\sqrt{34}} \approx -0.857$; $\cos \theta = \frac{3}{\sqrt{34}} \approx 0.514$;

$\tan \theta = -\frac{5}{3} \approx -1.666$; $\csc \theta = -\frac{\sqrt{34}}{5} \approx -1.166$;

$\sec \theta = \frac{\sqrt{34}}{3} \approx 1.944$; $\cot \theta = -\frac{3}{5} = -0.6$

7. $\sin \theta = \frac{4}{5}$; $\cos \theta = -\frac{3}{5}$; $\tan \theta = -\frac{4}{3}$;

$\csc \theta = \frac{5}{4}$; $\sec \theta = -\frac{5}{3}$; $\cot \theta = -\frac{3}{4}$

8. $\sin \theta = -\frac{4}{5}$; $\cos \theta = \frac{3}{5}$; $\tan \theta = -\frac{4}{3}$;

$\csc \theta = -\frac{5}{4}$; $\sec \theta = \frac{5}{3}$; $\cot \theta = -\frac{3}{4}$

9. $\sin \theta = 1$; $\cos \theta = 0$; $\tan \theta$ undefined;
$\csc \theta = 1$; $\sec \theta$ undefined; $\cot \theta = 0$

10. $\sin \theta = -1$; $\cos \theta = 0$; $\tan \theta$ undefined;
$\csc \theta = -1$; $\sec \theta$ undefined; $\cot \theta = 0$

11. $\sin \theta = 1$; $\cos \theta = 0$; $\tan \theta$ undefined;
$\csc \theta = 1$; $\sec \theta$ undefined; $\cot \theta = 0$

12. $\sin \theta = 1$; $\cos \theta = 0$; $\tan \theta$ undefined;
$\csc \theta = 1$; $\sec \theta$ undefined; $\cot \theta = 0$

13. 1   **14.** 0   **15.** 1

16. 1   **17.** 0   **18.** 0

4.4 Graphs of Sine and Cosine: Sinusoids

1. amplitude: 1/2; The graph of y_2 is a vertical shrink of the graph of y_1 by a factor of 1/2.

2. amplitude:3; The graph of y_2 is a vertical stretch of the graph of y_1 by a factor of 3.

3. amplitude: 1/2; The graph of y_2 is a vertical shrink of the graph of y_1 by a factor of 1/2.

4. amplitude: 4; The graph of y_2 is a vertical stretch of the graph of y_1 by a factor of 4, and a reflection of the result across the x-axis.

5. amplitude: 1/4; The graph of y_2 is a vertical shrink of the graph of y_1 by a factor of 1/4.

6. amplitude: 1; The graph of y_2 is a reflection of the graph of y_1 across the x-axis.

7. period: 4π; The graph of y_2 is a horizontal stretch of the graph of y_1 by a factor of 2 and a vertical shrink by a factor of 1/2.

8. period: $2\pi/3$; The graph of y_2 is a horizontal shrink of the graph of y_1 by a factor of 1/3 and a reflection across the x-axis.

9. period: 4π; The graph of y_2 is a horizontal stretch of the graph of y_1 by a factor of 2 and a vertical stretch by a factor of 2.

10. period: $\pi/2$; The graph of y_2 is a horizontal shrink of the graph of y_1 by a factor of 1/4.

11. 6π; The graph of y_2 is a horizontal stretch of the graph of y_1 by a factor of 3 and a vertical stretch by a factor of 3.

12. π; The graph of y_2 is a horizontal shrink of the graph of y_1 by a factor of 1/2, a vertical stretch by a factor of 5, and a reflection across the x-axis.

13. frequency: $\dfrac{1}{4\pi}$; The graph completes 1 full cycle per interval of length 4π.

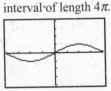

$[-2\pi, 2\pi]$ by $[-4, 4]$

14. frequency: $\dfrac{1}{\pi}$; The graph completes 1 full cycle per interval of length π.

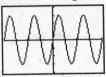

$[-2\pi, 2\pi]$ by $[-4, 4]$

15. frequency: $\dfrac{1}{5\pi}$; The graph completes 1 full cycle per interval of length 5π.

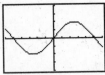

$[-3\pi, 3\pi]$ by $[-4, 4]$

16. frequency: $\dfrac{5}{4\pi}$; The graph completes 1 full cycle per interval of length $\dfrac{4}{5}\pi$.

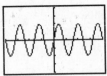

$[-2\pi, 2\pi]$ by $[-4, 4]$

17. frequency: $\dfrac{3}{4\pi}$; The graph completes 1 full cycle per interval of length $\dfrac{4}{3}\pi$.

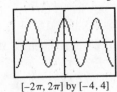

$[-2\pi, 2\pi]$ by $[-4, 4]$

18. frequency: $\dfrac{1}{3\pi}$; The graph completes 1 full cycle per interval of length 3π.

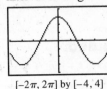

$[-2\pi, 2\pi]$ by $[-4, 4]$

19. $y = -2 \cos\left(\dfrac{\pi x}{6}\right) + 6$ **20.** $y = -2 \cos\left(\dfrac{\pi x}{4}\right) + 3$

21. $y = -3 \cos\left(\dfrac{\pi x}{5}\right) + 6$ **22.** $y = -\cos\left(\dfrac{\pi x}{2}\right) + 7$

23. $y = -3 \cos\left(\dfrac{\pi x}{6}\right) + 8$ **24.** $y = -4 \cos\left(\dfrac{\pi x}{5}\right) + 4$

4.5 Graphs of Tangent, Cotangent, Secant, and Cosecant

1. The graph is the reflection of the graph of $y = \tan x$ across the x-axis.; period: π; asymptotes: $x = \pi n/2$, where n is an odd integer.

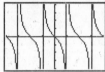

$[-2\pi, 2\pi]$ by $[-4, 4]$

2. The effect of the 2 is a horizontal shrink of the graph of $y = \tan x$ by a factor of 1/2; period: $\pi/2$; asymptotes: $x = \pi n/4$, where n is an odd integer.

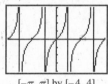

$[-\pi, \pi]$ by $[-4, 4]$

3. The effect of the 2 is a horizontal shrink of the graph of $y = \tan x$ by a factor of 1/2 and a reflection across the x-axis; period: $\pi/2$; asymptotes: $x = \pi n/4$, where n is an odd integer.

$[-\pi, \pi]$ by $[-4, 4]$

4. The effect of the 4 is a horizontal shrink of the graph of $y = \tan x$ by a factor of 1/4 and a reflection across the x-axis; period: $\pi/4$; asymptotes: $x = \pi n/8$, where n is an odd integer.

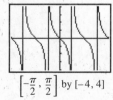

$\left[-\dfrac{\pi}{2}, \dfrac{\pi}{2}\right]$ by $[-4, 4]$

5. The effect of the 3 is a horizontal shrink of the graph of $y = \tan x$ by a factor of 1/3; period: $\pi/3$; asymptotes: $x = \pi n/6$, where n is an odd integer.

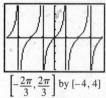

$\left[-\dfrac{2\pi}{3}, \dfrac{2\pi}{3}\right]$ by $[-4, 4]$

6. The effect of the 4 is a horizontal shrink of the graph of $y = \tan x$ by a factor of 1/4; period: $\pi/4$; asymptotes: $x = \pi n/8$, where n is an odd integer.

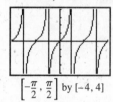

$\left[-\dfrac{\pi}{2}, \dfrac{\pi}{2}\right]$ by $[-4, 4]$

7. asymptotes: $x = \pi n$, where n is an integer; the graph is obtained from the graph of $y = \cot x$ by a vertical translation up one unit.

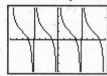

$[-2\pi, 2\pi]$ by $[-4, 4]$

8. asymptotes: $x = \pi n$, where n is an integer; the graph is obtained from the graph of $y = \cot x$ by a vertical translation down two units.

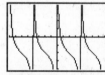

$[-2\pi, 2\pi]$ by $[-4, 4]$

9. asymptotes: $x = \pi n$, where n is an integer; the graph is obtained from the graph of $y = \cot x$ by a reflection across the x-axis.

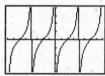

$[-2\pi, 2\pi]$ by $[-4, 4]$

10. asymptotes: $x = \pi n$, where n is an integer; the graph is obtained from the graph of $y = \cot x$ by a vertical stretch by a factor of 2 and a reflection across the x-axis.

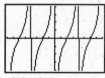

$[-2\pi, 2\pi]$ by $[-4, 4]$

11. asymptotes: $x = \pi n$, where n is an even integer; the graph is obtained from the graph of $y = \cot x$ by a horizontal stretch by a factor of 2 and a vertical translation up one unit.

$[-4\pi, 4\pi]$ by $[-4, 4]$

12. asymptotes: $x = \pi n$, where n is an integer; the graph is obtained from the graph of $y = \cot x$ by a horizontal shrink by a factor of 2, a vertical stretch by a factor of 2, and a vertical translation up one unit.

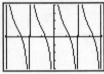

$[-\pi, \pi]$ by $[-4, 4]$

13. $\dfrac{\pi}{3}$ or $60°$ 14. $\dfrac{3\pi}{4}$ or $135°$

15. $\dfrac{7\pi}{4}$ or $315°$ 16. $\dfrac{3\pi}{4}$ or $135°$

17. $\dfrac{3\pi}{4}$ or $135°$ 18. $\dfrac{7\pi}{6}$ or $210°$ and $\dfrac{11\pi}{6}$ or $330°$

4.6 Graphs of Composite Functions

1. $f(x + 2\pi) = (\sin 3(x + 2\pi))^2 = (\sin 3x)^2 = f(x)$; period: $\pi/3$

2. $f(x + 2\pi) = (\sin 2(x + 2\pi))^2 = (\sin 2x)^2 = f(x)$; period: $\pi/2$

3. $f(x + 2\pi) = (\cos 2(x + 2\pi))^2 = (\cos 2x)^2 = f(x)$; period: $\pi/2$

4. $f(x + 2\pi) = (\cos 4(x + 2\pi))^2 = (\cos 4x)^2 = f(x)$; period: π

5. $f(x + 2\pi) = (-\cos (x + 2\pi))^2 = (-\cos x)^2 = f(x)$; period: π

6. $f(x + 2\pi) = (-\sin (x + 2\pi))^2 = (-\sin x)^2 = f(x)$; period: π

7. domain: all real numbers; range: $0 \le y \le 4$; period: π

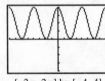

$[-2\pi, 2\pi]$ by $[-4, 4]$

8. domain: all real numbers; range: $0 \le y \le 1$; period: π

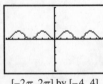

$[-2\pi, 2\pi]$ by $[-4, 4]$

9. domain: all real numbers; range: $0 \le y \le 4$; period: π

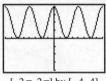

$[-2\pi, 2\pi]$ by $[-4, 4]$

10. domain: all real numbers; range: $-0.25 \le y \le 0$; period: π

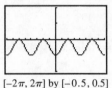

$[-2\pi, 2\pi]$ by $[-0.5, 0.5]$

11. domain: all real numbers except odd multiples of $\pi/2$; range: all nonnegative real numbers; period: π

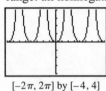

$[-2\pi, 2\pi]$ by $[-4, 4]$

12. domain: all real numbers except multiples of π; range: all nonnegative real numbers; period: π

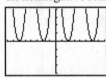

$[-2\pi, 2\pi]$ by $[-4, 4]$

13. period: 2π; $y = 1.41 \sin (x + 0.78)$

14. period: 2π; $y = 5 \sin (x + 0.64)$

15. period: 2π; $y = 2.24 \sin (x + 0.46)$

16. period: 2π; $y = 2.82 \sin (x + 0.79)$

17. period: 2π; $y = 1.41 \sin (x - 0.79)$

18. period: 2π; $y = 3.16 \sin (x + 0.32)$

4.7 Inverse Trigonometric Functions

1. $-45°$ 2. $-60°$ 3. does not exist

4. $\dfrac{\pi}{11}$ 5. $\dfrac{\pi}{3}$ 6. $\dfrac{\pi}{6}$

7. -0.535 8. 1.414 9. -1.098

10. -1.414 11. -0.212 12. 0.691

13. $\dfrac{5\pi}{6}$ or $150°$ 14. $\dfrac{3\pi}{4}$ or $135°$ 15. $-\dfrac{\pi}{4}$ or $-45°$

16. 0 17. 0.6 18. 1.8

4.8 Solving Problems with Trigonometry

1. about 596 ft   **2.** about 218 ft

3. about 584 ft   **4.** about 312 ft

5. about 2030 ft   **6.** about 668 ft

7. The boat's bearing from port is approximately 99.0°. They are about 102 miles out.

8. The boat's bearing from port is approximately 111.0°. They are about 102 miles out.

9. The boat's bearing from port is approximately 89.0°. They are about 102 miles out.

10. The boat's bearing from port is approximately 117.0°. They are about 102 miles out.

11. The boat's bearing from port is approximately 93.1°. They are about 150 miles out.

12. The boat's bearing from port is approximately 95.1°. They are about 150 miles out.

13. The boat's bearing from port is approximately 83.1°. They are about 150 miles out.

14. The boat's bearing from port is approximately 111.1°. They are about 150 miles out.

15. $d = 4 \sin\left(\dfrac{\pi}{4}t\right)$ **16.** $d = 7 \sin\left(\dfrac{\pi}{3}t\right)$

17. $d = 2 \sin\left(\dfrac{\pi}{4}t\right)$ **18.** $d = \sin 2\pi t$

19. $d = 10 \sin\left(\dfrac{2\pi}{5}t\right)$ **20.** $d = 16 \sin\left(\dfrac{\pi}{5}t\right)$

Chapter 5
Analytic Trigonometry

5.1 Fundamental Identities

1. $\sin\theta = -\dfrac{1}{\sqrt{2}}$ and $\cos\theta = -\dfrac{1}{\sqrt{2}}$

2. $\sin\theta = \dfrac{4}{\sqrt{17}}$ and $\cos\theta = \dfrac{1}{\sqrt{17}}$

3. $\sin\theta = \dfrac{3}{\sqrt{10}}$ and $\cos\theta = -\dfrac{1}{\sqrt{10}}$

4. $\sin\theta = -\dfrac{3}{\sqrt{10}}$ and $\cos\theta = \dfrac{1}{\sqrt{10}}$

5. $\sin\theta = -\dfrac{2}{\sqrt{5}}$ and $\cos\theta = \dfrac{1}{\sqrt{5}}$

6. $\sin\theta = \dfrac{5}{\sqrt{26}}$ and $\cos\theta = \dfrac{1}{\sqrt{26}}$

7. 1   **8.** $\tan^2 x \cos^4 x$

9. $\cos x + \sin^2 x$ **10.** 2

11. $2\cos x \sin^2 x$ **12.** $\sin^2 x \tan^2 x$

13. $\dfrac{\pi}{3}, \dfrac{\pi}{2}, \dfrac{3\pi}{2}, \dfrac{5\pi}{3}$ **14.** $0, \dfrac{\pi}{4}, \pi, \dfrac{5\pi}{4}$

15. $\dfrac{\pi}{3}, \dfrac{2\pi}{3}, \dfrac{4\pi}{3}, \dfrac{5\pi}{3}$ **16.** $\dfrac{\pi}{3}, \dfrac{2\pi}{3}, \dfrac{4\pi}{3}, \dfrac{5\pi}{3}$

17. $0, \pi$ **18.** $\dfrac{\pi}{2}, \dfrac{3\pi}{2}$

5.2 Proving Trigonometric Identities

1. $(1 - \cos x)^2 + 2\cos x = 1 - 2\cos x + \cos^2 x + 2\cos x$
$$= 1 + \cos^2 x$$
$$= (\sin^2 x + \cos^2 x) + \cos^2 x$$
$$= \sin^2 x + 2\cos^2 x$$

2. $(\sec x \div \csc x) \cot x = \left(\dfrac{1}{\cos x} \div \dfrac{1}{\sin x}\right) \cdot \cot x$
$$= \left(\dfrac{1}{\cos x} \cdot \sin x\right) \cdot \cot x$$
$$= \dfrac{\sin x}{\cos x} \cdot \dfrac{1}{\tan x}$$
$$= \tan x \cdot \dfrac{1}{\tan x}$$
$$= 1$$

3. $\dfrac{1 - \cos^2 x}{1 - \csc^2 x} = \dfrac{\sin^2 x}{-\cot^2 x}$
$$= -\sin^2 x \tan^2 x$$
$$= -\sin^2 x \cdot \dfrac{\sin^2 x}{\cos^2 x}$$
$$= -\dfrac{\sin^4 x}{\cos^2 x}$$

4. $\cos x + \sec x = \cos x + \dfrac{1}{\cos x} = \dfrac{\cos^2 x}{\cos x} + \dfrac{1}{\cos x}$
$$= \dfrac{1 + \cos^2 x}{\cos x}$$
$$= \dfrac{1 + (1 - \sin^2 x)}{\cos x}$$
$$= \dfrac{2 - \sin^2 x}{\cos x}$$

5. $1 + \tan^2 x = \sec^2 x = \dfrac{1}{\cos^2 x} = \dfrac{1}{1 - \sin^2 x}$

6. $\dfrac{\sin^2 x \csc^2 x}{\csc^2 x - \cot^2 x} = \dfrac{\sin^2 x \cdot \dfrac{1}{\sin^2 x}}{\csc^2 x - \cot^2 x}$

$= \dfrac{1}{\csc^2 x - \cot^2 x}$

$= \dfrac{1}{\csc^2 x - (\csc^2 x - 1)} = 1$

7. $\dfrac{1}{1 - \sin t} = \dfrac{1}{1 - \sin t} \cdot \dfrac{1 + \sin t}{1 + \sin t} = \dfrac{1 + \sin t}{(1 - \sin t)(1 + \sin t)}$

$= \dfrac{1 + \sin t}{1 - \sin^2 t}$

$= \dfrac{1 + \sin t}{\cos^2 t}$

8. $\dfrac{1}{1 - \cos t} = \dfrac{1}{1 - \cos t} \cdot \dfrac{1 + \cos t}{1 + \cos t} = \dfrac{1 + \cos t}{(1 - \cos t)(1 + \cos t)}$

$= \dfrac{1 + \cos t}{1 - \cos^2 t}$

$= \dfrac{1 + \cos t}{\sin^2 t}$

9. $\dfrac{1 + \cos t}{1 - \cos t} = \dfrac{1 + \cos t}{1 - \cos t} \cdot \dfrac{1 + \cos t}{1 + \cos t} = \dfrac{(1 + \cos t)^2}{1 - \sin^2 t}$

$= \dfrac{(1 + \cos t)^2}{\cos^2 t}$

10. $\dfrac{1 - \sin t}{1 + \sin t} = \dfrac{1 - \sin t}{1 + \sin t} \cdot \dfrac{1 - \sin t}{1 - \sin t} = \dfrac{(1 - \sin t)^2}{1 - \sin^2 t}$

$= \dfrac{(1 - \sin t)^2}{\cos^2 t}$

11. $\dfrac{\cos t - \sin t}{\cos t + \sin t} = \dfrac{\cos t - \sin t}{\cos t + \sin t} \cdot \dfrac{\cos t - \sin t}{\cos t - \sin t}$

$= \dfrac{(\cos t - \sin t)^2}{\cos^2 t - \sin^2 t}$

$= \dfrac{(\cos t - \sin t)^2}{(1 - \sin^2 t) - \sin^2 t}$

$= \dfrac{(\cos t - \sin t)^2}{1 - 2\sin^2 t}$

12. $\dfrac{1}{\sec t + \tan t} = \dfrac{1}{\sec t + \tan t} \cdot \dfrac{\sec t - \tan t}{\sec t - \tan t} = \dfrac{\sec t - \tan t}{\sec^2 t - \tan^2 t}$

$= \sec t - \tan t$

13. $\dfrac{\cos^2 x}{1 + \sin x} = \dfrac{1 - \sin^2 x}{1 + \sin x} = \dfrac{(1 - \sin x)(1 + \sin x)}{1 + \sin x}$

$= 1 - \sin x$

$= \dfrac{(1 - \sin x)(1 + \cos x)}{1 + \cos x}$

$= \dfrac{1 - \sin x \cos x - \sin x + \cos x}{1 + \cos x}$

14. $\dfrac{\cos^2 x - \sin^2 x}{\cos x + \sin x} = \dfrac{(\cos x - \sin x)(\cos x + \sin x)}{\cos x + \sin x}$

$= \cos x - \sin x$

$= \dfrac{1}{\sec x} - \dfrac{1}{\csc x}$

$= \dfrac{\csc x - \sec x}{\sec x \csc x}$

15. $\dfrac{\sec^2 x - \tan^2 x}{\sec x + \tan x} = \dfrac{(\sec x - \tan x)(\sec x + \tan x)}{\sec x + \tan x}$

$= \sec x - \tan x$

$= \dfrac{1}{\cos x} - \dfrac{\sin x}{\cos x}$

$= \dfrac{1 - \sin x}{\cos x}$

16. $\dfrac{\csc^2 x - \cot^2 x}{\csc x + \cot x} = \dfrac{(\csc x - \cot x)(\csc x + \cot x)}{\csc x + \cot x}$

$= \csc x - \cot x$

$= \dfrac{1}{\sin x} - \dfrac{\cos x}{\sin x}$

$= \dfrac{1 - \cos x}{\sin x}$

17.

$\dfrac{1}{\cos x - \sin x} + \dfrac{1}{\cos x + \sin x} = \dfrac{(\cos x + \sin x) + (\cos x - \sin x)}{\cos^2 x - \sin^2 x}$

$= \dfrac{2 \cos x}{\cos^2 x - \sin^2 x}$

$\dfrac{2}{\sec x(1 - 2\sin^2 x)} = \dfrac{2}{\sec x} \cdot \dfrac{1}{1 - 2\sin^2 x}$

$= 2 \cos x \cdot \dfrac{1}{(\cos^2 x + \sin^2 x) - 2\sin^2 x}$

$= 2 \cos x \cdot \dfrac{1}{\cos^2 x - \sin^2 x}$

$= \dfrac{2 \cos x}{\cos^2 x - \sin^2 x}$

18. $\dfrac{1}{\sec x - 1} + \dfrac{1}{\sec x + 1} = \dfrac{(\sec x - 1) + (\sec x + 1)}{\sec^2 x - 1}$

$= \dfrac{2 \sec x}{\sec^2 x - 1}$

$= \dfrac{2 \sec x}{\tan^2 x}$

$= \dfrac{2}{\cos x} \cdot \dfrac{1}{\tan^2 x}$

$= \dfrac{2}{\cos x \tan^2 x}$

5.3 Sum and Difference Identities

1. $\dfrac{\sqrt{6} + \sqrt{2}}{4}$ **2.** $\dfrac{1}{2}$ **3.** $\dfrac{\sqrt{6} - \sqrt{2}}{4}$

4. $\dfrac{\sqrt{6} + \sqrt{2}}{4}$ **5.** $\dfrac{\sqrt{6} - \sqrt{2}}{4}$ **6.** $\dfrac{-\sqrt{6} + \sqrt{2}}{4}$

7. $\sin 58°$ **8.** $\sin 41°$ **9.** $\cos \dfrac{\pi}{6}$

10. $\sin \dfrac{\pi}{20}$ **11.** $-\cos 11x$ **12.** $\cos 3x$

13. $\sin 55°$ **14.** $\cos \dfrac{\pi}{6}$

15. $\cos\left(x-\dfrac{3\pi}{2}\right)=\cos x\cos\dfrac{3\pi}{2}+\sin x\sin\dfrac{3\pi}{2}$
$$= \cos x\,(0)+\sin x\,(-1)$$
$$= -\sin x$$

16. $\sin\left(x-\dfrac{3\pi}{2}\right)=\sin x\cos\dfrac{3\pi}{2}-\cos x\sin\dfrac{3\pi}{2}$
$$= \sin x\,(0)-\cos x\,(-1)$$
$$= \cos x$$

17. $\sin(2\pi-x)=\sin 2\pi\cos x-\sin x\cos 2\pi$
$$= (0)\cos x-\sin x\,(1)$$
$$= -\sin x$$

18. $\sin\left(\dfrac{3\pi}{2}-x\right)=\sin\dfrac{3\pi}{2}\cos x-\cos\dfrac{3\pi}{2}\sin x$
$$= (-1)\cos x-(0)\sin x$$
$$= -\cos x$$

19. $\sin(x+\pi)=\sin x\cos\pi+\cos x\sin\pi$
$$= \sin x\,(-1)+\cos x\,(0)$$
$$= -\sin x$$

20. $\cos(x+\pi)=\cos x\cos\pi-\sin x\sin\pi$
$$= \cos x\,(-1)-\sin x\,(0)$$
$$= -\cos x$$

5.4 Multiple-Angle Identities

1. $\cos 4x=\cos(2x+2x)$
$$= \cos 2x\cos 2x-\sin 2x\sin 2x$$
$$= \cos^2 2x-\sin^2 2x$$

2. $\sin 4x=\sin(2x+2x)$
$$= \sin 2x\cos 2x+\cos 2x\sin 2x$$
$$= 2\sin 2x\cos 2x$$

3. $\cos 6x=\cos(3x+3x)$
$$= \cos 3x\cos 3x-\sin 3x\sin 3x$$
$$= \cos^2 3x-\sin^2 3x$$

4. $\sin 6x=\sin(3x+3x)$
$$= \sin 3x\cos 3x+\cos 3x\sin 3x$$
$$= 2\sin 3x\cos 3x$$

5. $\cos nx$
$$= \cos\left(\dfrac{n}{2}x+\dfrac{n}{2}x\right)$$
$$= \cos\left(\dfrac{n}{2}x\right)\cos\left(\dfrac{n}{2}x\right)-\sin\left(\dfrac{n}{2}x\right)\sin\left(\dfrac{n}{2}x\right)$$
$$= \cos^2\left(\dfrac{n}{2}x\right)-\sin^2\left(\dfrac{n}{2}x\right)$$

6. $\sin nx$
$$= \sin\left(\dfrac{n}{2}x+\dfrac{n}{2}x\right)$$
$$= \sin\left(\dfrac{n}{2}x\right)\cos\left(\dfrac{n}{2}x\right)+\cos\left(\dfrac{n}{2}x\right)\sin\left(\dfrac{n}{2}x\right)$$
$$= 2\sin\left(\dfrac{n}{2}x\right)\cos\left(\dfrac{n}{2}x\right)$$

7. $x=0,\dfrac{2\pi}{3},\dfrac{4\pi}{3}$ **8.** $x=0,\pi$

9. $x=\dfrac{\pi}{6},\dfrac{5\pi}{6},\dfrac{3\pi}{2}$ **10.** $x=0,\dfrac{\pi}{3},\dfrac{2\pi}{3},\dfrac{4\pi}{3},\dfrac{5\pi}{3},\pi$

11. $x=\dfrac{\pi}{3},\pi,\dfrac{5\pi}{3}$ **12.** $x=\dfrac{\pi}{3},\dfrac{5\pi}{3}$

13. $x=\dfrac{\pi}{3}+2\pi n,\ \dfrac{5\pi}{3}+2\pi n,\ n=0,\pm1,\pm2,\ldots$

14. $x=\dfrac{\pi}{3}+2\pi n,\ \dfrac{5\pi}{3}+2\pi n,\ n=0,\pm1,\pm2,\ldots$

15. $x=\dfrac{\pi}{3}+2\pi n,\dfrac{5\pi}{3}+2\pi n,\ n=0,\pm1,\pm2,\ldots$

16. $x=\dfrac{\pi}{3}+2\pi n,\dfrac{5\pi}{3}+2\pi n,\ \pi+2\pi n,\ n=0,\pm1,\pm2,\ldots$

17. $x=\dfrac{\pi}{2}+2\pi n,\dfrac{3\pi}{2}+2\pi n,\ n=0,\pm1,\pm2,\ldots$

18. $x=\dfrac{\pi}{2}+2\pi n,\ n=0,\pm1,\pm2,\ldots$

5.5 The Law of Sines

1. $A=40°$, $a=10$, $B=70°$, $b=14.6$, $C=70°$, and $c=14.6$

2. $A=45°$, $a=12$, $B=70°$, $b=15.9$, $C=65°$, and $c=15.4$

3. $A=100°$, $a=18$, $B=20°$, $b=6.3$, $C=60°$, and $c=15.8$

4. $A=33°$, $a=30$, $B=71°$, $b=52.1$, $C=76°$, and $c=53.4$

5. $A=100°$, $a=8$, $B=21°$, $b=16.5$, $C=149°$, and $c=23.7$

6. $A=68°$, $a=34$, $B=15°$, $b=9.5$, $C=97°$, and $c=36.4$

7. $A=41°$, $a=10$, $B=41°$, $b=10$, $C=98°$, and $c=15.1$

8. $A=35°$, $a=9$, $B=45°$, $b=11$, $C=100°$, and $c=15.4$

9. $A=23°$, $a=15$, $B=18°$, $b=12$, $C=139°$, and $c=25.3$

10. $A=32°$, $a=18$, $B=32°$, $b=18$, $C=116°$, and $c=30.5$

11. $A = 54°$, $a = 20$, $B = 43°$, $b = 17$, $C = 83°$, and $c = 24.5$

12. $A = 50°$, $a = 19$, $B = 47°$, $b = 18$, $C = 83°$, and $c = 24.6$

13. $A = 34°$, $a = 9$, $B = 43°$, $b = 11$, $C = 103°$, and $c = 15.7$
    $A = 34°$, $a = 9$, $B = 137°$, $b = 11$, $C = 9°$, and $c = 2.5$

14. $A = 20°$, $a = 15$, $B = 33°$, $b = 24$, $C = 127°$, and $c = 35.1$
    $A = 20°$, $a = 15$, $B = 147°$, $b = 24$, $C = 13°$, and $c = 9.9$

15. $A = 31°$, $a = 13$, $B = 49°$, $b = 19$, $C = 100°$, and $c = 24.8$
    $A = 31°$, $a = 13$, $B = 131°$, $b = 19$, $C = 18°$, and $c = 7.8$

16. $A = 48°$, $a = 25$, $B = 63°$, $b = 30$, $C = 69°$, and $c = 31.4$
    $A = 48°$, $a = 25$, $B = 117°$, $b = 30$, $C = 15°$, and $c = 8.8$

17. $A = 27°$, $a = 8$, $B = 43°$, $b = 12$, $C = 110°$, and $c = 16.6$
    $A = 27°$, $a = 8$, $B = 137°$, $b = 12$, $C = 16°$, and $c = 4.8$

18. $A = 29°$, $a = 13$, $B = 37°$, $b = 16$, $C = 114°$, and $c = 24.4$
    $A = 29°$, $a = 13$, $B = 143°$, $b = 16$, $C = 7°$, and $c = 3.7$

5.6 The Law of Cosines

1. $A = 115.0°$, $a = 15$, $B = 25.0°$, $b = 7$, $C = 40°$, and $c = 10.6$
2. $A = 72.5°$, $a = 20$, $B = 72.5°$, $b = 20$, $C = 35°$, and $c = 12.0$
3. $A = 29.5°$, $a = 6$, $B = 80.5°$, $b = 12$, $C = 70°$, and $c = 11.4$
4. $A = 60°$, $a = 1$, $B = 60°$, $b = 1$, $C = 60°$, and $c = 1$
5. $A = 48.0°$, $a = 15$, $B = 82.0°$, $b = 20$, $C = 50°$, and $c = 15.5$
6. $A = 55.2°$, $a = 35$, $B = 44.8°$, $b = 30$, $C = 80°$, and $c = 42.0$
7. $A = 41.8°$, $B = 91.0°$, and $C = 47.2°$
8. $A = 82.8°$, $B = 55.8°$, and $C = 41.4°$
9. $A = 27.7°$, $B = 68.2°$, and $C = 84.1°$
10. $A = 36.9°$, $B = 53.1°$, and $C = 90°$
11. $A = 27.3°$, $B = 27.3°$, and $C = 125.4°$
12. $A = 24.1°$, $B = 54.9°$, and $C = 101.0°$
13. about 103.6 square units
14. about 332.8 square units
15. about 149.3 square units
16. 6 square units
17. about 73.2 square units
18. about 1139.9 square units

Chapter 6
Applications of Trigonometry

6.1 Vectors in the Plane

1. $\langle 3, 2 \rangle$ 2. $\langle 3, 0 \rangle$ 3. $\langle -3, 0 \rangle$ 4. $\langle 8, 3 \rangle$

5. $\langle 0, 6 \rangle$ 6. $\langle 3, 6 \rangle$ 7. $\langle -1, 8 \rangle$ 8. $\langle 0, -6 \rangle$

9. $\left\langle \dfrac{2}{\sqrt{5}}, -\dfrac{1}{\sqrt{5}} \right\rangle$ 10. $\left\langle \dfrac{2}{\sqrt{13}}, \dfrac{3}{\sqrt{13}} \right\rangle$

11. $\dfrac{3}{\sqrt{13}}\mathbf{i} + \dfrac{2}{\sqrt{13}}\mathbf{j}$ 12. $-\dfrac{1}{\sqrt{2}}\mathbf{i} - \dfrac{1}{\sqrt{2}}\mathbf{j}$

13. $\left\langle \dfrac{4}{5}, \dfrac{3}{5} \right\rangle$ 14. $\left\langle -\dfrac{1}{\sqrt{10}}, -\dfrac{3}{\sqrt{10}} \right\rangle$

15. $\left\langle -\dfrac{4}{5}, \dfrac{3}{5} \right\rangle$ 16. $\dfrac{12}{\sqrt{153}}\mathbf{i} - \dfrac{3}{\sqrt{153}}\mathbf{j}$

17. $\langle -2.071, 7.727 \rangle$ 18. $\langle 1.941, 7.244 \rangle$

19. $\langle -3.532, -1.878 \rangle$ 20. $\langle -8, 13.856 \rangle$

21. $\langle 17.050, 18.284 \rangle$ 22. $\langle -16.168, -5.253 \rangle$

23. $\sqrt{5}$; $333.43°$ 24. 5; $36.87°$

25. $\sqrt{53}$; $254.05°$ 26. $\sqrt{2}$; $225°$

27. $\sqrt{4.25}$; $284.04°$ 28. $\sqrt{31.25}$; $116.57°$

29. $\sqrt{52}$ or $2\sqrt{13}$; $326.31°$ 30. $\sqrt{8.5}$; $59.04°$

6.2 Dot Product of Vectors

1. 13   2. 8   3. 0

4. -15 5. 22.5   6. 1.6

7. 2   8. -9 9. about $50°$

10. about $7°$ 11. $90°$ 12. about $162°$

13. about $7°$ 14. about $139°$ 15. about $164°$

16. $135°$

17. $\text{proj}_v\,\mathbf{u} = \left\langle -\dfrac{84}{17}, -\dfrac{21}{17} \right\rangle$; $\mathbf{u} = \left\langle -\dfrac{84}{17}, -\dfrac{21}{17} \right\rangle + \left\langle -\dfrac{18}{17}, \dfrac{72}{17} \right\rangle$

18. $\text{proj}_v\,\mathbf{u} = \left\langle \dfrac{248}{73}, -\dfrac{93}{73} \right\rangle$; $\mathbf{u} = \left\langle \dfrac{248}{73}, -\dfrac{93}{73} \right\rangle + \left\langle \dfrac{117}{73}, \dfrac{312}{73} \right\rangle$

19. $\text{proj}_v\,\mathbf{u} = \left\langle \dfrac{40}{73}, \dfrac{15}{73} \right\rangle$; $\mathbf{u} = \left\langle \dfrac{40}{73}, \dfrac{15}{73} \right\rangle + \left\langle -\dfrac{186}{73}, \dfrac{496}{73} \right\rangle$

20. $\text{proj}_v\,\mathbf{u} = \langle -1, 2 \rangle$; $\mathbf{u} = \langle -1, 2 \rangle + \langle 4, 2 \rangle$

21. $\text{proj}_v\,\mathbf{u} = \langle -3, 0 \rangle$; $\mathbf{u} = \langle -3, 0 \rangle + \langle 0, 2 \rangle$

22. $\text{proj}_v\,\mathbf{u} = \langle 2, 2 \rangle$; $\mathbf{u} = \langle 2, 2 \rangle + \langle 5, -5 \rangle$

23. $\text{proj}_v\,\mathbf{u} = \left\langle \dfrac{4}{5}, -\dfrac{12}{5} \right\rangle$; $\mathbf{u} = \left\langle \dfrac{4}{5}, -\dfrac{12}{5} \right\rangle + \left\langle -\dfrac{9}{5}, -\dfrac{3}{5} \right\rangle$

24. $\text{proj}_v\,\mathbf{u} = \left\langle -\dfrac{9}{2}, -\dfrac{3}{2} \right\rangle$; $\mathbf{u} = \left\langle -\dfrac{9}{2}, -\dfrac{3}{2} \right\rangle + \left\langle \dfrac{1}{2}, -\dfrac{3}{2} \right\rangle$

6.3 Parametric Equations and Motion

1. $y = -2x + 5$; line through $(0, 5)$ and $\left(\dfrac{5}{2}, 0\right)$

2. $y = \dfrac{x}{3} + \dfrac{10}{3}$; line through $(-10, 0)$ and $\left(0, \dfrac{10}{3}\right)$

3. $y = -x - 1$; line through $(0, -1)$ and $(-1, 0)$

4. $y = 4x - 9$; line through $(0, -9)$ and $\left(\dfrac{9}{4}, 0\right)$

5. $y = 1 - x^2$; parabola that opens downward, vertex at $(0, 1)$

6. $y = 1 + 2(x + 2)^2$; parabola that opens upward, vertex at $(-2, 1)$

7. $y = \dfrac{(x - 1)^2}{2} + 1$; parabola that opens upward, vertex at $(-1, 1)$

8. $x = (2 - y)^2$; parabola that opens to the right, vertex at $(0, 2)$

9. $x^2 + y^2 = 36$; circle centered at the origin with radius 6

10. $x^2 + y^2 = 6.25$; circle centered at the origin with radius 2.5

11. $x^2 + y^2 = 4$; circle centered at the origin with radius 2

12. $x^2 + y^2 = 16$; semicircle with center at the origin and radius 4 in first and fourth quadrants

13. $x^2 + y^2 = 9$; circle centered at the origin with radius 3, traced twice

14. $x^2 + y^2 = 25$; circular arc with center at the origin and radius 5, central angle $120°$, and initial side on the x-axis

15. $x = 2 + 4t,\ y = 1,\ 0 \le t \le 1$

16. $x = 1 - 3t,\ y = -1 + 8t,\ 0 \le t \le 1$

17. $x = 12 - 8t,\ y = -2 + 5t,\ 0 \le t \le 1$

18. $x = 1 + 2t,\ y = 1 + 2t,\ 0 \le t \le 1$

19. $x = 5 - 9t,\ y = 7 - 8t,\ 0 \le t \le 1$

20. $x = t,\ y = 1 - t,\ 0 \le t \le 1$

21. $x = t,\ y = -16t^2 + 25t + 4.5$; max height ≈ 14.27

22. $x = t,\ y = -16t^2 + 14t + 6$; max height $= 9.0625$

23. $x = t,\ y = -16t^2 + 9t + 1$; max height ≈ 2.266

24. $x = t,\ y = -16t^2 + 100$; $t = 2.5$ sec

25. $x = t,\ y = -16t^2 + 200$; $t \approx 3.536$ sec

26. $x = t,\ y = -16t^2 + 500$; $t \approx 5.59$ sec

6.4 Polar Coordinates

1. $(-2.12, 2.12)$ **2.** $(-1.73, -1)$ **3.** $(-2.60, -1.50)$

4. $(-1.91, -4.62)$ **5.** $(-1.64, 1.15)$ **6.** $(-0.69, -3.94)$

7. $(1.04, 5.91)$ **8.** $(6, -10.39)$

9. $\left(3\sqrt{2}, \dfrac{3\pi}{4}\right),\ \left(-3\sqrt{2}, \dfrac{-\pi}{4}\right)$

10. $\left(\sqrt{29}, 1.19\right),\ \left(-\sqrt{29}, -1.95\right)$

11. $\left(\sqrt{34}, -1.03\right),\ \left(-\sqrt{34}, 2.11\right)$

12. $\left(\sqrt{10}, -0.32\right),\ \left(-\sqrt{10}, 2.82\right)$

13. $\left(2\sqrt{2}, \dfrac{-3\pi}{4}\right),\ \left(-2\sqrt{2}, \dfrac{\pi}{4}\right)$

14. $\left(\sqrt{116}, 1.95\right),\ \left(-\sqrt{116}, -1.19\right)$

15. $(-13, -1.17),\ (13, 1.97)$

16. $\left(2, \dfrac{\pi}{4}\right),\ \left(-2, -\dfrac{3\pi}{4}\right)$

17. horizontal line

$y = 4$

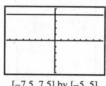

$[-7.5, 7.5]$ by $[-5, 5]$

18. circle

$(x - 1)^2 + \left(y - \dfrac{1}{2}\right)^2 = \dfrac{5}{4}$

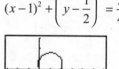

$[-3, 6]$ by $[-3, 3]$

19. circle

$x^2 + (y + 3)^2 = 9$

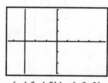

$[-6, 6]$ by $[-6, 2]$

20. vertical line

$x = -3$

$[-4.5, 4.5]$ by $[-3, 3]$

21. circle

$(x + 2.5)^2 + (y - 1.5)^2 = 8.5$

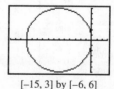

$[-6, 6]$ by $[-3, 5]$

22. circle

$(x + 6)^2 + y^2 = 36$

$[-15, 3]$ by $[-6, 6]$

23. $r = 8 \sin \theta$

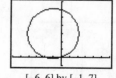

$[-7.5, 7.5]$ by $[-1, 9]$

24. $r = -2 \cos \theta + 6 \sin \theta$

$[-6, 6]$ by $[-1, 7]$

25. $r = 6\cos\theta + 8\sin\theta$

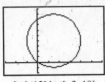

[−6, 12] by [−2, 10]

26. $r = 3\csc\theta$

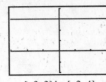

[−3, 3] by [−2, 4]

27. $r = \dfrac{3}{\cos\theta + 5\sin\theta}$

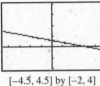

[−4.5, 4.5] by [−2, 4]

28. $r = \dfrac{3}{3\sin\theta - 2\cos\theta}$

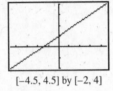

[−4.5, 4.5] by [−2, 4]

6.5 Graphs of Polar Equations

1. symmetric about the x-axis

2. symmetric about the x-axis

3. symmetric about the y-axis

4. all three symmetries

5. symmetric about the y-axis

6. x-axis

7. maximum r-value $= 5$ when $\theta = \dfrac{n\pi}{4}$ for any integer n.

8. maximum r-value $= 2$ when $\theta = \dfrac{n\pi}{10}$ for any odd integer n.

9. maximum r-value $= 2$ when $\theta = \dfrac{n\pi}{8}$ for any odd integer n.

10. maximum r-value $= 4$ when $\theta = 0$

11. maximum r-value $= 4$ when $\theta = \pi$

12. maximum r-value $= 7$ when $\theta = \dfrac{\pi}{2}$

13. Domain: All reals.
Range: [−4, 4]
Continuous
Symmetry: y-axis
Bounded
Maximum r-value: 4
No asymptotes

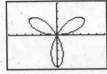

[−6, 6] by [−4, 4]

14. Domain: All reals.
Range: [−5, 5]
Continuous
Symmetry: x-axis, y-axis
origin
Bounded
Maximum r-value: 5
No asymptotes

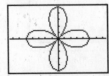

[−7.5, 7.5] by [−5, 5]

15. Domain: All reals.
Range: [−4, 4]
Continuous
Symmetry: x-axis, y-axis
origin
Bounded
Maximum r-value: 4
No asymptotes

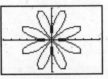

[−6, 6] by [−4, 4]

16. Domain: All reals.
Range: [−3, 3]
Continuous
Symmetry: x-axis
Bounded
Maximum r-value: 3
No asymptotes

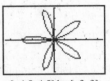

[−4.5, 4.5] by [−3, 3]

17. Domain: All reals.
Range: [−2, 2]
Continuous
Symmetry: x-axis, y-axis
origin
Bounded
Maximum r-value: 2
No asymptotes

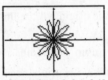

[−4.5, 4.5] by [−3.1, 3.1]

18. Domain: All reals.
Range: [−5, 5]
Continuous
Symmetry: x-axis
Bounded
Maximum r-value: 5
No asymptotes

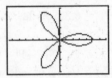

[−7.5, 7.5] by [−5, 5]

19. Domain: All reals.
Range: [−1, 3]
Continuous
Symmetry: x-axis
Bounded
Maximum r-value: 3
No asymptotes

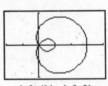

[−2, 4] by [−2, 2]

20. Domain: All reals.
Range: [−2, 6]
Continuous
Symmetry: y-axis
Bounded
Maximum r-value: 6
No asymptotes

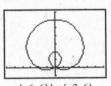

[−6, 6] by [−2, 6]

21. Domain: All reals.
Range: [1, 9]
Continuous
Symmetry: x-axis
Bounded
Maximum r-value: 9
No asymptotes

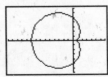

[−14, 7] by [−7, 7]

22. Domain: All reals.
Range: [5, 7]
Continuous
Symmetry: y-axis
Bounded
Maximum r-value: 7
No asymptotes

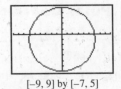

[−9, 9] by [−7, 5]

23. Domain: All reals.
Range: $[-3, 9]$
Continuous
Symmetry: x-axis
Bounded
Maximum r-value: 9
No asymptotes

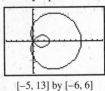

$[-5, 13]$ by $[-6, 6]$

24. Domain: All reals.
Range: $[0, 6]$
Continuous
Symmetry: y-axis
Bounded
Maximum r-value: 6
No asymptotes

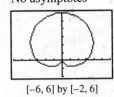

$[-6, 6]$ by $[-2, 6]$

6.6 De Moivre's Theorem and nth Roots

1. $3\sqrt{2}\left(\cos\dfrac{\pi}{4} + i\sin\dfrac{\pi}{4}\right)$

2. $6\left(\cos\dfrac{3\pi}{2} + i\sin\dfrac{3\pi}{2}\right)$

3. $\dfrac{2}{3}\left(\cos\dfrac{11\pi}{6} + i\sin\dfrac{11\pi}{6}\right)$

4. $4\sqrt{13}\left(\cos 1.85 + i\sin 1.85\right)$

5. $\sqrt{53}\left(\cos 4.43 + i\sin 4.43\right)$

6. $\sqrt{193}\left(\cos 2.10 + i\sin 2.10\right)$

7. $33(\cos 80° + i\sin 80°)$

8. $\dfrac{\sqrt{5}}{2}(\cos 90° + i\sin 90°)$

9. $14(\cos 142° + i\sin 142°)$

10. $\sqrt{6}\left(\cos\dfrac{17\pi}{12} + i\sin\dfrac{17\pi}{12}\right)$

11. $10\left(\cos\dfrac{\pi}{2} + i\sin\dfrac{\pi}{2}\right)$

12. $\dfrac{1}{2}\left(\cos\dfrac{19\pi}{12} + i\sin\dfrac{19\pi}{12}\right)$

13. $\dfrac{3}{2}(\cos 20° + i\sin 20°)$

14. $\dfrac{1}{4}(\cos 125° + i\sin 125°)$

15. $4\sqrt{2}\left(\cos(-200°) + i\sin(-200°)\right)$

16. $3(\cos(-\pi) + i\sin(-\pi))$

17. $\dfrac{3}{2}\left(\cos\dfrac{\pi}{6} + i\sin\dfrac{\pi}{6}\right)$

18. $\cos\dfrac{\pi}{6} + i\sin\dfrac{\pi}{6}$

19. $\cos\dfrac{\pi}{2} + i\sin\dfrac{\pi}{2},\ \cos\dfrac{5\pi}{6} + i\sin\dfrac{5\pi}{6},\ \cos\dfrac{7\pi}{6} + i\sin\dfrac{7\pi}{6},$

$\cos\dfrac{3\pi}{2} + i\sin\dfrac{3\pi}{2},\ \cos\dfrac{11\pi}{6} + i\sin\dfrac{11\pi}{6},$

$\cos\dfrac{13\pi}{6} + i\sin\dfrac{13\pi}{6}$

20. $2\left(\cos\dfrac{\pi}{8}\,i\sin\dfrac{\pi}{8}\right),\ 2\left(\cos\dfrac{5\pi}{8}\,i\sin\dfrac{5\pi}{8}\right),$

$2\left(\cos\dfrac{9\pi}{8}\,i\sin\dfrac{9\pi}{8}\right),\ 2\left(\cos\dfrac{13\pi}{8}\,i\sin\dfrac{13\pi}{8}\right)$

21. trig form: $z = \cos\pi + i\sin\pi$;

$\cos\dfrac{\pi}{5} + i\sin\dfrac{\pi}{5},\ \cos\dfrac{3\pi}{5} + i\sin\dfrac{3\pi}{5},\ \cos\pi + i\sin\pi,$

$\cos\dfrac{7\pi}{5} + i\sin\dfrac{7\pi}{5},\ \cos\dfrac{9\pi}{5} + i\sin\dfrac{9\pi}{5}$

22. trig form: $z = \cos\dfrac{\pi}{2} + i\sin\dfrac{\pi}{2}$;

$\cos\dfrac{\pi}{8}\,i\sin\dfrac{\pi}{8},\ \cos\dfrac{5\pi}{8}\,i\sin\dfrac{5\pi}{8},\ \cos\dfrac{9\pi}{8}\,i\sin\dfrac{9\pi}{8},$

$\cos\dfrac{13\pi}{8}\,i\sin\dfrac{13\pi}{8}$

23. trig form: $z = \cos\dfrac{11\pi}{6} + i\sin\dfrac{11\pi}{6}$;

$\left(\cos\dfrac{11\pi}{18} + i\sin\dfrac{11\pi}{18}\right),\ \left(\cos\dfrac{23\pi}{18} + i\sin\dfrac{23\pi}{18}\right),$

$\left(\cos\dfrac{35\pi}{18} + i\sin\dfrac{35\pi}{18}\right)$

24. trig form: $z = 4\sqrt{2}\left(\cos\dfrac{\pi}{4} + i\sin\dfrac{\pi}{4}\right)$;

$4\sqrt{2}\left(\cos\dfrac{\pi}{24} + i\sin\dfrac{\pi}{24}\right);\ 4\sqrt{2}\left(\cos\dfrac{3\pi}{8} + i\sin\dfrac{3\pi}{8}\right),$

$4\sqrt{2}\left(\cos\dfrac{17\pi}{24} + i\sin\dfrac{17\pi}{24}\right),\ 4\sqrt{2}\left(\cos\dfrac{25\pi}{24} + i\sin\dfrac{25\pi}{24}\right),$

$4\sqrt{2}\left(\cos\dfrac{11\pi}{8} + i\sin\dfrac{11\pi}{8}\right),\ 4\sqrt{2}\left(\cos\dfrac{41\pi}{24} + i\sin\dfrac{41\pi}{24}\right)$

Chapter 7
Systems and Matrices

7.1 Solving Systems of Two Equations

1. $(-2, 3)$

2. $(2, -1)$

3. $\left(\dfrac{21}{11}, \dfrac{8}{11}\right)$

4. $(5, 3)$

5. $(2, 3)$

6. $(-3, 1)$

7. $(5, 10)$ and $(22.5, -25)$

8. $(3, 9)$ and $(-5, 25)$

9. $(9, 0)$ and $(2, 7)$

10. $(0.5, -1.5)$ and $(6, 4)$

11. $(6, 8)$ and $(10, 0)$ **12.** $(2, 1)$ and $\left(\dfrac{20}{3}, -\dfrac{25}{3}\right)$

13. $(5, -3)$ **14.** $(1, -1)$

15. $(1, 4)$ **16.** $(3, 12)$

17. $(-6, -1)$ **18.** $(0, 2)$

19. no solution   **20.** no solution

21. infinitely many solutions

22. infinitely many solutions

23. no solution   **24.** no solution

25. infinitely many solutions

26. infinitely many solutions

7.2 Matrix Algebra

1. $\begin{bmatrix} 2 & 3 \\ 5 & 0 \\ 2 & 6 \end{bmatrix}; \begin{bmatrix} -4 & 1 \\ -5 & -2 \\ 2 & 2 \end{bmatrix}$ **2.** $\begin{bmatrix} -4 \\ 2 \\ 3 \end{bmatrix}; \begin{bmatrix} 2 \\ -2 \\ 3 \end{bmatrix}$

3. $\begin{bmatrix} 6 & -1 \\ 3 & 4 \end{bmatrix}; \begin{bmatrix} 16 & -1 \\ -3 & 2 \end{bmatrix}$

4. $\begin{bmatrix} -5 & 1 & -5 & 7 \\ 1 & 2 & -6 & -4 \end{bmatrix}; \begin{bmatrix} -9 & 1 & 5 & 1 \\ -1 & 6 & -6 & -2 \end{bmatrix}$

5. $\begin{bmatrix} 2 & 2 & 3 \\ -1 & 2 & 4 \\ -2 & -1 & 2 \end{bmatrix}; \begin{bmatrix} 0 & 2 & 3 \\ 1 & 0 & 4 \\ 2 & 1 & 0 \end{bmatrix}$

6. $\begin{bmatrix} 2 & 5 & 1 & 5 \\ 0 & 7 & 2 & 10 \\ 0 & 0 & 9 & -5 \end{bmatrix}; \begin{bmatrix} 4 & -1 & 1 & -5 \\ -4 & 3 & -2 & 4 \\ 0 & -2 & -5 & 3 \end{bmatrix}$

7. $\begin{bmatrix} 1 & 5 \\ 5 & -1 \\ 4 & 10 \end{bmatrix}; \begin{bmatrix} -11 & 1 \\ -15 & -5 \\ 4 & 2 \end{bmatrix}$ **8.** $\begin{bmatrix} -5 \\ 2 \\ 6 \end{bmatrix}; \begin{bmatrix} 7 \\ -6 \\ 6 \end{bmatrix}$

9. $\begin{bmatrix} 17 & -2 \\ 3 & 7 \end{bmatrix}; \begin{bmatrix} 37 & -2 \\ -9 & 3 \end{bmatrix}$

10. $\begin{bmatrix} -12 & 2 & -5 & 11 \\ 1 & 6 & -12 & -7 \end{bmatrix}; \begin{bmatrix} -20 & 2 & 15 & -1 \\ -3 & 14 & -12 & -3 \end{bmatrix}$

11. $\begin{bmatrix} 3 & 4 & 6 \\ -1 & 3 & 8 \\ -2 & -1 & 3 \end{bmatrix}; \begin{bmatrix} -1 & 4 & 6 \\ 3 & -1 & 8 \\ 6 & 3 & -1 \end{bmatrix}$

12. $\begin{bmatrix} 5 & 7 & 2 & 5 \\ -2 & 12 & 2 & 17 \\ 0 & -1 & 11 & -6 \end{bmatrix}; \begin{bmatrix} 9 & -5 & 2 & -15 \\ -10 & 4 & -6 & 5 \\ 0 & -5 & -17 & 10 \end{bmatrix}$

13. $\begin{bmatrix} -18 & -1 \\ -7 & 1 \end{bmatrix}; \begin{bmatrix} -15 & 5 \\ 11 & -2 \end{bmatrix}$ **14.** $\begin{bmatrix} -2 & -3 \\ 6 & 7 \end{bmatrix}; \begin{bmatrix} 1 & 0 \\ 3 & 4 \end{bmatrix}$

15. $\begin{bmatrix} 0 & -9 \\ 26 & 2 \end{bmatrix}; \begin{bmatrix} -3 & 20 & 11 & 1 \\ -7 & 4 & 15 & -19 \\ 0 & -8 & -2 & -4 \\ 1 & 0 & -2 & 3 \end{bmatrix}$

16. $\begin{bmatrix} 3 & 7 \\ -16 & 40 \end{bmatrix}; \begin{bmatrix} 1 & 5 & 10 \\ -1 & 2 & 5 \\ -17 & 13 & 40 \end{bmatrix}$

17. $\begin{bmatrix} -7 & -1 & 3 \\ -9 & -3 & 4 \\ -2 & -1 & 1 \end{bmatrix}; \begin{bmatrix} 1 & 2 & 3 \\ -1 & -1 & 1 \\ -2 & -5 & -9 \end{bmatrix}$

18. $\begin{bmatrix} 6 & -1 & -3 \\ 16 & 0 & -1 \\ -5 & 4 & 6 \end{bmatrix}; \begin{bmatrix} 5 & -7 & -1 \\ 2 & 7 & 1 \\ 3 & 11 & 0 \end{bmatrix}$ **19.** $\begin{bmatrix} 0.6 & -0.2 \\ -0.2 & 0.4 \end{bmatrix}$

20. no inverse   **21.** $\begin{bmatrix} 0.5 & 0.5 \\ 0.75 & 1.25 \end{bmatrix}$

22. no inverse   **23.** no inverse

24. $\begin{bmatrix} -0.1 & -0.4 & 0.6 \\ 0.1 & -0.6 & 0.4 \\ 0.4 & 0.6 & -0.4 \end{bmatrix}$

7.3 Multivariate Linear Systems and Row Operations

1. $\left(\dfrac{10}{3}, \dfrac{1}{2}, 1\right)$ **2.** $(1, -3, 3)$ **3.** $\left(\dfrac{1}{2}, 4, 2\right)$

4. $(2, 2, -2)$ **5.** $\left(1, \dfrac{2}{5}, -1\right)$ **6.** $(1, -4, 0)$

7. $(1, 2, -1)$ **8.** $\left(\dfrac{5}{13}, \dfrac{10}{13}, \dfrac{74}{13}\right)$ **9.** $(1, 2, 3)$

10. $(3, 1, 1)$ **11.** $\left(\dfrac{69}{20}, -\dfrac{3}{20}, \dfrac{67}{20}\right)$ **12.** $(0, 1, 2)$

13. $(-2z - 1, z + 3, z)$ **14.** $(3 - z, 0.5z - 2.5, z)$

15. no solution   **16.** $\left(-1 - 8z, 1 - \dfrac{2}{3}z, z\right)$

17. $(5 - z, 7 - z, z)$ **18.** $(1 - z, 2 - 2z, z)$

19. $(-71, 30)$ **20.** $(-12.5, -8.5)$

21. $(3, 0)$ **22.** $(2, 1, -1)$

23. $(-25.29, -13, -12.57)$ **24.** $(-3.2, 2.4, -1)$

7.4 Partial Fractions

1. $\dfrac{A_1}{x+2} + \dfrac{A_2}{(x+2)^2} + \dfrac{B_1}{3x-2} + \dfrac{B_2}{(3x-2)^2}$

2. $3 + \dfrac{A_1}{x} + \dfrac{A_2}{x^2} + \dfrac{B_1}{x-1}$

3. $x - 1 + \dfrac{A_1}{x} + \dfrac{B_1}{x-1} + \dfrac{C_1}{x+2}$

4. $\dfrac{A_1}{x} + \dfrac{A_2}{x^2} + \dfrac{A_3}{x^3} + \dfrac{B_1}{2x+3} + \dfrac{C_1 x + D_1}{x^2+9}$

5. $\dfrac{A_1 x + B_1}{x^2+3x+3} + \dfrac{C_1}{x+1} + \dfrac{C_2}{(x+1)^2} + \dfrac{D_1}{2x-1} + \dfrac{D_2}{(2x-1)^2}$

6. $\dfrac{A_1}{x} + \dfrac{A_2}{x^2} + \dfrac{A_3}{x^3} + \dfrac{B_1}{2x+3} + \dfrac{C_1}{x-3} + \dfrac{D_1}{x+3}$

7. $\dfrac{3}{2x+1} + \dfrac{2}{x-1}$ **8.** $\dfrac{1}{x+5} + \dfrac{-1}{x+4}$

9. $\dfrac{2}{x-2} + \dfrac{-3}{2x-3}$ **10.** $\dfrac{1}{3x-2} + \dfrac{3}{x-1}$

11. $\dfrac{2}{x-2} + \dfrac{5}{x-5}$ **12.** $\dfrac{1}{3x-2} + \dfrac{1}{3x+2}$

13. $\dfrac{3}{x} + \dfrac{1}{x^2} + \dfrac{1}{x+2}$ **14.** $\dfrac{-1}{x} + \dfrac{4}{x-1} + \dfrac{-1}{(x-1)^2}$

15. $\dfrac{-2}{x} + \dfrac{1}{x^2} + \dfrac{1}{2x+3}$ **16.** $\dfrac{-3}{x} + \dfrac{1}{x^2} + \dfrac{-2}{x+5}$

17. $\dfrac{-1}{x} + \dfrac{2}{x+3} + \dfrac{3}{(x+3)^2}$ **18.** $\dfrac{1}{x-1} + \dfrac{-1}{(x-1)^2} + \dfrac{2}{x+2}$

19. $\dfrac{2}{x+1} + \dfrac{-1}{x^2-x+1}$ **20.** $\dfrac{-2}{x-2} + \dfrac{7x+1}{x^2-2x+2}$

21. $\dfrac{4}{x^2+1} + \dfrac{-2}{x+1}$ **22.** $\dfrac{5}{x-1} + \dfrac{x-1}{x^2+4}$

23. $\dfrac{3x-1}{x^2+2x+4} + \dfrac{-2}{x-2}$ **24.** $\dfrac{2x}{x^2-x+1} + \dfrac{1}{x+1}$

7.5 Systems of Inequalities in Two Variables

1.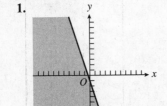

2.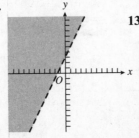

3.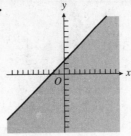

4.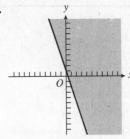

5.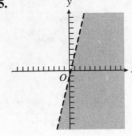

6.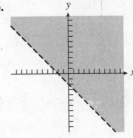

7.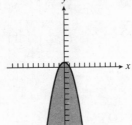

8.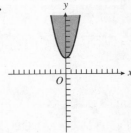

9.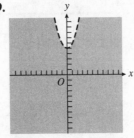

10.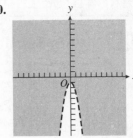

11.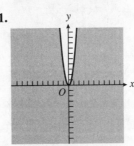

12.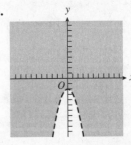

13.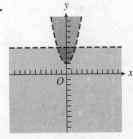

14.

15.

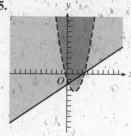

16.

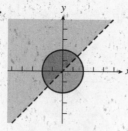

17.

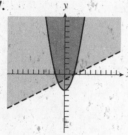

18.

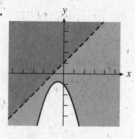

19. 37, 0; (4, 3); (0, 0) **20.** 12.5, –4; (4.5, 0); (0, 3)
21. 27, –5; (0, 12); (0, –4) **22.** 9, –3; (6, 3); (0, –3)
23. 8, –13; (4, 0); (1, 5) **24.** 7, –15; (0, 2; (6, 0)

Chapter 8
Analytic Geometry in Two and Three Dimensions

8.1 Conic Sections and Parabolas

1. $y^2 = 20x$ **2.** $x^2 = -16y$ **3.** $y^2 = -4x$
4. $x^2 = -40y$ **5.** $y^2 = -36x$ **6.** $x^2 = 28y$
7. $y^2 = 8x$ **8.** $(x-3)^2 = 16y$ **9.** $(x-1)^2 = 8(y-2)$
10. $(y-5)^2 = 24(x-6)$
11. $(y-6)^2 = 32(x-8)$
12. $x^2 = 8(y+5)$

13.

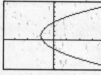

$[-3.8, 3.8]$ by $[-6, 8]$

14.

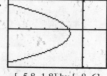

$[-5.8, 1.8]$ by $[-8, 6]$

15.

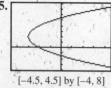

$[-4.5, 4.5]$ by $[-4, 8]$

16.

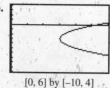

$[0, 6]$ by $[-10, 4]$

17.

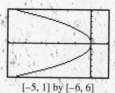

$[-5, 1]$ by $[-6, 6]$

18.

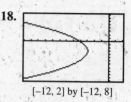

$[-12, 2]$ by $[-12, 8]$

8.2 Ellipses

1. vertices: $(0, \pm5)$; foci: $\left(0, \pm\sqrt{21}\right)$

2. vertices: $(0, \pm3)$; foci: $\left(0, \pm\sqrt{5}\right)$

3. vertices: $(\pm2, 0)$; foci: $\left(\pm\sqrt{3}, 0\right)$

4. vertices: $(\pm6, 0)$; foci: $\left(\pm\sqrt{32}, 0\right)$

5. vertices: $(0, \pm2)$; foci: $\left(0, \pm\sqrt{3}\right)$

6. vertices: $(\pm5, 0)$; foci: $\left(\pm4, 0\right)$

7. $\frac{x^2}{16} + \frac{y^2}{2.25} = 1$ **8.** $\frac{x^2}{16} + \frac{(y-2)^2}{4} = 1$

9. $\frac{x^2}{4} + \frac{(y-1)^2}{25} = 1$ **10.** $\frac{(x-10)^2}{9} + \frac{y^2}{4} = 1$

11. $\frac{(x-6)^2}{0.25} + \frac{(y-2)^2}{4} = 1$ **12.** $\frac{(x+5)^2}{9} + \frac{(y+2)^2}{2.25} = 1$

13. center: $(-1, 1)$; vertices: $(-1, 6)$ and $(-1, -4)$; foci: $(-1, 4)$ and $(-1, -2)$

14. center: $(-5, 6)$; vertices: $(-5, 9)$ and $(-5, 3)$; foci: $\left(-5, 6\pm\sqrt{5}\right)$

15. center: $(3, 3)$; vertices: $(3, 9)$ and $(3, -3)$; foci: $\left(3, 3\pm4\sqrt{2}\right)$

16. center: $(2, -5)$; vertices: $(2, 2)$ and $(2, -12)$; foci: $\left(2, 2\pm2\sqrt{6}\right)$

17. center: $(3, 3)$; vertices: $(11, 3)$ and $(-5, 3)$; foci: $\left(3\pm\sqrt{15}, 3\right)$

18. center: $(-5, 7)$; vertices: $(8, 7)$ and $(-18, 7)$; foci: $(0, 7)$ and $(-10, 7)$

8.3 Hyperbolas

1. vertices: $(\pm5, 0)$; foci: $\left(\pm\sqrt{26}, 0\right)$

2. vertices: $(\pm5, 0)$; foci: $\left(\pm\sqrt{29}, 0\right)$

3. vertices: ($\pm$2, 0); foci: ($\pm 2\sqrt{10}$, 0)

4. vertices: ($\pm$10, 0); foci: ($\pm\sqrt{104}$, 0)

5. vertices: ($\pm$3, 0); foci: ($\pm\sqrt{109}$, 0)

6. vertices: ($\pm$5, 0); foci: ($\pm 5\sqrt{2}$, 0)

7. $\dfrac{y^2}{32} - \dfrac{x^2}{4} = 1$

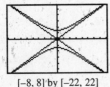

[–8, 8] by [–22, 22]

8. $\dfrac{x^2}{12} - \dfrac{y^2}{4} = 1$

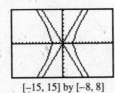

[–15, 15] by [–8, 8]

9. $\dfrac{x^2}{128} - \dfrac{y^2}{16} = 1$

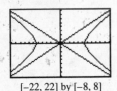

[–22, 22] by [–8, 8]

10. $\dfrac{y^2}{15} - \dfrac{x^2}{1} = 1$

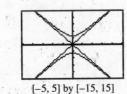

[–5, 5] by [–15, 15]

11. $\dfrac{y^2}{16} - \dfrac{x^2}{9} = 1$

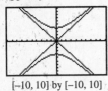

[–10, 10] by [–10, 10]

12. $\dfrac{x^2}{72} - \dfrac{y^2}{9} = 1$

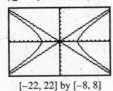

[–22, 22] by [–8, 8]

13. center: (1, –2); vertices: (4, –2) and (–2, –2);
    foci: $(1 \pm \sqrt{34}, -2)$

14. center: (–2, –1); vertices: (2, –1) and (–6, –1);
    foci: $(-2 \pm \sqrt{52}, -1)$

15. center: (4, –1); vertices: (4, 1) and (4, –3);
    foci: $(4, -1 \pm \sqrt{5})$

16. center: (–1, –1); vertices: (5, –1) and (–7, –1);
    foci: $(-1 \pm 3\sqrt{5}, -1)$

17. center: (0, –2); vertices: (9, –2) and (–9, –2);
    foci: $(\pm\sqrt{130}, -2)$

18. center: (4, 0); vertices: (4, 10) and (4, –10);
    foci: $(4, \pm\sqrt{149})$

8.4 Translation and Rotation of Axes

1. $y = 1 \pm \dfrac{1}{3}\sqrt{36 - 4x^2}$

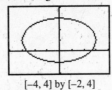

[–4, 4] by [–2, 4]

2. $y = \pm\dfrac{2}{3}\sqrt{x^2 + 2x - 8}$

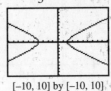

[–10, 10] by [–10, 10]

3. $y = 0.25(x^2 - 4x - 7)$

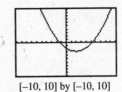

[–10, 10] by [–10, 10]

4. $y = -1 \pm \dfrac{1}{3}\sqrt{-x^2 + 2x + 8}$

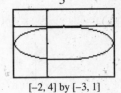

[–2, 4] by [–3, 1]

5. $y = 1 \pm 2.5\sqrt{-x^2 - 2x + 3}$

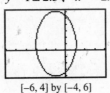

[–6, 4] by [–4, 6]

6. $y = 2 \pm 2.5\sqrt{x^2 - 4x + 8}$

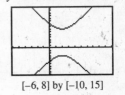

[–6, 8] by [–10, 15]

7. $y = \dfrac{1}{x}$

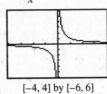

[–4, 4] by [–6, 6]

8. $y = -\dfrac{3}{2x}$

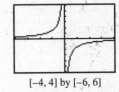

[–4, 4] by [–6, 6]

9. $y = -\dfrac{3}{x}$

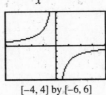

[–4, 4] by [–6, 6]

10. $y = \dfrac{3}{x}$

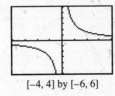

[–4, 4] by [–6, 6]

11. $y = \dfrac{5}{2x}$

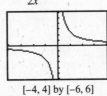

[–4, 4] by [–6, 6]

12. $y = -\dfrac{1}{4x}$

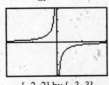

[–2, 2] by [–3, 3]

13. ellipse; $\dfrac{(x-2)^2}{9} + \dfrac{(y-3)^2}{16} = 1$; $\dfrac{(x')^2}{9} + \dfrac{(y')^2}{16} = 1$

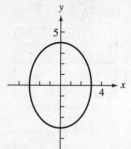

14. parabola; $(y-3)^2 = -8(x+4)$; $(y')^2 = -8x'$

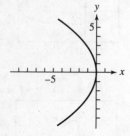

15. hyperbola; $\dfrac{(y+1)^2}{4} - \dfrac{(x+1)^2}{49} = 1$; $\dfrac{(y')^2}{4} - \dfrac{x^2}{49} = 1$

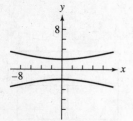

16. parabola; $(x-1)^2 = -4y$; $(x')^2 = -4y'$

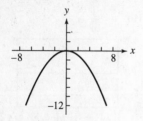

17. hyperbola; $\dfrac{(x+1)^2}{4} - \dfrac{(y-1)^2}{25} = 1$; $\dfrac{(x')^2}{4} - \dfrac{(y')^2}{25} = 1$

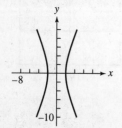

18. ellipse; $\dfrac{(x-2)^2}{4} + \dfrac{(y-2)^2}{25} = 1$; $\dfrac{(x')^2}{4} + \dfrac{(y')^2}{25} = 1$

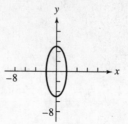

19. hyperbola; $\dfrac{(x')^2}{5} - \dfrac{(y')^2}{5} = 1$

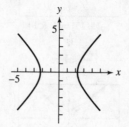

20. ellipse; $\dfrac{(x')^2}{2} + \dfrac{(y')^2}{2/3} = 1$

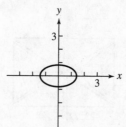

21. ellipse; $\dfrac{(x')^2}{4} + \dfrac{(y')^2}{9} = 1$

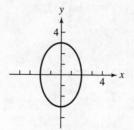

22. hyperbola; $\dfrac{(y')^2}{9} - \dfrac{(x')^2}{4} = 1$

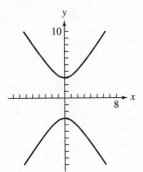

23. hyperbola; $\dfrac{(x')^2}{1} - \dfrac{(y')^2}{1/2} = 1$

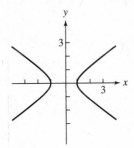

24. parabola; $4(x')^2 + 16y' = 0$

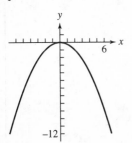

8.5 Polar Equations of Conics

1. $r = \dfrac{3}{1 + 1.5 \sin \theta}$

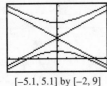

[−5.1, 5.1] by [−2, 9]

2. $r = \dfrac{0.8}{1 + 0.4 \cos \theta}$

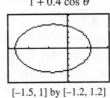

[−1.5, 1] by [−1.2, 1.2]

3. $r = \dfrac{1}{1 + 0.5 \cos \theta}$

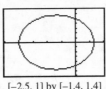

[−2.5, 1] by [−1.4, 1.4]

4. $r = \dfrac{4}{1 + 2 \sin \theta}$

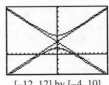

[−12, 12] by [−4, 10]

5. $r = \dfrac{2}{5 + \cos \theta}$

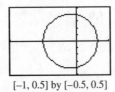

[−1, 0.5] by [−0.5, 0.5]

6. $r = \dfrac{4}{1 + \sin \theta}$

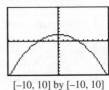

[−10, 10] by [−10, 10]

7. eccentricity: 2.5; type of conic: hyperbola; directrix: $x = 1.8$

8. eccentricity: 3; type of conic: hyperbola; directrix: $x = 7/15$

9. eccentricity: 2/3; type of conic: ellipse; directrix: $y = -3/2$

10. eccentricity: 7/9; type of conic: ellipse; directrix: $y = -1$

11. eccentricity: 0.5; type of conic: ellipse; directrix: $y = -2$

12. eccentricity: 1; type of conic: parabola; directrix: $x = 2$

13. $e = 2$, directrix $y = 1$, vertices: $\left(\dfrac{2}{3}, \dfrac{\pi}{2}\right)$ and $\left(-2, \dfrac{3\pi}{2}\right)$,

$a = -\dfrac{2}{3}$, $c = -\dfrac{4}{3}$, $b = \dfrac{2\sqrt{3}}{2}$; $\dfrac{\left(y - \dfrac{4}{3}\right)^2}{4/9} - \dfrac{x^2}{12/9} = 1$

14. $e = 1$, $x = 2$, vertex: $(1, 0)$, $p = 1$; $y^2 = -4(x - 1)$

15. $e = 2$, $k = -2$, vertices: $(-4, 0)$ and $\left(\dfrac{4}{3}, \pi\right)$,

$a = -\dfrac{4}{3}$, $c = -\dfrac{8}{3}$, $b = \dfrac{4\sqrt{3}}{3}$; $\dfrac{\left(x + \dfrac{8}{3}\right)^2}{16/9} - \dfrac{y^2}{48/9} = 1$

16. $e = 0.75$, directrix $y = -\dfrac{8}{3}$, vertices: $\left(8, \dfrac{\pi}{2}\right)$ and

$\left(\dfrac{8}{7}, \dfrac{3\pi}{2}\right)$, $a = \dfrac{32}{7}$, $c = \dfrac{24}{7}$, $b = \dfrac{8\sqrt{7}}{7}$;

$\dfrac{\left(y - \dfrac{24}{7}\right)^2}{\left(\dfrac{32}{7}\right)^2} + \dfrac{x^2}{\left(\dfrac{8\sqrt{7}}{7}\right)^2} = 1$

17. $e = \dfrac{2}{3}$, directrix $y = 3$, vertices: $\left(\dfrac{6}{5}, \dfrac{\pi}{2} \right)$ and

$\left(6, \dfrac{3\pi}{2} \right)$, $a = \dfrac{18}{5}$, $c = \dfrac{12}{5}$, $b = \dfrac{6\sqrt{5}}{5}$;

$$\dfrac{\left(y + \dfrac{12}{5} \right)^2}{\left(\dfrac{18}{5} \right)^2} + \dfrac{x^2}{\left(\dfrac{6\sqrt{5}}{5} \right)^2} = 1$$

18. $e = 1.5$, directrix $x = 0.5$, vertices: $\left(\dfrac{6}{5}, 0 \right)$ and $(-6, \pi)$,

$a = -\dfrac{12}{5}$, $c = -\dfrac{18}{5}$, $b = \dfrac{6\sqrt{5}}{5}$;

$$\dfrac{\left(x - \dfrac{18}{5} \right)^2}{\left(\dfrac{12}{5} \right)^2} - \dfrac{y^2}{\left(\dfrac{6\sqrt{5}}{5} \right)^2} = 1$$

8.6 Three-Dimensional Cartesian Coordinate System

1. distance: $\sqrt{65}$; midpoint: $(3, -1, -2.5)$

2. distance: $3\sqrt{5}$; midpoint: $(5, 1, -2.5)$

3. distance: $\sqrt{26}$; midpoint: $(3.5, 0, 1.5)$

4. distance: $\sqrt{221}$; midpoint: $(3.5, 4, -1)$

5. distance: $6\sqrt{3}$; midpoint: $(0, 0, 0)$

6. distance: 9; midpoint: $(-1, -0.5, 2)$

7.

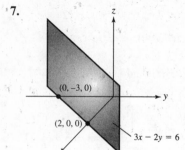

8.

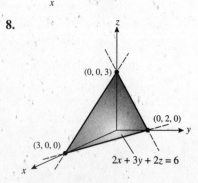

9.

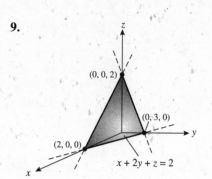

10.

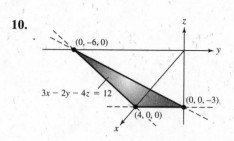

11.

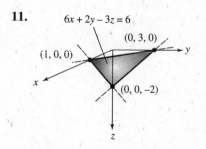

12.

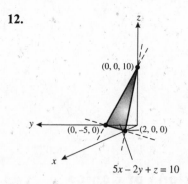

13. $\left\langle 4, -6, -14 \right\rangle$ **14.** $\left\langle 1, 3, -3 \right\rangle$ **15.** $\left\langle -1, 9, -3 \right\rangle$

16. $3\sqrt{3}$ **17.** $\left\langle 3, 0, 0 \right\rangle$ **18.** 0

19. 5   **20.** 3

21. $x\mathbf{i} + y\mathbf{j} + z\mathbf{k} = (t)\mathbf{i} + (t)\mathbf{j} + (t)\mathbf{k}$;

$x = t$, $y = t$, and $z = t$

22. $x\mathbf{i} + y\mathbf{j} + z\mathbf{k} = (1 + 2t)\mathbf{i} + (2 - t)\mathbf{j} + (-2 + t)\mathbf{k}$;

$x = 1 + 2t$, $y = 2 - t$, and $z = -2 + t$

23. $x\mathbf{i} + y\mathbf{j} + z\mathbf{k} = (-4 + t)\mathbf{i} + (2 - t)\mathbf{j} + (-1)\mathbf{k}$;

$x = -4 + t$, $y = 2 - t$, and $z = -1$

24. $x\mathbf{i} + y\mathbf{j} + z\mathbf{k} = (t)\mathbf{i} + (2 - 2t)\mathbf{j} + (3t)\mathbf{k}$;

$x = t$, $y = 2 - 2t$, and $z = 3t$

25. $x\mathbf{i} + y\mathbf{j} + z\mathbf{k} = (-3 + 7t)\mathbf{i} + (5 - 6t)\mathbf{j} + (2 - 2t)\mathbf{k}$;

$x = -3 + 7t$, $y = 5 - 6t$, and $z = 2 - 2t$

26. $x\mathbf{i} + y\mathbf{j} + z\mathbf{k} = (8 - 6t)\mathbf{i} + (-9 + 8t)\mathbf{j} + (-1 + 3t)\mathbf{k}$;

$x = 8 - 6t$, $y = -9 + 8t$, and $z = -1 + 3t$

Chapter 9
Discrete Mathematics

9.1 Basic Combinatorics

1. 175,760,000 possible license plates
2. 196,560,000 possible license plates
3. 26,000,000 possible license plates
4. 3,315,312,000 possible license plates
5. 3,089,157,760 possible license plates
6. 88,583,040 possible license plates
7. 720 distinguishable permutations
8. 720 distinguishable permutations
9. 360 distinguishable permutations
10. 360 distinguishable permutations
11. 360 distinguishable permutations
12. 180 distinguishable permutations
13. 180 distinguishable permutations
14. 180 distinguishable permutations
15. 20,160   16. 60,480   17. 60
18. 840   19. 3,628,800
20. $n(n-1)(n-2)(n-3)(n-4)(n-5)(n-6)(n-7)$
21. $n(n-1)(n-2)(n-3)(n-4)(n-5)$
22. 937,845,656,300 ways   23. 119,759,850 ways
24. 17,310,309,460,000 ways   25. 2,042,975 ways
26. 386,206,920 ways   27. 91,390 ways
28. 67,525 ways   29. 1,225 ways

9.2 The Binomial Theorem

1. $a^3 + 3a^2b + 3ab^2 + b^3$
2. $a^4 + 4a^3b + 6a^2b^2 + 4ab^3 + b^4$
3. $a^5 + 5a^4b + 10a^3b^2 + 10a^2b^3 + 5ab^4 + b^5$
4. $a^6 + 6a^5b + 15a^4b^2 + 20a^3b^3 + 15a^2b^4 + 6ab^5 + b^6$

5. $a^8 + 8a^7b + 28a^6b^2 + 56a^5b^3 + 70a^4b^4$
   $+ 56a^3b^5 + 28a^2b^6 + 8ab^7 + b^8$
6. $a^9 + 9a^8b + 36a^7b^2 + 84a^6b^3 + 126a^5b^4$
   $+ 126a^4b^5 + 84a^3b^6 + 36a^2b^7 + 9ab^8 + b^9$
7. 105   8. 680   9. 2,422,500
10. 8,008   11. −54   12. 15
13. −77,520   14. −5,940
15. $125x^3 - 225x^2y^2 + 135xy^4 - 27y^6$
16. $x^{10} - 5x^8y^2 + 10x^6y^4 - 10x^4y^6 + 5x^2y^8 - y^{10}$
17. $10,000x^4 - 60,000x^3y + 135,000x^2y^2$
    $- 135,000xy^3 + 50,625y^4$
18. $16x^{12} + 64x^9y^2 + 96x^6y^4 + 64x^3y^6 + 16y^8$
19. $3,125x^{15} - 15,265x^{12}y^3 + 31,250x^9y^6$
    $- 31,250x^6y^9 + 15,265x^3y^{12} - 3,125y^{15}$
20. $27x^9 - 27x^6y + 9x^3y^2 - y^3$
21. $x^3 - 9x^2y^3 + 27xy^6 - 27y^9$
22. $x^{25} - 5x^{20}y^5 + 10x^{15}y^{10} - 10x^{10}y^{15} + 5x^5y^{20} - y^{25}$

9.3 Probability

1. $\dfrac{1}{10,626}$ 2. $\dfrac{1}{29,716}$ 3. $\dfrac{1}{81,719}$
4. $\dfrac{5}{138}$ 5. $\dfrac{1}{2,704,156}$ 6. $\dfrac{1}{6}$
7. (a) 48.57%   8. (a) 72.22%   9. (a) 50%
   (b) 0.23   (b) 0.175   (b) 0.43
10. (a) 41.88%   11. (a) 64%   12. (a) 34%
    (b) 0.045   (b) 0.48   (b) 0.23
13. 2/3   14. 1/4   15. 3/4
16. 1/3   17. 5/7   18. 5/21
19. 0.51291   20. 0.00003   21. 0.00107
22. 0.10468   23. 0.01495   24. 0.36636

9.4 Sequences

1. diverges   2. diverges   3. converges to 0
4. diverges   5. converges to 0   6. converges to −2
7. (a) 5   8. (a) 2
   (b) $a_{10} = 76$ (b) $a_{10} = 3$
   (c) $a_n = a_{n-1} + 5$ (c) $a_n = a_{n-1} + 3$
   (d) $a_n = 5n + 26$ (d) $a_n = 2n - 17$

9. (a) 8
   (b) $a_{10} = 63$
   (c) $a_n = a_{n-1} + 8$
   (d) $a_n = 8n - 17$

10. (a) 10
    (b) 91
    (c) $a_n = a_{n-1} + 10$
    (d) $a_n = 10n - 9$

11. (a) 12
    (b) 110
    (c) $a_n = a_{n-1} + 12$
    (d) $a_n = 12n - 10$

12. (a) 9
    (b) 89
    (c) $a_n = a_{n-1} + 9$
    (d) $a_n = 9n - 1$

13. (a) 15
    (b) 139
    (c) $a_n = a_{n-1} + 15$
    (d) $a_n = 15n - 11$

14. (a) 3
    (b) 133,781
    (c) $a_1 = 7$ and $a_n = 3a_{n-1}$
    (d) $a_n = 7 \cdot 3^{n-1}$

15. (a) 0.2
    (b) 0.0128
    (c) $a_1 = 2500$ and $0.2a_{n-1}$ for $n \geq 2$
    (d) $a_n = 2500 \cdot 0.2^{n-1}$

16. (a) 2
    (b) 5,120
    (c) $a_1 = 10$ and $a_n = 5a_{n-1}$ for $n \geq 2$
    (d) $a_n = 10 \cdot 2^{n-1}$

17. (a) –3
    (b) –59,049
    (c) $a_1 = 3$ and $a_n = -3a_{n-1}$ for $n \geq 2$
    (d) $a_n = -3 \cdot (-3)^{n-1} = 3^n$

18. (a) 6
    (b) 70,543,872
    (c) $a_1 = 7$ and $a_n = 6a_{n-1}$ for $n \geq 2$
    (d) $a_n = 7 \cdot 6^{n-1}$

19. (a) –2
    (b) $a_{10} = -9 \cdot 2^{10-1} = -9 \cdot 2^9 = 4{,}608$
    (c) $a_1 = -9$ and $a_n = -2a_{n-1}$ for $n \geq 2$
    (d) $a_n = -9 \cdot 2^{n-1}$

20. $a_n = a_{n-1} + 2$ and $a_1 = -5$

21. $a_n = a_{n-1} + 13$ and $a_1 = -17$

22. $a_n = a_{n-1} + 7.2$ and $a_1 = -23.8$

23. $a_n = 7(3)^{n-1}$ and $a_1 = 7$

24. $a_n = 2(5)^{n-1}$ and $a_1 = 2$

25. $a_n = 4(3)^{n-1}$ and $a_1 = 4$

9.5 Series

1. 280	**2.** 3,990	**3.** 366
4. 273	**5.** 429	**6.** 92
7. ≈ 8.75	**8.** 177,146	**9.** 511
10. 1.19997	**11.** 7.875	**12.** 7.19985

13. converges, 16/3
14. converges, 40
15. diverges
16. diverges
17. converges, 3/4
18. converges, 2/3

9.6 Mathematical Induction

1. $P_n: 2 + 4 + 6 + \ldots + 2n = n(n+1)$. P_1 is true:
   $2 = 1(1 + 1)$. Now assume P_k is true:
   $2 + 4 + 6 + \ldots + 2k = k(k+1)$.
   Add $2(k + 1)$ to both sides:
   $2 + 4 + 6 + \ldots + 2k + 2(k+1) = k(k+1) + 2(k+1)$
   $= (k+1)(k+2)$, so P_{k+1} is true. Therefore, P_n is true
   for all $n \geq 1$.

2. $P_n: 1 + 4 + 7 + \ldots + (3n - 2) = \dfrac{n(3n-1)}{2}$. P_1 is true:

   $1 = \dfrac{1(3 \cdot 1 - 1)}{2}$. Now assume P_k is true:

   $1 + 4 + 7 + \ldots + (3k - 2) = \dfrac{k(3k-1)}{2}$.

   Add $3(k+1) - 2$ to both sides:

   $1 + 4 + 7 + \ldots + (3k - 2) + \big(3(k+1) - 2\big)$

   $= \dfrac{k(3k-1)}{2} + 2\big(3(k+1) - 2\big) = \dfrac{k(3k-1)}{2} + \dfrac{3(k+1)}{2}$

   $= \dfrac{(k+1)(3(k+1) - 1)}{2}$ so P_{k+1} is true. Therefore,

   P_n is true for all $n \geq 1$.

3. $P_n: 1 + 7 + 13 + \ldots + (6n - 5) = n(3n - 2)$. P_1 is true:
   $1 = 1(3 \cdot 1 - 2)$. Now assume P_k is true:
   $1 + 7 + 13 + \ldots + (6k - 5) = k(3k - 2)$. Add $6(k + 1) - 5$
   to both sides: $1 + 7 + 13 + \ldots + (6k - 5) + 6(k + 1) - 5$
   $= k(3k - 2) + 6(k + 1) - 5 = (k + 1)(3(k + 1) - 2)$
   so P_{k+1} is true. Therefore, P_n is true for all $n \geq 1$.

4. $P_n : 1 + 5 + 5^2 + \ldots + 5^{n-1} = \frac{1}{4}\left(5^n - 1\right)$. P_1 is true:

$1 = \frac{1}{4}\left(5^1 - 1\right)$. Now assume P_k is true:

$1 + 5 + 5^2 + \ldots + 5^{k-1} = \frac{1}{4}\left(5^k - 1\right)$. Add 5^k to both sides:

$1 + 5 + 5^2 + \ldots + 5^{k-1} + 5^k = \frac{1}{4}\left(5^k - 1\right) + 5^k = \frac{1}{4}\left(5^{k+1} - 1\right)$

so P_{k+1} is true. Therefore, P_n is true for all $n \geq 1$.

5. $P_n : 1 \bullet 2 + 2 \bullet 3 + 3 \bullet 4 + \ldots + n(n+1) =$

$\frac{n(n+1)(n+2)}{3}$. P_1 is true: $1 \bullet 2 = \frac{1(1+1)(1+2)}{3}$.

Now assume P_k is true: $1 \bullet 2 + 2 \bullet 3 + 3 \bullet 4 + \ldots + k(k+1)$

$= \frac{k(k+1)(k+2)}{3}$. Add $(k+1)(k+2)$ to both sides:

$1 \bullet 2 + 2 \bullet 3 + 3 \bullet 4 + \ldots + k(k+1) + (k+1)(k+2)$

$= \frac{k(k+1)(k+2)}{3} + (k+1)(k+2) = \frac{(k+1)(k+2)(k+3)}{3}$

so P_{k+1} is true. Therefore, P_n is true for all $n \geq 1$.

6. $P_n : 5 + 7 + 9 + \ldots + [5 + 2(n-1)] = \frac{n(8 + 2n)}{2}$.

P_1 is true: $5 = \frac{1(8 + 2(1))}{2}$. Now assume P_k is true:

$5 + 7 + 9 + \ldots + [5 + 2(k-1)] = \frac{k(8 + 2k)}{2}$.

Add $5 + 2((k+1) - 1) = (5 + 2k)$ to both sides:

$5 + 7 + 9 + \ldots + (5 + 2(k-1)) + (5 + 2k)$

$= \frac{k(8 + 2k)}{2} + (5 + 2k) = \frac{(k+1)(8 + 2(k+1))}{2}$

so P_{k+1} is true. Therefore, P_n is true for all $n \geq 1$.

7. $P_n : 3 + 9 + 15 + \ldots + (6n - 3) = 3n^2$. P_1 is true: $3 = 3(1)^2$.

Now assume P_k is true: $3 + 9 + 15 + \ldots + (6k - 3) = 3k^2$.

Add $6(k+1) - 3 = (6k+3)$ to both sides:

$3 + 9 + 15 + \ldots + (6k - 3) + (6k + 3) = 3k^2 + (6k + 3)$

$= 3(k+1)^2$ so P_{k+1} is true. Therefore, P_n is true

for all $n \geq 1$.

8. $P_n : 3 + 8 + 13 + \ldots + (5n - 2) = \frac{n}{2}(5n + 1)$.

P_1 is true: $3 = \frac{1}{2}(5(1) + 1)$. Now assume P_k is true:

$3 + 8 + 13 + \ldots + (5k - 2) = \frac{k}{2}(5k + 1)$.

Add $5(k+1) - 2 = (5k + 3)$ to both sides:

$3 + 8 + 13 + \ldots + (5k - 2) + (5k + 3)$

$= \frac{k}{2}(5k + 1) + (5k + 3) = \frac{k+1}{2}(5(k+1) + 1)$ so P_{k+1} is true.

Therefore, P_n is true for all $n \geq 1$.

9. $P_n : 7 + 11 + 15 + \ldots + (4n + 3) = n(2n + 5)$.

P_1 is true: $7 = 1(2 \bullet 1 + 5)$. Now assume P_k is true:

$7 + 11 + 15 + \ldots + (4k + 3) = k(2k + 5)$.

Add $4(k+1) + 3 = (4k + 7)$ to both sides:

$7 + 11 + 15 + \ldots + (4k + 3) + (4k + 7) = k(2k + 5) + (4k + 7)$

$= (k+1)\left(2(k+1) + 5\right)$ so P_{k+1} is true. Therefore,

P_n is true for all $n \geq 1$.

10. $P_n : 1 + 6 + 6^2 + \ldots + 6^{n-1} = \frac{1}{5}\left(6^n - 1\right)$.

P_1 is true: $1 = \frac{1}{5}\left(6^1 - 1\right)$. Now assume P_k is true:

$1 + 6 + 6^2 + \ldots + 6^{k-1} = \frac{1}{5}\left(6^k - 1\right)$. Add 6^k to both sides:

$1 + 6 + 6^2 + \ldots + 6^{k-1} + 6^k = \frac{1}{5}\left(6^k - 1\right) + 6^k = \frac{1}{5}\left(6^{k+1} - 1\right)$

so P_{k+1} is true. Therefore, P_n is true for all $n \geq 1$.

11. $P_n : \left(1 + \frac{1}{1}\right) + \left(1 + \frac{1}{2}\right) + \left(1 + \frac{1}{3}\right) + \ldots + \left(1 + \frac{1}{n}\right) = n + 1$.

P_1 is true: $\left(1 + \frac{1}{1}\right) = 1 + 1$. Now assume P_k is true:

$\left(1 + \frac{1}{1}\right) + \left(1 + \frac{1}{2}\right) + \left(1 + \frac{1}{3}\right) + \ldots + \left(1 + \frac{1}{k}\right) = k + 1$.

Add $\left(1 + \frac{1}{k+1}\right)$ to both sides:

$\left(1 + \frac{1}{1}\right) + \left(1 + \frac{1}{2}\right) + \left(1 + \frac{1}{3}\right) + \ldots + \left(1 + \frac{1}{k}\right) + \left(1 + \frac{1}{k+1}\right)$

$= k + 1 + \left(1 + \frac{1}{k+1}\right)$, so P_{k+1} is true. Therefore,

P_n is true for all $n \geq 1$.

12. P_n : $n! > 2^n$, for $n \geq 4$. P_4 is true: $4! > 2^4$. Now assume P_k is true: $k! > 2^k$, for $k \geq 4$. Since $k \geq 4$, $k+1 \geq 5 > 2$, multiplying the left hand side by $k + 1$ and multiplying the right hand side by 2 preserves the inequality. Hence, P_{k+1} is true. Therefore, P_n is true for all $n \geq 1$.

13. P_n : $18^n - 1$ is evenly divisible by 17. (That is, 17 is a factor of $18^n - 1$.) P_1 is true: 17 is a factor of $18^1 - 1 = 17$. Now, assume P_k is true: 17 is a factor of $18^k - 1$. Then, $18^{k+1} - 1 = 18(18^k - 1) + 17$. Since 17 is a factor of both terms of this sum, it is a factor of the sum, so P_{k+1} is true. Therefore, P_n is true for all $n \geq 1$.

14. P_n : $n(n^2 + 5)$ is evenly divisible by 6. (That is, 6 is a factor of $n(n^2 + 5)$.) P_1 is true: 6 is a factor of $1(1^2 + 5)$. Now, assume P_k is true: 6 is a factor of $k(k^2 + 5)$. Then, $(k+1)\left[(k+1)^2 + 5\right] = (k+1)^3 + 5(k+1)$ $= k^3 + 3k^2 + 8k + 6 = (k^3 + 5k) + (3k^2 + 3k) + 6$ $= k(k^2 + 5) + 3k(k+1) + 6$. If k is either even or odd, $3k(k+1)$ is even and a multiple of 3; hence a multiple of 6. Since 6 is a factor of $k(k^2 + 5)$ and 6, it is a factor of all three terms of this sum, and it is a factor of the sum, so P_{k+1} is true. Therefore, P_n is true for all $n \geq 1$.

15. P_n : $11^n - 4^n$ is evenly divisible by 7. (That is, 7 is a factor of $11^n - 4^n$.) P_1 is true: 7 is a factor of $11^1 - 4^1 = 7$. Now, assume P_k is true: 7 is a factor of $11^k - 4^k$. Then, $11^{k+1} - 4^{k+1} = 11(11^k) - 4(4^k)$ $= 11(11^k) - (11 - 7)(4^k) = 11(11^k) - 11(4^k) + 7(4^k)$ $= 11(11^k - 4^k) + 7(4^k)$. Since 7 is a factor of both terms of this sum, it is a factor of the sum, so P_{k+1} is true. Therefore, P_n is true for all $n \geq 1$.

16. P_n : $n(n^2 + 3n + 8)$ is evenly divisible by 6. (That is, 6 is a factor of $n(n^2 + 3n + 8)$.) P_1 is true: 6 is a factor of $1(1^2 + 3 \cdot 1 + 8) = 12$. Now, assume P_k is true: 6 is a factor of $k(k^2 + 3k + 8)$. Then, $(k+1)\left((k+1)^2 + 3(k+1) + 8\right)$ $= (k+1)^3 + 3(k+1)^2 + 8(k+1)$ $= (k^3 + 3k^2 + 8k) + (3k^2 + 9k + 12)$ $= k(k^2 + 3k + 8) + 3(k^2 + 3k + 4)$. Now, 6 is a factor of $k^2 + 3k + 4$ regardless of whether k is even or odd. Since 6 is a factor of both terms of this sum, it is a factor of the sum, so P_{k+1} is true. Therefore, P_n is true for all $n \geq 1$.

17. P_n : $2^{5n} - 1$ is evenly divisible by 31. (That is, 31 is a factor of $2^{5n} - 1$.) P_1 is true: 31 is a factor of $2^{5(1)} - 1 = 31$. Now, assume P_k is true: 31 is a factor of $2^{5k} - 1$. Then, $2^{5(k+1)} - 1 = 2^5\left(2^{5k} - 1\right) + 31$. Since 31 is a factor of both terms of this sum, it is a factor of the sum, so P_{k+1} is true. Therefore, P_n is true for all $n \geq 1$.

18. P_n : $(n+1)(n+2)...(2n)$ is evenly divisible by 2^n. (That is, 2^n is a factor of $(n+1)(n+2)...(2n)$.) P_1 is true: $2^1 = 2$ is a factor of $(1+1) = 2$. Now, assume P_k is true: 2^k is a factor of $(k+1)(k+2)...(2k)$. Then, multiplying the first term of $((k+1)+1)((k+1)+2)...(2(k+1)) =$ $(k+2)(k+3)...(2k)(2(k+1))$ by $(k+1)$ and dividing the last term by the same quantity gives $(k+1)(k+2)...(2k)(2)$, which is divisible by 2^k, so P_{k+1} is true. Therefore, P_n is true for all $n \geq 1$.

9.7 Statistics and Data (Graphical)

1.

Carbohydrates		Fat
	1	0 4 5 9
6 8 8 8 9 9	2	5 8
0	3	1 3 8
6 6 8 8	4	8
	5	5

The bacon double cheeseburger would be your best option.

2.

Protein		Fat
4 4 6 6	1	0 4 5 9
7 7	2	5 8
0 0 2	3	1 3 8
6	4	8
1	5	5

Answers may vary.

3.

Protein		Carbohydrates
4 4 6 6	1	
7 7	2	6 8 8 8 9 9
0 0 2	3	0
6	4	6 6 8 8
1	5	

The bacon double cheeseburger deluxe would be your best option.

4.

National		American
	2	2
1 6 6 7 7 8 8	3	2 2 2 3 6 7 9 9 9
0 0 0 4 5 8 8 8	4	0 1 3 4 5 6
2	5	

5. The National League champion tends to hit more home runs.

6. 52 home runs by George Foster in the National League in 1977, and 22 home runs by 4 players in 1981 in the American League seem to be outliers. Explanations may vary.

7.
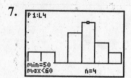
[10, 80] by [−1, 5]

The data does not necessarily appear normal.

8.
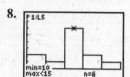
[0, 25] by [−1, 8]

The data does not necessarily appear normal.

9.
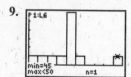
[0, 50] by [−1, 6]

Yes, there is an outlier of 45 field goal attempts. See graph.

10.
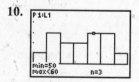
[10, 80] by [−1, 5]

The data does not necessarily appear normal.

11.
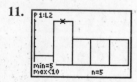
[0, 25] by [−1, 7]

The data does not necessarily appear normal.

12.
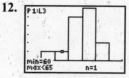
[50, 85] by [−1, 6]

There appear to be two low outliers, at 57% and 63% proficiency.

13.
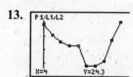

Window: [0, 10], where x represents decades since 1900, and [22.2, 26.7], where y represents age.

14.
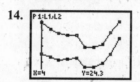

Window: [0, 10], where x represents decades since 1900, and [19.3, 27.1], where y represents age. The higher set of data represents the median age for the males, while the lower set represents the median age for the females.

15. Between 1940 to 1950, the median age of first marriage decreased most rapidly for both sets. This could be attributed to World War II and/or the Korean War.

16.
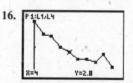

Window: [0, 10], where x represents decades since 1900, and [1.9, 4.3], where y represents the difference between the median ages of the males and the females. The difference in the median ages seem to be decreasing, except for the decade surrounding 1980.

17.
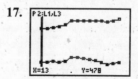

Window: [0, 14], where x represents years since 1980, and [412.5, 487.5], where y represents the average SAT scores. The mathematics data set appears above the verbal data set. Mathematics scores have increased more during the time period.

18.

Window: [0, 14], where x represents years since 1980, and [38.8, 56.2], where y represents the difference between the average math and verbal scores.

19. Over time, the difference between the average math and verbal scores have generally increased.

20. The slope of the trend line through the first and last points is 12/13. Through the time period from 1980 to 1993, the gap between the average math and verbal scores have been increasing at the rate of 12/13 point per year.

9.8 Statistics and Data (Algebraic)

1. 8.05   **2.** 8.05   **3.** 8.15

4. 8.55   **5.** 8.1   **6.** 5.8

7. The mean score for males is higher than the mean score for females.

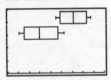

[90, 180] by [0, 5]

8. The mean score for males is higher than the mean score for females.

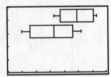

[90, 160] by [0, 5]

9. The mean score for females is higher than the mean score for males

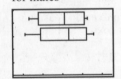

[135, 185] by [0, 5]

10. The mean score for males is higher than the mean score for females

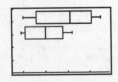

[75, 125] by [0, 5]

11. The mean score for males is higher than the mean score for females

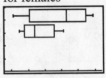

[12, 185] by [0, 5]

12. The mean score for females is higher than the mean score for males

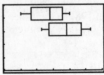

[140, 200] by [0, 5]

13. (a) 85.7   **14.** (a) 84.81

    (b) 13.865424   (b) 2.99976

15. (a) $7.3\bar{3}$ **16.** (a) $17.3\bar{3}$

    (b) 2.121320344   (b) 1.8027756

17. (a) 3.4055   **18.** (a) 112.77

    (b) 0.36407798   (b) 13.98908

19. 2.5%   **20.** 16%   **21.** 2.5%

22. 16%   **23.** 2.5%   **24.** 2.5%

Chapter 10
An Introduction to Calculus

10.1 Limits and Motion: The Tangent Problem

1. 20 meters/second; 1200 meters/minute

2. 400 kilometers/hour   **3.** 20 yards/second

4. 230.77 feet/minute   **5.** 41.54 miles/hour

6. 147.37 kilometers/hour   **7.** 4

8. 0   **9.** 16

10. −4   **11.** 3

12. 3   **13.** 6

14. 4   **15.** $\dfrac{1}{6}$

16. 0   **17.** $-\dfrac{1}{20}$

18. 54   **19.** $f'(x) = -2$

20. $f'(x) = \dfrac{2}{3}x$ **21.** $f'(x) = \dfrac{1}{2\sqrt{x-7}}$

22. $\dfrac{dy}{dx} = -\dfrac{3}{(x-1)^2}$

23. $\dfrac{dy}{dx} = 0$

24. $\dfrac{dy}{dx} = -15x^2 + 7$

10.2 Limits and Motion: The Area Problem

1. 945 miles

2. 40.5 yards

3. 240 kilometers

4. 310 meters

5. 18 inches

6. 190 centimeters

7. 10.47

8. 2.461

9. 1.614

10. 11.0625

11. 6.146

12. 8.4375

13. 25

14. 24

15. 100

16. 4

17. 40.5

18. 20

19. 30 meters

20. 96 feet

21. 36 feet

22. 21 meters

23. 9π meters

24. 4π meters

10.3 More on Limits

1. 1/3

2. $\dfrac{5\sqrt{10}}{3}$

3. -4

4. $\dfrac{(\ln 4) - 8}{6}$

5. $\dfrac{\sqrt{2}}{6}$

6. $\dfrac{-3(-1 + \pi^3)}{\pi}$

7. $\lim\limits_{x \to 4^-} f(x) = 48, \quad \lim\limits_{x \to 4^+} f(x) = 33$

8. $\lim\limits_{x \to 0^-} f(x) = 1, \quad \lim\limits_{x \to 0^+} f(x) = 0$

9. $\lim\limits_{x \to e^-} f(x) = 1, \quad \lim\limits_{x \to e^+} f(x) = e$

10. $\lim\limits_{x \to 1^-} f(x) = e, \quad \lim\limits_{x \to 1^+} f(x) = e$

11. $\lim\limits_{x \to \pi^-} f(x) = -1, \quad \lim\limits_{x \to \pi^+} f(x) = 1$

12. $\lim\limits_{x \to \sqrt{5}^-} f(x) = 15, \quad \lim\limits_{x \to \sqrt{5}^+} f(x) = 25$

13. $\lim\limits_{x \to \infty} f(x) = 0, \quad \lim\limits_{x \to -\infty} f(x) = 0$

14. $\lim\limits_{x \to \infty} f(x) = \infty, \quad \lim\limits_{x \to -\infty} f(x) = 0$

15. $\lim\limits_{x \to \infty} f(x) = 0, \quad \lim\limits_{x \to -\infty} f(x) = -\infty$

16. $\lim\limits_{x \to \infty} f(x) = 4, \quad \lim\limits_{x \to -\infty} f(x) = \infty$

17. $\lim\limits_{x \to \infty} f(x) = \infty, \quad \lim\limits_{x \to -\infty} f(x) = 1$

18. $\lim\limits_{x \to \infty} f(x) = \dfrac{5}{3}, \quad \lim\limits_{x \to -\infty} f(x) = 0$

19. $\lim\limits_{x \to -\frac{1}{4}^-} f(x) = -\infty, \quad \lim\limits_{x \to -\frac{1}{4}^+} f(x) = -\infty$

20. $\lim\limits_{x \to 0^-} f(x) = -\infty, \quad \lim\limits_{x \to 0^+} f(x) = -\infty$

21. $\lim\limits_{x \to 1^-} f(x) = -\infty, \quad \lim\limits_{x \to 1^+} f(x) = \infty$

22. $\lim\limits_{x \to 0^-} f(x) = -\infty, \quad \lim\limits_{x \to 0^+} f(x) = \infty$

23. $\lim\limits_{x \to 0^-} f(x) = \infty, \quad \lim\limits_{x \to 0^+} f(x) = \infty$

24. $\lim\limits_{x \to 0^-} f(x) = -\infty, \quad \lim\limits_{x \to 0^+} f(x) = \infty$

10.4 Numerical Derivatives and Integrals

1. 108

2. 1/4

3. $-3/8$

4. 3

5. 3/2

6. -1

7. 323.333

8. 1

9. 1

10. 0.386

11. 4.5

12. 11.599

13. 44 miles

14. 100.625 miles

15. 1515 miles

16. 123.635 miles

17. 113.071 miles

18. 17.261 kilometers